Heath Mathematics

Walter E. Rucker
Clyde A. Dilley

D. C. Heath and Company
Lexington, Massachusetts Toronto

About the authors

Walter E. Rucker Former Specialist in Education with the Curriculum Laboratory of the University of Illinois, has taught mathematics in public schools and is a coauthor of successful mathematics programs for elementary and junior high schools.

Clyde A. Dilley Professor, University of Toledo, Toledo, Ohio, is teaching methods courses in elementary and secondary mathematics. He has taught mathematics in public schools and is a coauthor of successful mathematics programs for elementary and junior high schools.

Illustrations Mark Kelley/Sharon Kurlansky and Bill Morrison

Photography Jeff Albertson: 372, 372/Jonathan Barkan: 151/J. Berndt, Stock Boston: 88/Fredrik Bodin: 56, 63, 95, 190, 225, 266, 268, 301, 350/ Christian Delbert: 199, 259/Owen Franken: 360/Kevin Galvin: 27, 43, 76, 87, 175, 218, 219, 235, 292, 312, 313/Mike Malyszko: 352, 353, 364, 365/Terry McCoy: 1, 68, 279/Julie O'Neil: 38, 72, 73, 91, 130, 162, 168, 169, 235, 245/ Edward Pieratt: 356, 357/Richard Raphael: 82/Rick Rizzotto: 33/John Running: 370, 371/Deidra Delano Stead: Cover, 22, 24, 46, 47, 48, 49, 67, 106, 117, 121, 147, 192, 193, 221, 264, 297, 317/Bruce Stearns: 5

Published simultaneously in Canada.

Printed in the United States of America.

International Standard Book Number: 0-669-03427-4

Cont

1 Whole Numbers–Addition and Subtraction

Thousands

Did you ever stop to think about our place-value system? With only ten digits we can write about any whole number, no matter how large. Place value also makes our simple methods of computation possible. That is why you know more arithmetic than ancient Greek and Roman scholars!

Thousands					
Hundreds	Tens	Ones	Hundreds	Tens	Ones
3	6	5	2	4	8

Standard Numeral 3 6 5,2 4 8

- 8 ones = 8
- 4 tens = 40
- 2 hundreds = 200
- 5 thousands = 5000
- 6 ten thousands = 60,000
- 3 hundred thousands = 300,000

The value of a digit depends on both the digit and the value of its place in the numeral.

Words Three hundred sixty-five thousand, two hundred forty-eight

EXERCISES

In 953,674, which digit is in the

1. hundreds place?
2. hundred thousands place?
3. tens place?
4. ten thousands place?
5. thousands place?
6. ones place?

What does the red digit stand for? Give two answers.

7. 5938 (answer: 3 tens, 30)
8. 2651
9. 7402
10. 5763
11. 84,295
12. 63,171
13. 92,158
14. 76,053
15. 129,748
16. 638,415
17. 291,645
18. 378,905
19. 444,444
20. 444,444
21. 444,444
22. 444,444

Write the standard numeral.

23. nine thousand, four hundred twenty-six
24. fifty-three thousand, five hundred six
25. four hundred sixty-one thousand, five hundred eighty-five
26. seventy-three thousand, six hundred ninety-four
27. nine hundred thousand, nine hundred
28. eight hundred thousand, eight
29. eight hundred eight thousand
30. seven hundred thousand, seventy

Write in words.

31. 5618
32. 8940
33. 25,008
34. 403,752
35. 351,692
36. 834,162
37. 702,300
38. 202,020

Here is how a standard numeral is written as an expanded numeral.

Standard Numeral	*Expanded Numeral*
3 5 6 2 5 8	300,000 + 50,000 + 6000 + 200 + 50 + 8

Write as expanded numerals.

39. 382
40. 576
41. 1948
42. 3759
43. 26,534
44. 71,823
45. 591,648
46. 715,384

Comparing numbers

Lifetime Winnings of Some Race Horses	
Horse	*Amount won*
Bold Ruler	$764,204
Gun Bow	$798,722
Hasty Road	$541,402
Native Dancer	$785,240
Never Bend	$641,524
Ponder	$541,275
TV Lark	$902,194
Whirlaway	$561,161

Here is how to compare the winnings of Hasty Road and Whirlaway. Both numbers have the same number of digits, so we start at the left and compare the digits that are in the same place one by one.

541,402 < 561,161

is less than

4 < 6

Another example.

349,106 > 345,798

is greater than

9 > 5

How would you compare two numbers that have different numbers of digits?

EXERCISES

Give the correct sign.

1. 753 ● 754

2. 500 ● 498

3. 999 ● 1000

4. 906 ● 960

5. 5382 ● 5296

6. 4821 ● 865

7. 78,234 ● 81,639

8. 65,382 ● 65,379

9. 53,462 ● 71,893

10. 60,009 ● 59,999

11. 243,842 ● 243,851

12. 506,735 ● 506,673

13. 268,183 ● 267,997

14. 625,835 ● 625,853

15. 529,376 ● 530,500

16. 406,218 ● 406,128

17. 594,999 ● 595,000

18. 307,561 ● 300,861

Refer to the table for exercises 19 – 26.

19. Which horse won more, TV Lark or Bold Ruler?
20. Which horse won more, Gun Bow or Native Dancer?
21. Which horse won more, Never Bend or Ponder?
22. Which horse won less, Hasty Road or Ponder?
23. Which of the horses won the most?
24. Which of the horses won the least?
25. Which horse won five hundred sixty-one thousand, one hundred sixty-one dollars?
26. List the horses in order from least to greatest winnings.

Rounding

Sometimes approximate numbers are used. The exact price of the car is not \$6000, but \$5835. The exact price was **rounded to the nearest thousand.**

5835 rounded to the nearest thousand is 6000.

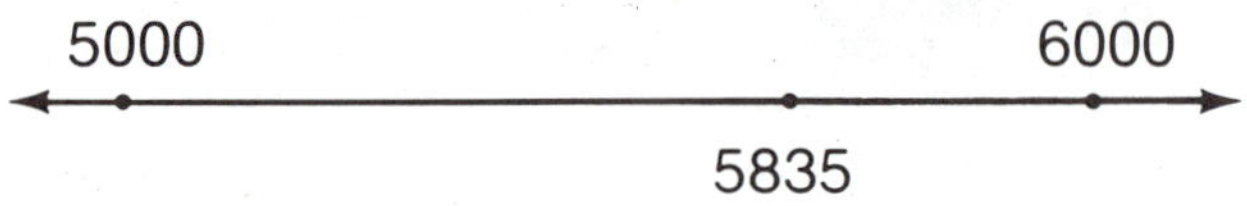

5835 rounded to the nearest hundred is 5800.

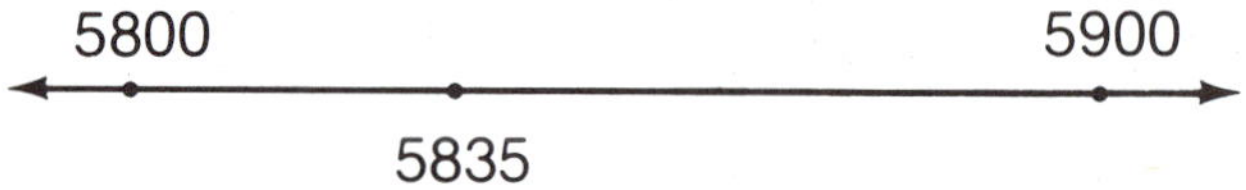

When the number is halfway between, round up.

5835 rounded to the nearest ten is 5840.

EXERCISES

Round to the nearest thousand.

1. 5826	**2.** 6374	**3.** 9028	**4.** 4691
5. 14,039	**6.** 28,243	**7.** 47,500	**8.** 39,800

Round to the nearest hundred.

9. 7346	**10.** 9214	**11.** 3560	**12.** 4938
13. 29,643	**14.** 75,385	**15.** 29,580	**16.** 86,950

Here is a shortcut for rounding.

Step 1. Find the place to which you are rounding.

Step 2. Look at the next digit to the right.

Step 3. If that digit is 5 or greater, round up. If that digit is less than 5, round down.

Round to the nearest thousand.

578,341 ↑

578,341 ↑ ↑

578,341 ↑ ↑

3 is less than 5.
Round down: 578,000.

Use the shortcut to round

17. 35,928 to the nearest ten.

18. 26,753 to the nearest thousand.

19. 84,219 to the nearest hundred.

20. 59,037 to the nearest thousand.

21. 482,961 to the nearest hundred.

22. 357,281 to the nearest ten.

23. 426,350 to the nearest hundred.

24. 291,784 to the nearest thousand.

25. 565,000 to the nearest ten thousand.

26. 399,999 to the nearest hundred thousand.

Pretend that you are a newspaper reporter. Write a headline for each fact. *Hint:* Round the numbers.

27. The city council adopted a budget of $598,382.

28. The Concorde reached 2015 kilometers per hour during a recent flight.

29. Last night 42,056 went to Municipal Stadium to see the Cleveland Indians win.

★ **30.** Some of the digits have been turned face down. Study the clues to find the number.

a. If you round to the nearest thousand or ten thousand, you get 70,000.

b. If you round to the nearest ten or hundred you get the same number.

c. There is no smaller number that rounds this way.

Millions and billions

Of course you have heard of "large" numbers such as a million or a billion. But do you really know how large a billion is? Such large numbers will never be as familiar as small numbers such as 5.

Pretend that you decided to count to a billion. If you counted a number every second, you would be over 40 years old by the time you finished.

A million is 1000 thousand. 1,000,000
A billion is 1000 million. 1,000,000,000

When writing a large number in standard form we start at the right and separate the periods of three digits with commas.

Billions	Millions	Thousands	
465,	009,	046,	581

Four hundred sixty-five billion, nine million, forty-six thousand, five hundred eighty-one

EXERCISES

Read aloud.

1. 2000
2. 7,000,000
3. 5,000,000,000
4. 358,219
5. 6,540,320
6. 58,473,000
7. 29,368,152
8. 591,346,782
9. 712,543,264
10. 5,368,934,156
11. 7,846,915,274
12. 3,258,265,192
13. 72,359,065,020
14. 95,004,290,000
15. 259,068,342,517
16. If you had \$1,000,000, you could spend more than \$20 every day for 100 years.
17. If your heart averages 72 beats a minute, it will beat 3,110,400 times in 30 days.
18. If your average pace is $2\frac{1}{2}$ feet, then 1,000,000,000 paces would take you 473,484 miles.
19. Traveling 24 hours a day, it would take you over 2000 years to travel 1,000,000,000 miles in a car.

Give the standard numeral.

20. 242 million **21.** 18 billion

22. ten million **23.** forty-two billion

24. fifty-eight billion, two hundred twelve million, nine hundred thousand

Solve.

25. Which planet is nearest to the sun?

26. Which planet is farthest from the sun?

27. How far is Jupiter from the sun?

Average distances of planets from the sun	
Planet	*Distance* (rounded to nearest thousand kilometers)
Mercury	57,937,000
Venus	108,260,000
Earth	149,668,000
Mars	228,048,000
Jupiter	778,695,000
Saturn	1,427,665,000
Uranus	2,872,285,000
Neptune	4,500,632,000
Pluto	5,905,534,000

28. Which planet is fifth farthest from the sun?

29. How many planets are farther from the sun than Earth?

30. How many planets are closer to the sun than Jupiter?

31. Round Mercury's distance to the nearest million kilometers.

32. Round Mars' distance to the nearest ten million kilometers.

1 trillion is 1000 billion. 1,000,000,000,000

Trillions	Billions	Millions	Thousands	
314,	926,	785,	642,	085

Read aloud.

33. 5,000,000,000,000

34. 36,450,000,000,000

35. 243,715,821,400,000

36. 791,365,425,381,742

Roman numerals

Ancient Romans wrote numerals that did *not* use place value. They combined the symbols shown below and usually added the values.

I	V	X	L	C	D	M
1	5	10	50	100	500	1000

Here are some examples of the addition rule.

VII = 7 (5 + 1 + 1)

XVI = 16 (10 + 5 + 1)

LXXV = 75 (50 + 10 + 10 + 5)

The Romans also had a subtraction rule.

IV = 4 (5 − 1)

IX = 9 (10 − 1)

XXIX = 29 (10 + 10 + (10 − 1))

XC = 90 (100 − 10)

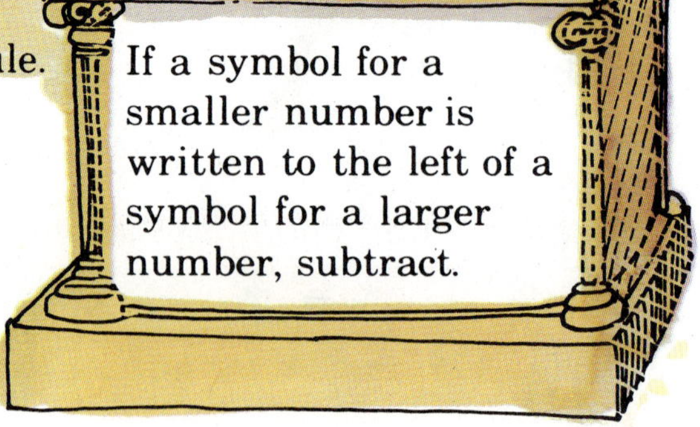

If a symbol for a smaller number is written to the left of a symbol for a larger number, subtract.

Study these examples.

CCCX = 310 MDC = 1600 CCLXVI = 266

CCXC = 290 MCD = 1400 CCLXIV = 264

EXERCISES

Write standard numerals.

1. III 2. VI 3. XIII 4. XXXVI
5. IV 6. IX 7. XIV 8. XXXIV
9. LX 10. XL 11. CX 12. XC
13. CD 14. DC 15. MC 16. XLIX
17. CMX 18. MXC 19. MCMLXXX 20. MDCCLXXVI

Write Roman numerals.

21. 10 22. 11 23. 12 24. 13 25. 14
26. 25 27. 29 28. 30 29. 40 30. 65
31. 267 32. 591 33. 658 34. 827 35. 927
36. 1027 37. 2568 38. 3006 39. 2573 40. 1981

Change the dates to standard numerals.

41.

Copyright
MCMLXXVI

42.

Columbus discovered America
MCDXCII

43.

Lincoln became president
MDCCCLXI

44.

Telescope invented
MDIX

Addition

The numbers you add are called **addends.** The answer is called the **sum.**

How well do you know your basic addition facts? You could find the answers by counting, but by now you should have all the facts memorized. It is important that you take the time to learn them well. In this lesson you will review and practice the basic addition facts. Before practicing the facts, let's quickly review two properties that could help you remember some facts.

The Adding 0 Property

The sum of any number and 0 is the number.

The Commutative (Order) Property of Addition

Changing the order of the addends does not change the sum.

Here is another property of addition.

The Associative (Grouping) Property of Addition

Changing the grouping of the addends does not change the sum.

$$(5 + 2) + 4 = 5 + (2 + 4)$$

Remember to work inside the grouping symbols first.

See how long it takes you to write these sums.

1. 6 +6	**2.** 3 +2	**3.** 8 +8	**4.** 8 +1	**5.** 9 +8	**6.** 6 +9	**7.** 5 +5	**8.** 8 +7
9. 2 +2	**10.** 7 +7	**11.** 9 +5	**12.** 9 +0	**13.** 7 +3	**14.** 9 +3	**15.** 7 +9	**16.** 5 +7
17. 6 +3	**18.** 8 +3	**19.** 5 +2	**20.** 9 +2	**21.** 6 +5	**22.** 8 +9	**23.** 9 +9	**24.** 4 +4
25. 6 +2	**26.** 6 +4	**27.** 5 +8	**28.** 9 +7	**29.** 1 +9	**30.** 8 +2	**31.** 9 +4	**32.** 7 +4
33. 3 +3	**34.** 8 +4	**35.** 0 +7	**36.** 7 +5	**37.** 5 +4	**38.** 7 +8	**39.** 7 +6	**40.** 5 +6
41. 6 +7	**42.** 4 +3	**43.** 8 +5	**44.** 6 +8	**45.** 5 +9	**46.** 8 +6	**47.** 9 +6	**48.** 5 +3

49. Check your work. The correct answers are given at the bottom of the page.

50. Make two bar graphs like the ones shown. Record your score on one graph and your time on the other.

51. Practice the facts that you missed.

52. Take the speed drill on six more days. Record your results on your graphs. On which graph should the bars get taller? Get shorter?

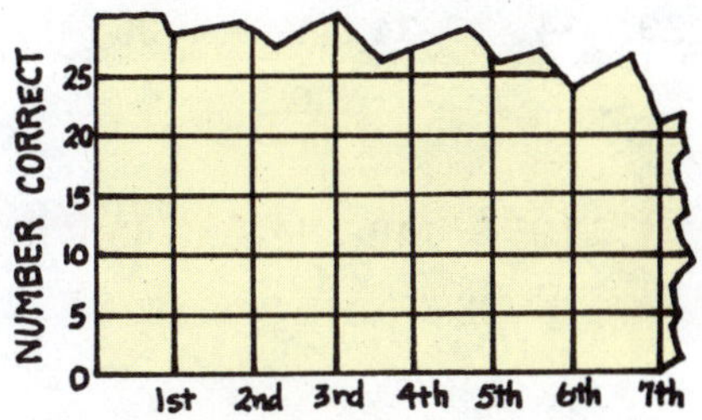

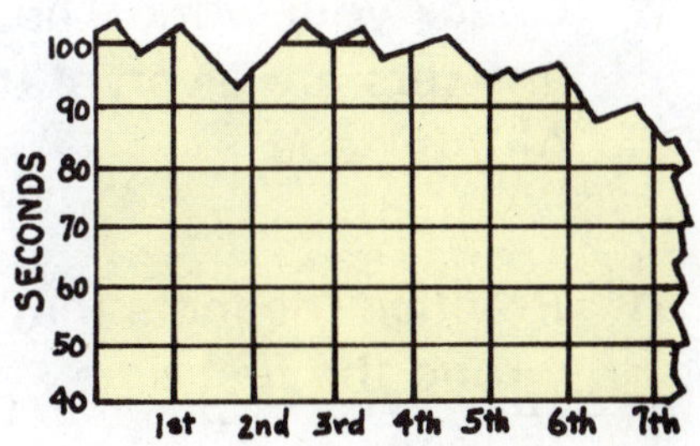

Answers.

1. 12 **2.** 5 **3.** 16 **4.** 9 **5.** 17 **6.** 15 **7.** 10 **8.** 15 **9.** 4 **10.** 14
11. 14 **12.** 9 **13.** 10 **14.** 12 **15.** 16 **16.** 12 **17.** 9 **18.** 11 **19.** 7 **20.** 11
21. 11 **22.** 17 **23.** 18 **24.** 8 **25.** 8 **26.** 10 **27.** 13 **28.** 16 **29.** 10 **30.** 10
31. 13 **32.** 11 **33.** 6 **34.** 12 **35.** 7 **36.** 12 **37.** 9 **38.** 15 **39.** 13 **40.** 11
41. 13 **42.** 7 **43.** 13 **44.** 14 **45.** 14 **46.** 14 **47.** 15 **48.** 8

Subtraction—basic facts

Remember: Subtracting is finding a missing addend.

$16 - 7 = 9$ $7 + 9 = 16$

The answer is called the **difference.**

See how long it takes you to write these differences.

1. 15 − 9	**2.** 12 − 4	**3.** 14 − 5	**4.** 10 − 8	**5.** 12 − 3	**6.** 13 − 9	**7.** 11 − 5	**8.** 12 − 5
9. 14 − 9	**10.** 10 − 6	**11.** 11 − 6	**12.** 13 − 4	**13.** 11 − 7	**14.** 17 − 9	**15.** 10 − 7	**16.** 15 − 6
17. 17 − 8	**18.** 9 − 4	**19.** 10 − 5	**20.** 9 − 3	**21.** 9 − 7	**22.** 12 − 7	**23.** 16 − 9	**24.** 15 − 7
25. 13 − 8	**26.** 9 − 5	**27.** 16 − 8	**28.** 12 − 9	**29.** 13 − 5	**30.** 8 − 0	**31.** 15 − 8	**32.** 10 − 2
33. 14 − 6	**34.** 11 − 8	**35.** 18 − 9	**36.** 12 − 6	**37.** 14 − 8	**38.** 10 − 4	**39.** 11 − 3	**40.** 10 − 9
41. 13 − 6	**42.** 11 − 9	**43.** 10 − 3	**44.** 13 − 7	**45.** 10 − 1	**46.** 14 − 7	**47.** 12 − 8	**48.** 16 − 7

49. Check your work. The correct answers are given at the bottom of this page.

50. Make two bar graphs for subtraction like those you made for addition on page 13.

51. Practice the facts that you got wrong the first time.

52. Take the speed drill on six more days. Keep a graph of your scores and times.

Answers.

1. 6 **2.** 8 **3.** 9 **4.** 2 **5.** 9 **6.** 4 **7.** 6 **8.** 7 **9.** 5 **10.** 4
11. 5 **12.** 9 **13.** 4 **14.** 8 **15.** 3 **16.** 9 **17.** 9 **18.** 5 **19.** 5 **20.** 6
21. 2 **22.** 5 **23.** 7 **24.** 8 **25.** 5 **26.** 4 **27.** 8 **28.** 3 **29.** 8 **30.** 8
31. 7 **32.** 8 **33.** 8 **34.** 3 **35.** 9 **36.** 6 **37.** 6 **38.** 6 **39.** 8 **40.** 1
41. 7 **42.** 2 **43.** 7 **44.** 6 **45.** 9 **46.** 7 **47.** 4 **48.** 9

Find the missing addends.

53. 6 + ? = 9
54. 7 + ? = 15
55. 8 + ? = 13
56. 7 + ? = 14
57. 9 + ? = 17
58. 6 + ? = 15
59. 8 + ? = 12
60. 8 + ? = 14
61. 7 + ? = 16
62. 7 + ? = 13
63. 8 + ? = 11
64. 6 + ? = 13
65. 6 + ? = 14
66. 8 + ? = 17
67. 8 + ? = 15
68. 8 + ? = 16
69. 7 + ? = 12
70. 7 + ? = 15
71. 9 + ? = 16
72. 8 + ? = 8
73. 9 + ? = 14
74. 9 + ? = 18
75. 9 + ? = 15
76. 6 + ? = 7

Compute. Watch the signs.

77. (5 + 6) + 2
78. 5 + (6 + 2)
79. (4 + 5) − 2
80. (8 − 5) − 2
81. 8 − (5 − 2)
82. (9 − 3) + 8
83. (6 − 2) + 5
84. (7 + 2) + 9
85. (8 − 2) + 9

Complete. Add across and add down.

86. (+ across, + down)

3	4	?
2	5	?
?	?	?

87. (+ across, + down)

6	?	6
3	?	8
?	?	?

88. (+ across, + down)

4	?	8
?	?	?
6	?	15

Add across and subtract down.

89. (+ across, − down)

8	2	?
4	2	?
?	?	?

90. (+ across, − down)

9	?	13
2	?	5
?	?	?

91. (+ across, − down)

?	5	12
?	?	?
5	?	7

Addition

Remember that you add in one column at a time. If the sum in a column is 10 or greater, you have to regroup.

Step 1. Add ones.

```
  57
+25
```

12 ones

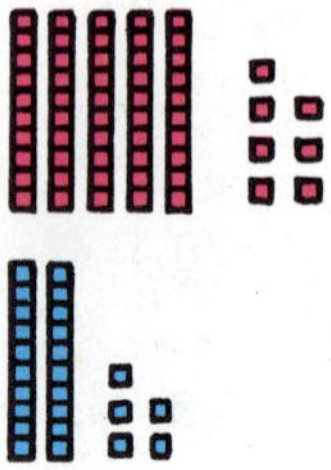

Step 2. Regroup.

```
  1
  57
+25
   2
```

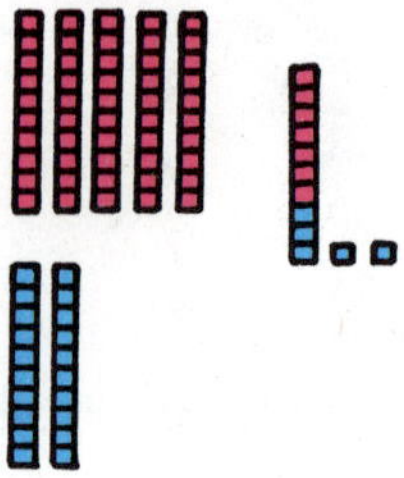

Step 3. Add tens.

```
  1
  57
+25
  82
```

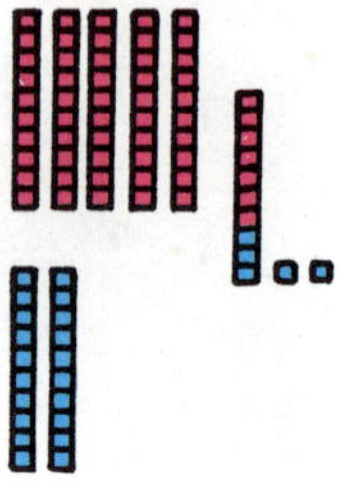

Study these examples.

```
  11
  594
+678
 1272
```

```
  111
  3978
+5697
  9675
```

```
  11  1
  286354
+379518
 665,872
```

EXERCISES

Add.

1. 72 + 15	2. 34 + 23	3. 52 + 56	4. 79 + 25	5. 94 + 37
6. 63 + 28	7. 29 + 47	8. 67 + 91	9. 86 + 56	10. 77 + 39
11. 857 + 462	12. 156 + 238	13. 437 + 286	14. 593 + 865	15. 297 + 486
16. 635 + 779	17. 835 + 768	18. 546 + 888	19. 357 + 758	20. 487 + 993
21. 48364 + 29541	22. 27835 + 58877	23. 29443 + 67581	24. 46358 + 2977	25. 86431 + 987

26. 3742 + 5916

27. 7418 + 9253

28. 5069 + 2788

29. 5349 + 2761

30. 2761 + 5349

31. 5271 + 935

32. 7819 + 599

33. 7438 + 2965

34. 5384 + 2999

35. 2999 + 5384

36. 48364 + 29541

37. 27835 + 58877

38. 29943 + 67581

39. 46358 + 2977

40. 86431 + 987

41. $4.67 + 2.93

42. $7.54 + .99

43. $83.56 + 42.77

44. $29.57 + 30.00

45. $86.83 + 68.08

Solve.

46. Miss Cooke spent $85.65 to get her car tuned up and $52.75 for a new tire. What was the total bill?

47. Miss Cooke drove her car 15,287 kilometers one year and 14,796 kilometers the next year. How far did she drive the car in the two years?

48. A car dealer wanted $2975 and Miss Cooke's car for a new model. If the dealer allowed $3245 for her old car, what was the total price of the new car?

Find the missing digits.

49. 7585 + ■34■ = 8■■7

50. ■865 + 4■7■ = 14,5■3

51. 89563 + ■■■■■ = 115,461

Addition—3 or more addends

When you add more than two numbers, you have to remember each sum as you add.

Step 1. Add 9 and 7. Remember the sum.

```
 9 .. (16)
 7
 1
+8
```

Step 2. Add 16 and 1. Remember the sum.

```
 9    16
 7   +1
 1 .. 17
+8
```

Step 3. Add 17 and 8. *Write* the sum.

```
 9    17
 7   +8
 1    25
+8 ..
25
```

Step 1. Add 9 and 1. Remember the sum.

```
 9 ..
 7    (10)
 1 ..
+8
```

Step 2. Add 7 and 8. Remember the sum.

```
 9
 7 ..
 1    (15)
+8 ..
```

Step 3. Add 10 and 15. Write the sum.

```
 9   (10)
 7
 1   (15)
+8
25
```

When a sum is 10 or greater, you have to regroup.

```
 2 ... (Regrouped 20 ones for 2 tens)
 35     13
 28    +9
+39
102
```

```
   22
  392
  614
  258
 +757
 2021
```

EXERCISES

Add. Look for sums of 10.

1.	2.	3.	4.	5.	6.
8	5	8	9	8	9
2	5	7	6	6	7
+7	+9	+3	+4	+2	+1

7.	8.	9.	10.	11.	12.
8	9	6	5	7	9
6	3	8	8	5	1
2	4	4	2	4	7
+5	+1	+7	+5	+3	+6

13.	14.	15.	16.	17.	18.
8	9	8	7	9	9
3	6	5	6	6	3
5	7	8	4	4	1
7	4	5	8	8	8
+2	+3	+2	+3	+1	+4

19.	20.	21.	★22.	★23.	★24.
57	58	883	6382	8426	923
38	83	596	5759	3928	529
23	42	741	1258	718	316
+41	+76	+97	+7367	556	741
				+8291	+89

Solve.

25. Who had the best (lowest score) first round?
26. Who had the worst fourth round?
27. What was Allen's total tournament score?
28. Who had the better (lower) score, Davis or Norwood?
29. Who won the tournament?

Elmwood Golf Tournament

	SCORES			
Name	*1st Round*	*2nd Round*	*3rd Round*	*4th Round*
Allen	82	85	79	86
Davis	91	82	80	85
Johnson	86	85	89	92
Jones	85	86	83	79
Norwood	83	92	91	81
Thomas	85	87	95	85

Keeping Skills Sharp

Less than (<) or greater than (>)?

1. 832 ● 833
2. 526 ● 518
3. 699 ● 700
4. 4821 ● 4852
5. 3627 ● 3672
6. 22,428 ● 31,002

Subtraction

Remember that you subtract in one column at a time. Sometimes you have to regroup.

Step 1.
Not enough ones.
Regroup 1 ten to 10 ones.

$$\begin{array}{r} \overset{3}{\cancel{4}}{}^{1}2 \\ -\ 17 \\ \hline \end{array}$$

Step 2.
Subtract ones.

$$\begin{array}{r} \overset{3}{\cancel{4}}{}^{1}2 \\ -\ 17 \\ \hline 5 \end{array}$$

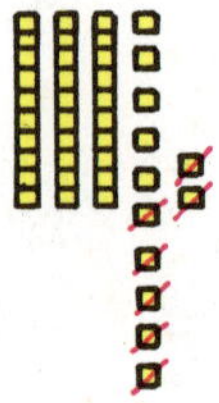

Step 3.
Subtract tens.

$$\begin{array}{r} \overset{3}{\cancel{4}}{}^{1}2 \\ -\ 17 \\ \hline 25 \end{array}$$

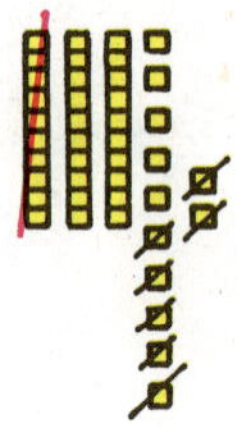

Remember that you can check subtraction by addition.

$$\begin{array}{r} \overset{6}{\cancel{7}}\overset{11}{\cancel{2}}{}^{1}1 \\ -\ 385 \\ \hline 336 \end{array} \qquad \begin{array}{r} \overset{1\,1}{336} \\ +\ 385 \\ \hline 721 \end{array}$$

EXERCISES

Subtract and check.

1. $85 - 24$
2. $57 - 16$
3. $35 - 17$
4. $46 - 28$
5. $94 - 57$
6. $93 - 48$
7. $97 - 53$
8. $85 - 68$
9. $98 - 39$
10. $85 - 65$
11. $463 - 267$
12. $584 - 297$
13. $643 - 581$
14. $391 - 293$
15. $674 - 188$
16. $235 - 167$
17. $851 - 276$
18. $377 - 86$
19. $983 - 642$
20. $754 - 186$

21. 5918 − 2653	22. 7419 − 2836	23. 6283 − 1946	24. 8372 − 2851	25. 4693 − 2582
26. 53891 − 18246	27. 76382 − 45968	28. 83274 − 58166	29. 92834 − 7165	30. 68345 − 7998
31. \$9.56 − 2.59	32. \$7.56 − 2.48	33. \$19.56 − 7.89	34. \$53.54 − 27.83	35. \$295.64 − 58.39
36. 91487 − 32608	37. 59638 − 14979	38. 65387 − 43659	39. 49736 − 6918	40. 88451 − 9377

Solve.

41. Jill bought a can of tennis balls for \$3.09. She gave the clerk \$3.50. How much change did she get?

42. Carol bowled a score of 172. Jim bowled 210. By how many points did Jim win?

43. Carl was a weight lifter. In one match he lifted 485 kilograms in one lift and 377 kilograms in another lift. What was the total weight lifted?

44. Five bowlers on a team made these scores in one game: 178, 235, 188, 196, and 202. What was the total team score for the game?

Who am I?

45. If you add me to 758, you get 2173.

46. If you subtract me from 4963, you get 2173.

47. If you subtract 2856 from me, you get 2743.

More about subtracting

There were 8005 beans in the jar. John's guess was 6879. We can subtract to see how close John's guess was to the actual number of beans.

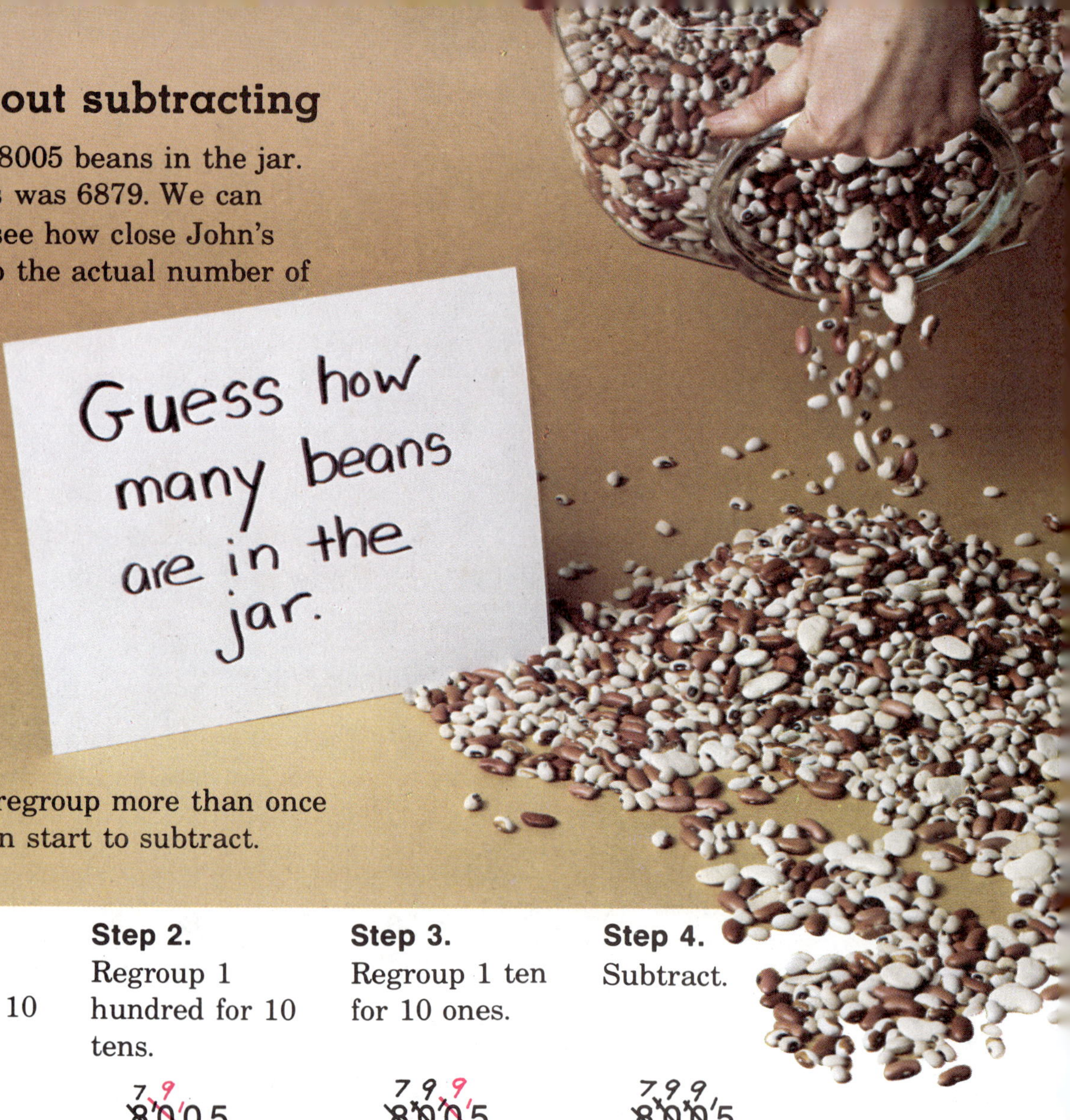

$$\begin{array}{r} 8005 \\ -\ 6879 \\ \hline \end{array}$$

We have to regroup more than once before we can start to subtract.

Step 1.
Regroup 1 thousand for 10 hundreds.

$$\begin{array}{r} \scriptstyle 7\ \ 10\ \ \ \ \ \ \\ \not{8}\,0\,0\,5 \\ -\ 6\,8\,7\,9 \\ \hline \end{array}$$

Step 2.
Regroup 1 hundred for 10 tens.

$$\begin{array}{r} \scriptstyle 7\ \ 9\ \ 10\ \ \ \\ \not{8}\,\not{0}\,0\,5 \\ -\ 6\,8\,7\,9 \\ \hline \end{array}$$

Step 3.
Regroup 1 ten for 10 ones.

$$\begin{array}{r} \scriptstyle 7\ \ 9\ \ 9\ \ 15 \\ \not{8}\,\not{0}\,\not{0}\,\not{5} \\ -\ 6\,8\,7\,9 \\ \hline \end{array}$$

Step 4.
Subtract.

$$\begin{array}{r} \scriptstyle 7\ \ 9\ \ 9\ \ 15 \\ \not{8}\,\not{0}\,\not{0}\,\not{5} \\ -\ 6\,8\,7\,9 \\ \hline 1\,1\,2\,6 \end{array}$$

EXERCISES

Subtract.

1. 600 − 231	**2.** 503 − 276	**3.** 901 − 463	**4.** 305 − 197	**5.** 804 − 438
6. 900 − 528	**7.** 800 − 574	**8.** 600 − 159	**9.** 700 − 347	**10.** 500 − 256

11. 5012 − 1378
12. 5006 − 3561
13. 3509 − 1254
14. 7400 − 3561
15. 8000 − 7483

16. 9603 − 4538
17. 8291 − 2714
18. 7406 − 863
19. 5930 − 95
20. 8706 − 588

21. 92008 − 7461
22. 50074 − 3996
23. 10000 − 9684
24. 253015 − 84976
25. 700000 − 156834

Here are the six winners of the guess-the-number-of-beans contest.

How close to the number of beans (8005) was

26. Sue Davis?
27. Robert Garcia?
28. Who was closer, Bob Bender or Jane Wilson?
29. Who was the closest of the winners?
30. Who came in second?

4000
− 1673

★ **Use the shortcut to find these differences.**

31. 6000 − 4136 (5999 − 4135)
32. 9000 − 5866
33. 50000 − 29584

34. 802 − 467 (799 − 464)
35. 401 − 167
36. 2003 − 896

Keeping Skills Sharp

Round to the nearest ten.

1. 71
2. 78
3. 45
4. 156
5. 302
6. 1672

Round to the nearest hundred.

7. 359
8. 416
9. 550
10. 1745
11. 2863
12. 5987

Estimating and practice

Sometimes Andy uses a calculator. When he does, he generally estimates each answer first. Estimating helps Andy catch errors caused by punching the wrong keys. One way to estimate is to round.

This difference is near 1000.

$$\begin{array}{r} 3004 \\ -\ 1885 \\ \hline \end{array} \quad \begin{array}{r} 3000 \\ -2000 \\ \hline 1000 \end{array}$$

This sum is near 180.

$$\begin{array}{r} 68 \\ 34 \\ 21 \\ +\ 57 \\ \hline \end{array} \quad \begin{array}{r} 70 \\ 30 \\ 20 \\ +60 \\ \hline 180 \end{array}$$

EXERCISES

First estimate the answer. Then compute.

1. 78 + 29	2. 82 + 49	3. 90 + 42	4. 78 + 91	5. 63 + 82
6. 29 + 36 + 40	7. 78 + 29 + 65	8. 39 + 42 + 58	9. 67 + 75 + 91	10. 82 + 36 + 74
11. 594 − 220	12. 703 − 141	13. 619 − 512	14. 879 − 196	15. 958 − 449
16. 8012 − 6957	17. 5974 − 3102	18. 9113 − 6957	19. 8146 − 5857	20. 6007 − 3818

Estimate and compute.

21. 59 + 72
22. 68 + 91
23. (36 + 25) + 19
24. (46 + 82) + 41
25. (59 − 23) + 74
26. (56 + 39) − 23
27. (75 − 32) − 10
28. 75 − (32 − 10)
29. (259 + 40) − 38
30. (561 − 59) + 152
31. (3765 + 1992) − 2026
32. (6247 − 849) + 2694

33. Here is a student's paper. Estimate each answer. Which answers do your estimates tell you are wrong?

★34. Estimating will not find all mistakes. The answers to exercises 4 and 8 are also wrong. Why didn't estimating help find those errors?

Problem solving

Some of the problems in this lesson are two-step problems. Often such problems can be solved in more than one way. Compare Bob's way with Jean's way.

Bob's Way

1. Read the problem and find the question.

   Erica bought some hiking boots for \$29.75 and some wool socks for \$3.19. She gave the clerk \$40. How much change should she get?

2. What are the facts?

   had \$40
   spent \$29.75
   spent \$3.19

3. Decide what to do.

   First add to find the money spent. Then subtract to find the amount of change.

4. Answer the question.

   $$\begin{array}{r} \$29.75 \\ +\ 3.19 \\ \hline \$32.94 \end{array} \quad \text{amount spent}$$

   $$\begin{array}{r} \$\ 40.00 \\ -\ 32.94 \\ \hline \$\ \ 7.06 \end{array} \quad \text{change}$$

5. Estimate! Does your answer seem right?

   Spent \$30 + \$3 = \$33
   Have left \$40 − \$33 = \$7

   The answer seems right.

Jean's Way

1. Read the problem and find the question.

   Erica bought some hiking boots for \$29.75 and some wool socks for \$3.19. She gave the clerk \$40. How much change should she get?

2. What are the facts?

   had \$40
   spent \$29.75
   spent \$3.19

3. Decide what to do.

   First subtract to find how much is left after buying the boots. Then subtract to find how much is left after buying the socks.

4. Answer the question.

   $$\begin{array}{r} \$40.00 \\ -\ 29.75 \\ \hline \$10.25 \\ -\ 3.19 \\ \hline \$\ \ 7.06 \end{array} \quad \begin{array}{l} \\ \\ \text{left after buying boots} \\ \\ \text{left after buying socks} \end{array}$$

5. Estimate! Does your answer seem right?

   \$40 − \$30 = \$10
   \$10 − \$3 = \$7

   The answer seems right.

EXERCISES
Solve

1. Luisa and her parents went backpacking on the Appalachian Trail. Luisa carried 12 kilograms of supplies and each of her parents carried 22 kg. What was the total?

2. They planned to hike a total of 200 kilometers. They hiked 32 km the first day, 35 km the second day, and 29 km the third day. How many km did they have left to hike?

3. One day they climbed a mountain. They started at an elevation of 259 meters above sea level. When they reached the top, they had climbed 578 meters. What was the elevation of the peak?

4. One day they hiked 2 hours and 30 minutes in the morning and 1 hour and 45 minutes in the afternoon. How many minutes did they hike?

5. After hiking for 4 days, they stopped for supplies. They spent $12.65 for food, $1.69 for candles, $1.75 for flashlight batteries, and $4.29 for a new compass. How much did they spend in all?

6. At the end of the fifth day they had hiked a total of 148 kilometers. They planned to hike 28 kilometers the next day. How many kilometers would they have left then? (See exercise 2.)

CHAPTER CHECKUP

[pages 2–3, 8–9]

In 59,634,275, which digit is in the

1. ten thousands place? 2. millions place? 3. hundred thousands place?

Write the standard numeral. [pages 2–3, 8–9]

4. seven hundred thousand 5. eighty-five thousand, two hundred forty-five
6. nine million, two hundred fifty thousand 7. two billion

Less than (<) or greater than (>)? [pages 4–5]

8. 7000 ◯ 6987 9. 954,621 ◯ 956,421 10. 286,538 ◯ 39,786

Round to the nearest thousand. [pages 6–7]

11. 8246 12. 9875 13. 25,672 14. 46,500

Round to the nearest million. [pages 8–9]

15. 5,926,384 16. 7,286,095 17. 15,354,296 18. 245,500,000

Add. [pages 12–13, 16–19, 24–25]

19. $\begin{array}{r} 67 \\ +38 \\ \hline \end{array}$ 20. $\begin{array}{r} 834 \\ +177 \\ \hline \end{array}$ 21. $\begin{array}{r} 835416 \\ +278887 \\ \hline \end{array}$ 22. $\begin{array}{r} 293 \\ 377 \\ +689 \\ \hline \end{array}$

Subtract. [pages 14–15, 20–25]

23. $\begin{array}{r} 93 \\ -\ 57 \\ \hline \end{array}$ 24. $\begin{array}{r} 725 \\ -\ 548 \\ \hline \end{array}$ 25. $\begin{array}{r} 68732 \\ -\ 51965 \\ \hline \end{array}$ 26. $\begin{array}{r} 6002 \\ -\ \ 489 \\ \hline \end{array}$

Solve. [pages 4–5, 26–27]

27. Which state is largest (ranks first in area)?
28. Which state ranks sixth?
29. What is the total area of the two largest states?
30. California is how much larger than New Mexico?

The Six Largest States	
State	Area in square miles
Alaska	586,400
Arizona	113,909
California	158,693
Montana	147,138
New Mexico	121,666
Texas	267,339

Project

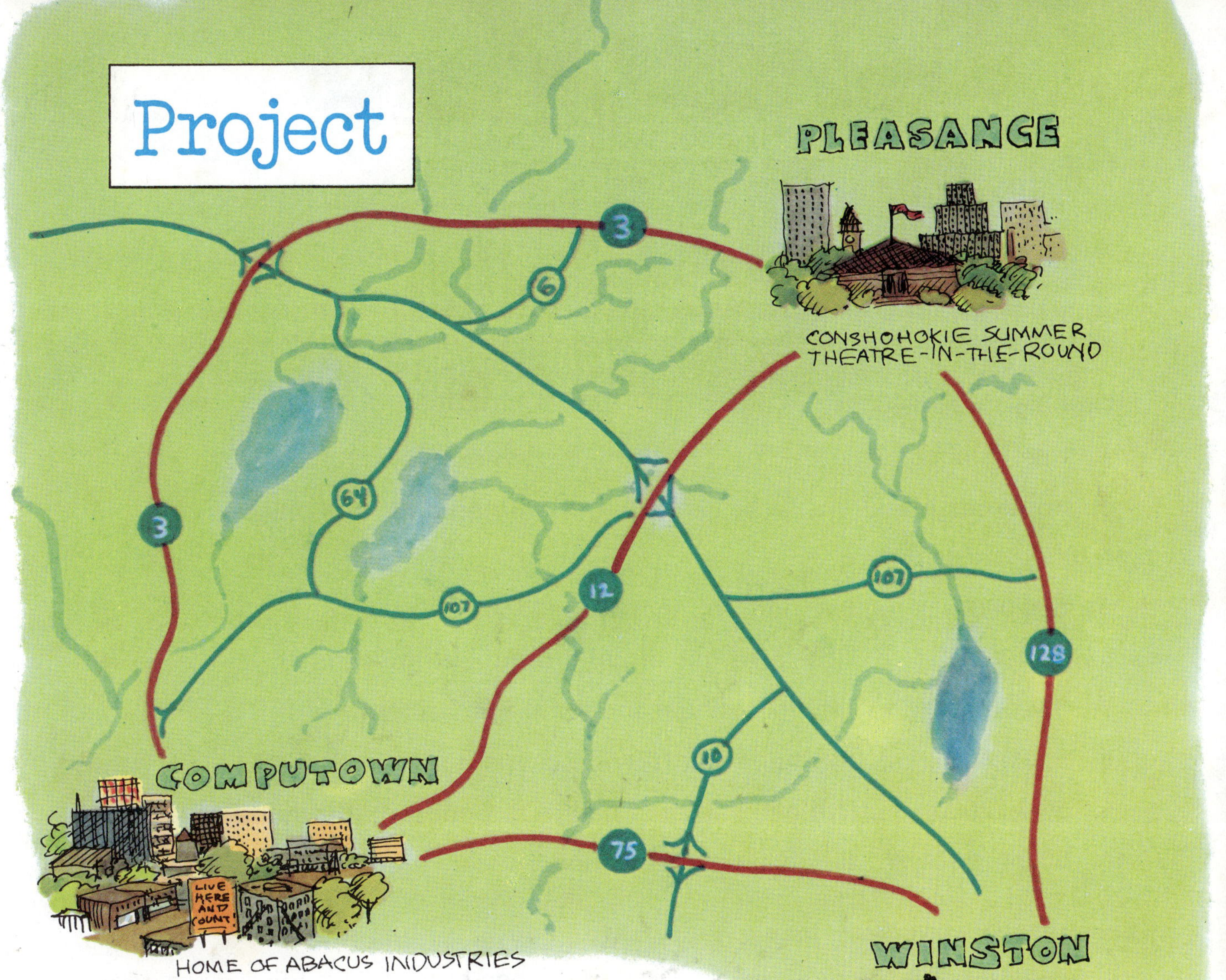

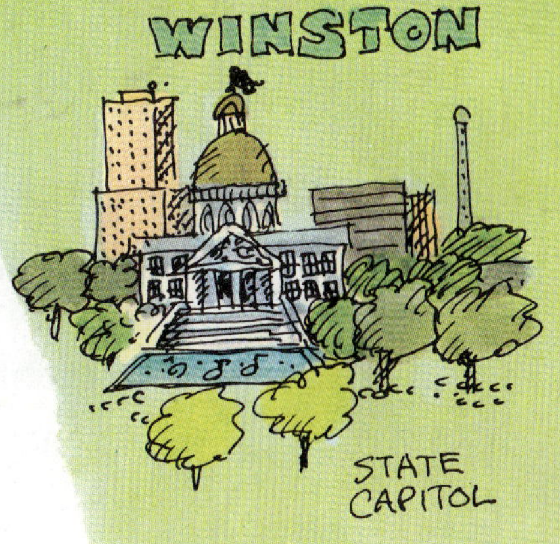

1. List two cities that you would like to visit. Use a map or a mileage chart to compute the total number of miles that you would have to drive to visit the two cities and return home.
2. Plan a three-city trip. List all possible orders in which you could visit the three cities. Which order gives the shortest route? How long is the shortest route?
3. Plan a four-city trip. List all possible orders in which you could visit the four cities. How long is the shortest route?

CHAPTER REVIEW

Billions			Millions			Thousands					
Hundreds	Tens	Ones	Hundreds	Tens	Ones	Hundreds	Tens	Ones	Hundreds	Tens	Ones
5	7	3	4	9	1	6	8	2	6	0	4

What digit is in the

1. one millions place?
2. ten thousands place?
3. hundreds place?
4. hundred billions place?

Match.

5. 530,040 — **a.** five hundred thirty-four thousand
6. 53,400 — **b.** fifty three million, four hundred thousand
7. 534,000 — **c.** fifty-three thousand, four hundred
8. 5,300,400 — **d.** five hundred thirty thousand, forty
9. 53,400,000 — **e.** five million, three hundred thousand, four hundred
10. 50,340,000 — **f.** fifty million, three hundred forty thousand

Less than (<) or greater than (>)?

11. 6 million ● 6 billion **12.** 599,999 ● 1,000,000 **13.** 67,423 ● 67,432

Round 24,965 to the nearest

14. ten. **15.** hundred. **16.** thousand. **17.** ten thousand.

$$\begin{array}{r} \overset{1}{5}\overset{1}{7}28 \\ +4936 \\ \hline 10,664 \end{array}$$

Add.

$$\textbf{18.}\ \begin{array}{r} 298 \\ +467 \\ \hline \end{array} \qquad \textbf{19.}\ \begin{array}{r} 8563 \\ +6899 \\ \hline \end{array} \qquad \textbf{20.}\ \begin{array}{r} 58 \\ 67 \\ 39 \\ +46 \\ \hline \end{array}$$

$$\begin{array}{r} 38\overset{7\,9}{\not{0}}4 \\ -1456 \\ \hline 2348 \end{array}$$

Subtract.

$$\textbf{21.}\ \begin{array}{r} 563 \\ -258 \\ \hline \end{array} \qquad \textbf{22.}\ \begin{array}{r} 63155 \\ -1867 \\ \hline \end{array} \qquad \textbf{23.}\ \begin{array}{r} 5004 \\ -2896 \\ \hline \end{array}$$

CHAPTER CHALLENGE

Computer circuits use a place-value system for representing numbers, but it is not a base ten system. It is a base two system. A base two system uses only the digits 0 and 1. Study these examples of base ten numerals. Notice that in a base two system we group by twos.

Base ten numeral	Eights	Fours	Twos	Ones	Base two numeral
1				1	1_{two}
2			1	0	10_{two}
3			1	1	11_{two}
4		1	0	0	100_{two}
5		1	0	1	101_{two}
6		1	1	0	110_{two}
7		1	1	1	111_{two}
8	1	0	0	0	1000_{two}

This shows the base

Read as "one zero base two"

Change to base two numerals.

1. 9 **2.** 10 **3.** 11 **4.** 12 **5.** 13 **6.** 14

7. 15 **8.** 16 **9.** 17 **10.** 18 **11.** 19 **12.** 20

13. 21 **14.** 22 **15.** 23 **16.** 24 **17.** 25 **18.** 26

You will need another place

19. Continue writing base two numerals. See if you can go to 60.

20. Write some number facts about yourself in base two. **Example.** I am 1100_{two} years old.

MAJOR CHECKUP

Standardized Format

Choose the correct letter.

1. In 8763 the 7 stands for

a. 7 tens
b. 7 hundreds
c. 7 thousands
d. none of these

2. In 954,638 the 9 stands for

a. 9 hundreds
b. 9 millions
c. 9 hundred thousands
d. none of these

3. The standard numeral for six million sixty thousand six hundred is

a. 6,600,060
b. 6,060,006
c. 60,060,600
d. none of these

4. Which is greater than 873,214?

a. 99,846
b. 1,000,000
c. 799,999
d. none of these

5. Which is less than 93,472,681?

a. 93,482,681
b. 93,476,281
c. 93,472,671
d. none of these

6. 86,345 rounded to the nearest thousand is

a. 86,000
b. 87,000
c. 90,000
d. none of these

7. 8,921,542 rounded to the nearest million is

a. 8,000,000
b. 9,000,000
c. 8,900,000
d. none of these

8. 5,742,934,650 rounded to the nearest billion is

a. 5,000,000,000
b. 5,743,000,000
c. 5,740,000,000
d. none of these

9. Add.

$$\begin{array}{r} 48632 \\ +29584 \\ \hline \end{array}$$

a. 78,116
b. 77,216
c. 78,216
d. none of these

10. Add.

$48 + 206 + 57$

a. 411
b. 671
c. 1256
d. none of these

11. Subtract.

$$\begin{array}{r} 8874 \\ -3775 \\ \hline \end{array}$$

a. 5109
b. 5099
c. 5101
d. none of these

12. Subtract.

$$\begin{array}{r} 30042 \\ -15673 \\ \hline \end{array}$$

a. 14,369
b. 14,379
c. 14,469
d. none of these

2
Multiplication

Multiplication–basic facts

$3 \times 5 = ?$

You can find the **product** of two **factors** by laying out objects and counting:

3 sets of 5

or by adding:

$5 + 5 + 5$

or by skip counting.

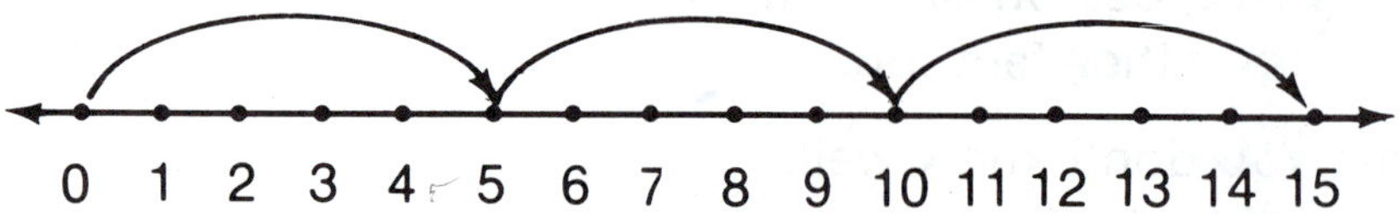

But if you want to do well in math, you must memorize the basic multiplication facts.

$3 \times 5 = ?$

These exercises can help you review and memorize the basic multiplication facts.

EXERCISES

1. Make a multiplication fact table.

X	0	1	2	3	4	5	6	7	8	9
0										
1										
2										
3						15				
4										
5										
6										
7										
8										
9										

The fact $3 \times 5 = 15$ has been written in the table.

2. Cross out all the products that you know *well*.
These properties help you to know some facts.

The Multiplying by 1 Property The product of any number and 1 is the number.	$6 \times 1 = 6$
The Multiplying by 0 Property The product of any number and 0 is 0.	$6 \times 0 = 0$

3. Now you should have just a few facts not crossed off. This property can also help.

The Commutative (Order) Property of Multiplication Changing the order of the factors does not change the product.	$3 \times 5 = 5 \times 3$

If you learn $7 \times 8 = 56$, you will also know $8 \times 7 = 56$. When you can cross off one of these facts, you can cross off the other fact too.

4. Study the few facts that you don't know well. Here is another property of multiplication which you will use later.

The Associative (Grouping) Property of Multiplication Changing the grouping of the factors does not change the product.	$(2 \times 3) \times 4 = 2 \times (3 \times 4)$

Multiply.

5. 7×7	**6.** 8×3	**7.** 9×3	**8.** 8×8	**9.** 7×6	**10.** 6×6	**11.** 9×8	**12.** 6×4
13. 8×2	**14.** 5×5	**15.** 8×0	**16.** 7×5	**17.** 9×9	**18.** 8×7	**19.** 7×2	**20.** 6×3
21. 9×4	**22.** 9×7	**23.** 9×1	**24.** 7×4	**25.** 7×3	**26.** 6×5	**27.** 9×5	**28.** 6×2
29. 4×4	**30.** 8×4	**31.** 5×4	**32.** 9×2	**33.** 9×6	**34.** 8×6	**35.** 5×3	**36.** 8×5

Multiplication—basic facts

See how long it takes you to write these products.

1. 7×5	2. 5×5	3. 8×6	4. 9×6	5. 5×3	6. 7×6	7. 7×8	8. 9×1
9. 6×6	10. 4×1	11. 8×5	12. 8×4	13. 6×2	14. 7×4	15. 9×7	16. 4×9
17. 7×2	18. 3×3	19. 9×8	20. 6×8	21. 3×9	22. 5×4	23. 9×5	24. 3×8
25. 3×2	26. 5×9	27. 4×3	28. 2×2	29. 3×7	30. 4×2	31. 9×0	32. 6×7
33. 6×3	34. 8×8	35. 4×4	36. 9×9	37. 8×2	38. 5×8	39. 9×2	40. 7×9
41. 5×2	42. 7×3	43. 9×4	44. 8×7	45. 8×3	46. 9×3	47. 6×5	48. 7×7
			49. 8×9	50. 6×4			

51. Check your work. The answers are on page 37.

52. Make two bar graphs like these. Record your score and your time.

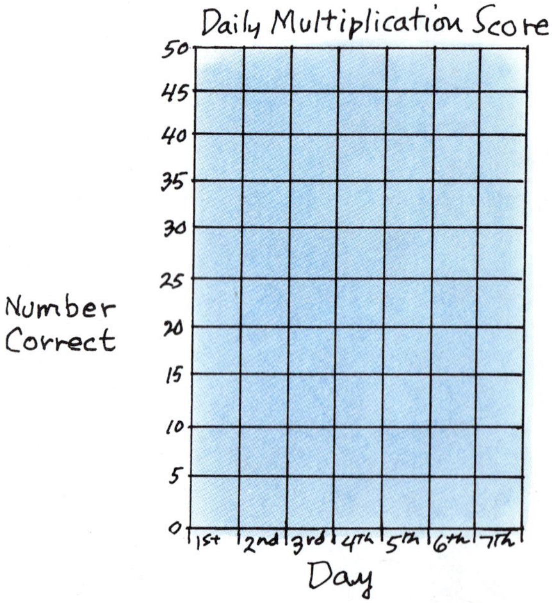

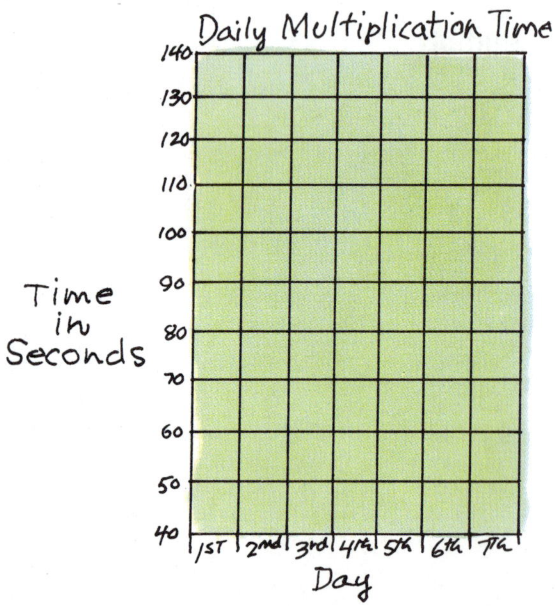

53. Study the facts that you got wrong.

54. Take the speed drill daily for several days.
Keep your results on your graphs.
See if you improve.

Find the missing factors.

55. $7 \times \underline{?} = 21$	56. $\underline{?} \times 8 = 56$	57. $\underline{?} \times 7 = 35$
58. $9 \times \underline{?} = 81$	59. $\underline{?} \times 7 = 42$	60. $9 \times \underline{?} = 63$
61. $6 \times \underline{?} = 36$	62. $4 \times \underline{?} = 16$	63. $\underline{?} \times 5 = 25$
64. $8 \times \underline{?} = 48$	65. $9 \times \underline{?} = 54$	66. $\underline{?} \times 8 = 64$
67. $7 \times \underline{?} = 28$	68. $5 \times \underline{?} = 20$	69. $8 \times \underline{?} = 32$

Compute. Watch the signs.

70. $(2 \times 3) \times 4$	71. $(2 + 3) \times 4$	72. $(2 \times 3) + 4$
73. $(5 + 3) \times 2$	74. $5 + (3 \times 2)$	75. $(5 - 3) \times 2$
76. $6 \times (4 - 3)$	77. $6 \times (4 + 3)$	78. $(6 \times 4) - 3$

Complete. Multiply across. Multiply down.

79.
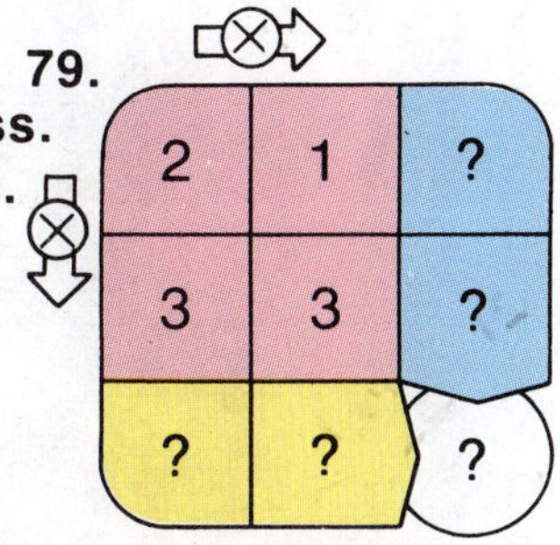

80.
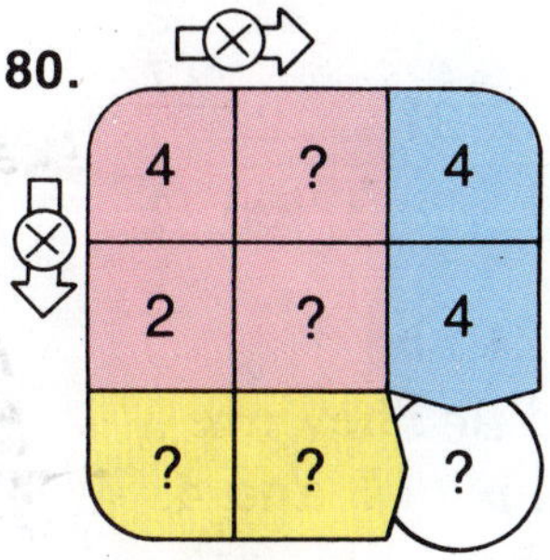

81.
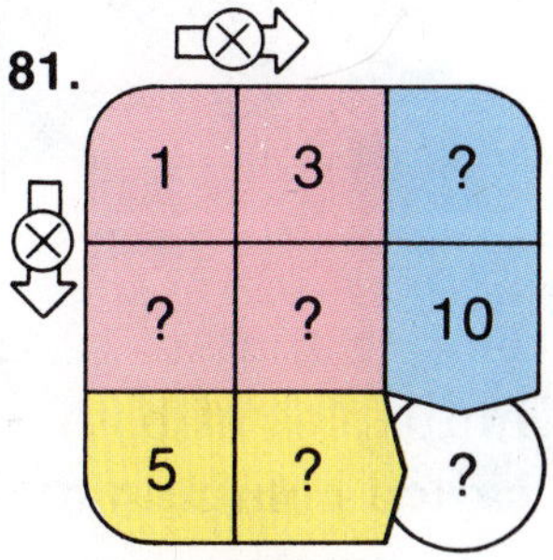

82.
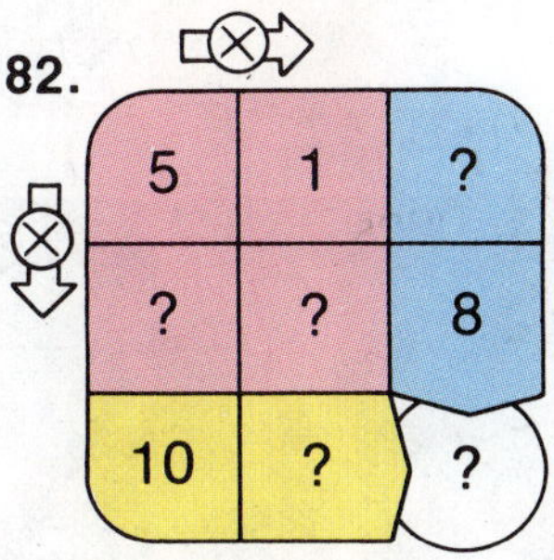

83.
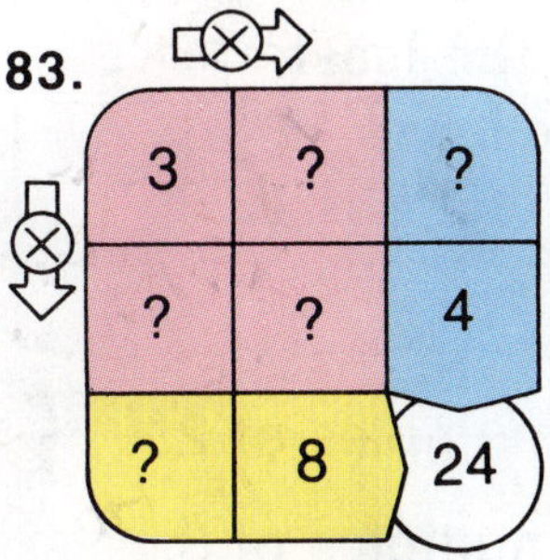

84.
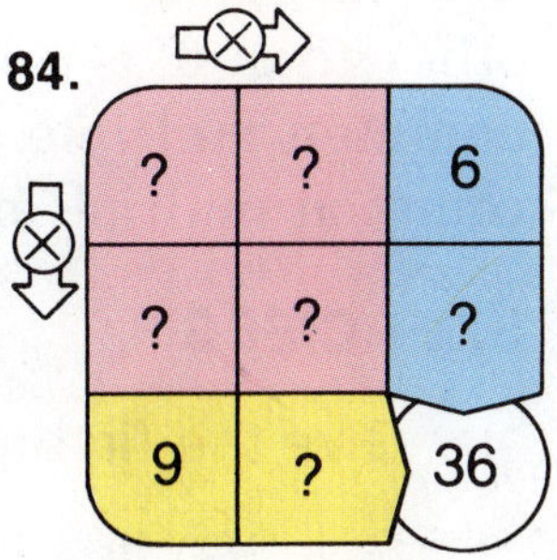

Answers.

1. 35	2. 25	3. 48	4. 54	5. 15	6. 42	7. 56	8. 9	9. 36	10. 4
11. 40	12. 32	13. 12	14. 28	15. 63	16. 36	17. 14	18. 9	19. 72	20. 48
21. 27	22. 20	23. 45	24. 24	25. 9	26. 45	27. 12	28. 4	29. 21	30. 8
31. 0	32. 42	33. 18	34. 64	35. 16	36. 81	37. 16	38. 40	39. 18	40. 63
41. 10	42. 21	43. 36	44. 56	45. 24	46. 27	47. 30	48. 49	49. 72	50. 24

Multiples and common multiples

The products shown in yellow are **multiples** of 3. 10 is not a multiple of 3 because there is no whole number that can be multiplied by 3 to give a product of 10.

X	0	1	2	3	4	5	6	7	8	9
0	0	0	0	0	0	0	0	0	0	0
1	0	1	2	3	4	5	6	7	8	9
2	0	2	4	6	8	10	12	14	16	18
3	0	3	6	9	12	15	18	21	24	27
4	0	4	8	12	16	20	24	28	32	36

Look at the multiplication table shown above. What multiples of 4 are listed in the table?

3	0	3	6	9	12	15	18	21	24	27
4	0	4	8	12	16	20	24	28	32	36

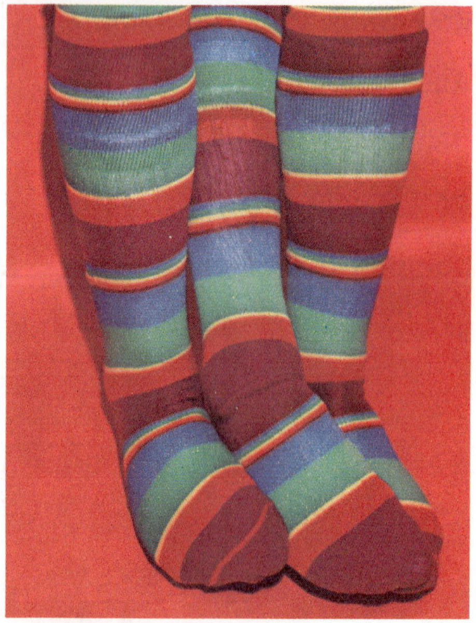

Notice that 0, 12, and 24 are multiples of both 3 and 4. They are called **common multiples** of 3 and 4. The smallest common multiple, other than 0, is called the **least common multiple.** The least common multiple of 3 and 4 is 12.

EXERCISES

1. Give the first ten multiples of 2.
2. Give the first ten multiples of 3.
3. Give the first two common multiples of 2 and 3.
4. Give the least common multiple of 2 and 3.

True or false?

5. 30 is a multiple of 5.

6. 30 is a multiple of 6.

7. 30 is a common multiple of 5 and 6.

8. 30 is the least common multiple of 5 and 6.

9. 7 is a multiple of 21.

10. 18 is a multiple of 3.

11. 18 is a multiple of 5.

12. 18 is a common multiple of 3 and 5.

13. 18 is the least common multiple of 3 and 5.

Give the least common multiple of each pair of numbers.

14. 3, 4 **15.** 3, 6 **16.** 4, 5 **17.** 4, 8 **18.** 2, 6 **19.** 2, 7 **20.** 3, 8 **21.** 7, 8

There are some simple rules that tell whether one number is a multiple of another number. You probably already know this one.

Multiples of 2

If the digit in the ones place is a multiple of 2, the number is a multiple of 2.

16 is a multiple of 2 because 6 is a multiple of 2.

58 is a multiple of 2 because 8 is a multiple of 2.

Are these numbers multiples of 2?

22. 10 **23.** 18 **24.** 37 **25.** 94 **26.** 625 **27.** 10034

28. What is a rule for multiples of 5?

Are these numbers multiples of 5?

29. 15 **30.** 40 **31.** 42 **32.** 85 **33.** 2760 **34.** 5043

★35. Here are some multiples of 3: 6, 15, 27, 81, 252, 681

For each number, add the digits. How are the sums related to 3? Make a rule for multiples of 3.

Multiplying by a 1-digit number

Remember that you multiply one place at a time.

Step 1.
Multiply to find how many ones.

$$\begin{array}{r} 24 \\ \times\ 3 \\ \hline \end{array}$$

12 ones

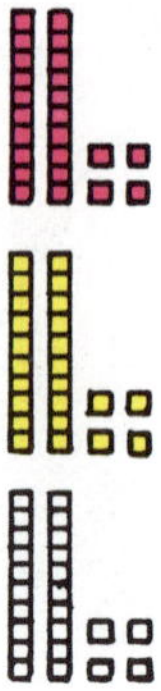

Step 2.
Regroup.

$$\begin{array}{r} ^{1} \\ 24 \\ \times\ 3 \\ \hline 2 \end{array}$$

12 ones = 1 ten + 2 ones

Step 3.
Multiply and add to find how many tens.

$$\begin{array}{r} ^{1} \\ 24 \\ \times\ 3 \\ \hline 72 \end{array}$$

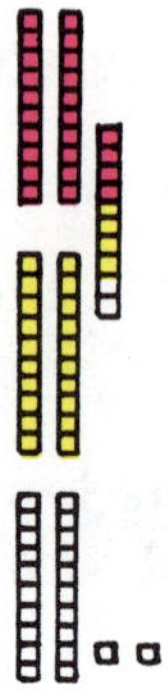

Other examples.

$$\begin{array}{r} ^{1} \\ 524 \\ \times 4 \\ \hline 2096 \end{array} \qquad \begin{array}{r} ^{23} \\ 735 \\ \times 6 \\ \hline 4410 \end{array} \qquad \begin{array}{r} ^{71} \\ 5821 \\ \times 9 \\ \hline 52,389 \end{array}$$

EXERCISES

Multiply.

1. 23 × 2	2. 41 × 2	3. 32 × 3	4. 22 × 4	5. 43 × 2
6. 522 × 3	7. 831 × 2	8. 942 × 2	9. 210 × 7	10. 401 × 9
11. 57 × 3	12. 85 × 4	13. 67 × 8	14. 95 × 6	15. 27 × 3
16. 48 × 2	17. 40 × 3	18. 97 × 8	19. 84 × 7	20. 28 × 5

21. 342 ×5

22. 235 ×6

23. 481 ×7

24. 305 ×6

25. 528 ×8

26. 673 ×9

27. 341 ×2

28. 408 ×9

29. 325 ×8

30. 679 ×8

31. \$5.98 ×3

32. \$6.05 ×4

33. \$2.24 ×7

34. \$2.25 ×4

35. \$8.98 ×6

36. 2531 ×3

37. 3817 ×4

38. 4928 ×6

39. 5873 ×7

40. 8314 ×9

41. 3826 ×8

42. 5934 ×5

43. 7568 ×6

44. 2351 ×9

45. 7482 ×7

46. 82106 ×7

47. 79238 ×9

48. 59382 ×5

49. 632541 ×6

50. 590367 ×8

Solve.

51. How many days in 24 weeks?

52. How many feet in 275 yards?

53. How many quarts in 1120 gallons?

54. How many pints in 1256 gallons?

Find the end number.

55.

Practice exercises

Multiply.

1. 133×3	2. 213×4	3. 308×6	4. 484×2	5. 467×8
6. 845×5	7. 691×3	8. 847×4	9. 378×6	10. 539×7
11. 5628×8	12. 1793×6	13. 2593×3	14. 6657×8	15. 3555×9
16. 4768×2	17. 5681×5	18. 8509×8	19. 3276×6	20. 7225×5
21. 14835×6	22. 28593×4	23. 49148×7	24. 78265×9	25. 95362×3
26. 236858×5	27. 392161×8	28. 774304×7	29. 259903×6	30. 560794×4

31. 794×5
32. 378×6
33. 943×9
34. 401×4
35. 299×7
36. 753×8
37. 8216×6
38. 4375×9
39. 6068×5
40. 3928×6
41. 7628×3
42. 5274×7
43. 2923×7
44. 8061×4
45. 6324×9
46. 29167×8
47. 38296×6
48. 59274×5
49. 13059×6
50. 68957×5
51. 39648×7
52. 431795×8
53. 708612×6
54. 259803×9

Remember that these steps can help you solve problems.

1. Read the problem and find the question.
2. What are the facts?
3. Decide what to do.
4. Answer the question.
5. Estimate! Does your answer seem right?

In the exercises that follow you will be asked to think through the first three steps and decide what to do.

EXERCISES

The numbers have been covered. Tell whether you would add, subtract, or multiply to solve the problem.

55. Carl ran ■ kilometers. Each kilometer took him ■ minutes. How many minutes did he run?

56. Julie ran ■ kilometers in the morning and ■ kilometers in the afternoon. How far did she run?

57. Angela ran 1 kilometer in ■ minutes and another kilometer in ■ minutes. How long did it take her to run the 2 kilometers?

58. Sam ran ■ kilometers and Luis ran ■ kilometers. How many kilometers farther did Sam run?

59. The school record for the high jump is ■ centimeters. Beth missed the record by ■ centimeters.. How high did Beth jump?

60. The old school record for the long jump was ■ centimeters. Juan bettered the old record by ■ centimeters. How far did Juan jump?

61. Jim ran his first dash in ■ seconds. He ran his second dash in ■ seconds. By how much did he improve his time in the second dash?

62. To warm up, Susan ran ■ laps. Each lap was ■ meters. How far did she run?

Multiplying by 10, 100, or 1000

There are 12 cards in each package. There are 10 packages. There are 120 cards in all.

$$\begin{array}{r} 12 \\ \times 10 \\ \hline 120 \end{array}$$

The same kind of shortcut works for multiplying by 100 and by 1000.

$$\begin{array}{r} 438 \\ \times 100 \\ \hline 43{,}800 \end{array}$$

$$\begin{array}{r} 596 \\ \times 1000 \\ \hline 596{,}000 \end{array}$$

EXERCISES

Multiply.

1. 57×10
2. 38×10
3. 26×10
4. 90×10
5. 83×10
6. 125×10
7. 256×10
8. 734×10
9. 825×10
10. 674×10
11. 36×100
12. 58×100
13. 92×100
14. 47×100
15. 85×100
16. 158×100
17. 275×100
18. 346×100
19. 468×100
20. 692×100

21. 57×1000 **22.** 82×1000 **23.** 93×1000 **24.** 60×1000 **25.** 94×1000

26. 772×1000 **27.** 864×1000 **28.** 953×1000 **29.** 675×1000 **30.** 558×1000

$$\begin{array}{r} 1000 \\ \times 100 \\ \hline 100{,}000 \end{array}$$

Same number of zeros

31. 10×10 **32.** 100×10 **33.** 1000×10

34. 100×100 **35.** 100×1000 **36.** 10000×100 **37.** 10000×1000 **38.** 100000×1000

Complete.

39. 1 centimeter = 10 millimeters

centimeters	8	17	58	154	793	1246
millimeters						

40. 1 kilogram = 1000 grams

kilograms	7	24	96	245	700	2425
grams						

Solve.

41. J. P. Morgan used to hand out dimes to poor people he met on the street. If he changed a $10 bill into dimes, how many dimes would he get?

42. Ron Laird once set an American record of 1 hour, 35 minutes, 25.8 seconds in the 20-kilometer walking race. A kilometer is 1000 meters. How many meters was the race?

Keeping Skills Sharp

1. $359 + 260$ **2.** $894 + 765$ **3.** $938 + 297$ **4.** $568 + 774$ **5.** $937 + 826$

6. $8592 + 7468 + 3142$ **7.** $5926 + 3174 + 2835$ **8.** $6934 + 1782 + 9346$ **9.** $5674 + 9283 + 1562$ **10.** $4695 + 3821 + 2746$

Multiplying by multiples of 10, 100, or 1000

You have learned these two multiplication skills.

Skill 1. Multiplying by a 1-digit number.

Skill 2. Multiplying by 10, 100, or 1000.

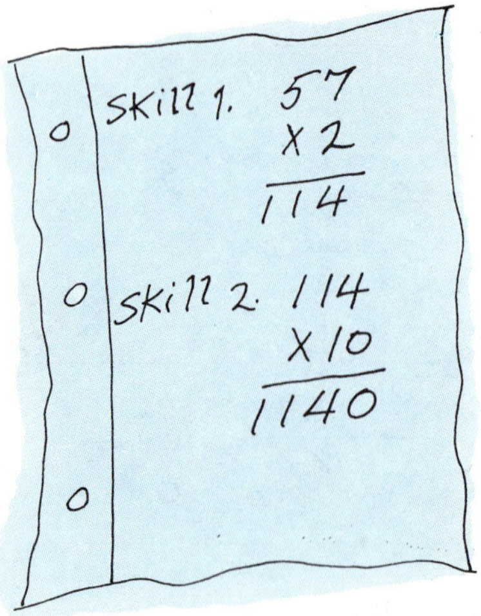

If you put these two skills together, you can multiply by multiples of 10, 100, or 1000.

To multiply a number by 20, you can first multiply the number by 2 and then multiply the answer by 10.

When you do this you are using the associative (grouping) property of multiplication.

$57 \times (2 \times 10) = (57 \times 2) \times 10$

Other examples.

$$\begin{array}{r} 621 \\ \times 30 \\ \hline 18{,}630 \end{array} \qquad \begin{array}{r} 422 \\ \times 300 \\ \hline 126{,}600 \end{array} \qquad \begin{array}{r} 573 \\ \times 8000 \\ \hline 4{,}584{,}000 \end{array}$$

EXERCISES

Multiply.

1. 58×30
2. 95×20
3. 146×50
4. 234×70
5. 861×90
6. 250×80
7. 347×20
8. 587×60
9. $\$9.84 \times 30$
10. $\$4.62 \times 50$
11. 727×200
12. 548×600
13. 239×800
14. 1346×500
15. 2372×2000

16. 30×50
17. 80×70
18. 700×30
19. 500×90
20. 600×300
21. 600×400
22. 8000×200
23. 9000×4000

Solve.

24. A huge omelet was made with 60 dozen eggs. How many eggs in 60 dozen?

25. 

How much for 15 dozen?

26. 

How much money will she get?

27. How many minutes are there in a day?
28. How many seconds are there in a day?

Multiplying by a 2-digit number

You can combine the skills that you have learned thus far to multiply by a 2-digit number.

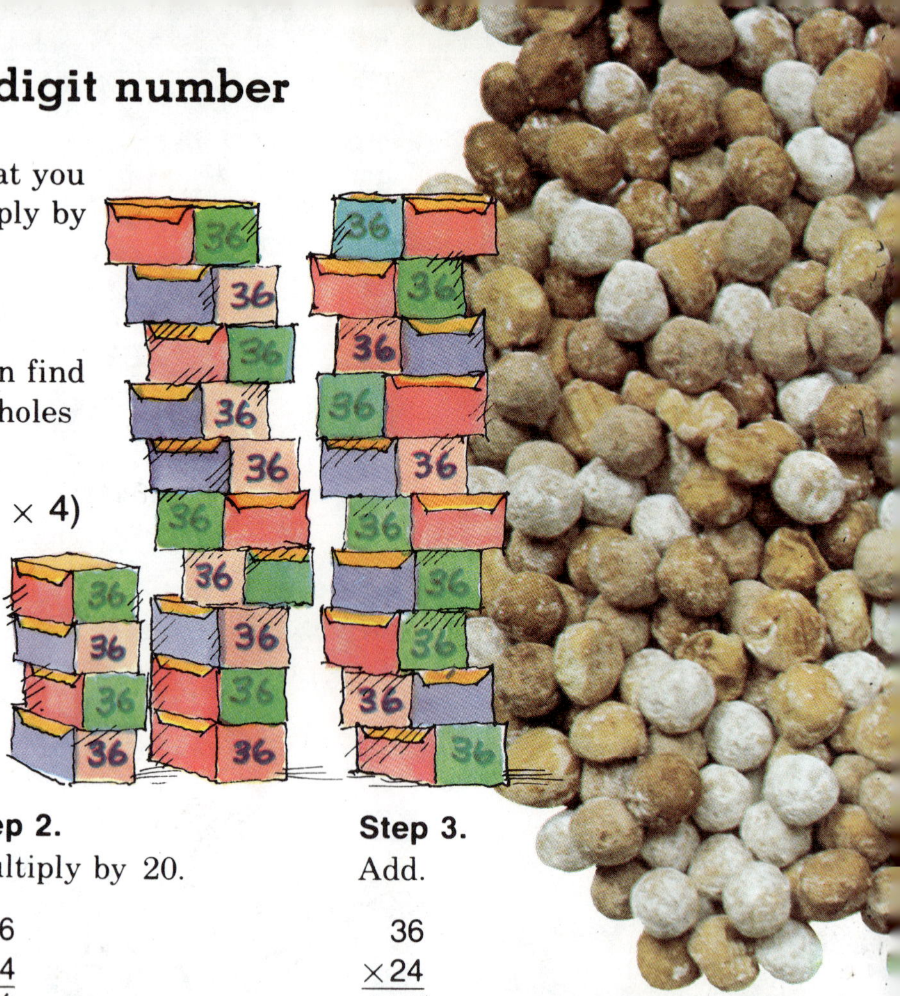

There are 24 boxes with 36 doughnut holes in each. We can find the total number of doughnut holes by multiplying.

$$36 \times 24 = (36 \times 20) + (36 \times 4)$$

[This is an example of the Distributive Property.]

Step 1.
Multiply by 4.

$$\begin{array}{r} 36 \\ \times 24 \\ \hline 144 \end{array}$$

144 ← doughnut holes in 4 boxes

Step 2.
Multiply by 20.

$$\begin{array}{r} 36 \\ \times 24 \\ \hline 144 \\ 720 \end{array}$$

720 ← doughnut holes in 20 boxes

Step 3.
Add.

$$\begin{array}{r} 36 \\ \times 24 \\ \hline 144 \\ 720 \\ \hline 864 \end{array}$$

864 ← doughnut holes in 24 boxes

Other examples.

$$\begin{array}{r} 374 \\ \times 53 \\ \hline 1122 \\ 18700 \\ \hline 19{,}822 \end{array} \qquad \begin{array}{r} 5936 \\ \times 74 \\ \hline 23744 \\ 415520 \\ \hline 439{,}264 \end{array}$$

EXERCISES

Multiply.

1. 78×23
2. 59×42
3. 64×36
4. 70×53
5. 93×40
6. 62×51
7. 58×46
8. 75×38
9. 86×69
10. 92×75

11. 246 × 23	**12.** 387 × 41	**13.** 168 × 22	**14.** 425 × 46	**15.** 238 × 57
16. 529 × 48	**17.** 3526 × 31	**18.** 684 × 13	**19.** 827 × 23	**20.** 596 × 42
21. 864 × 65	**22.** 298 × 56	**23.** 677 × 48	**24.** \$9.23 × 51	**25.** \$6.85 × 36
26. 2592 × 61	**27.** 3684 × 43	**28.** 5644 × 21	**29.** 2897 × 74	**30.** 4294 × 85
31. 3596 × 52	**32.** 7841 × 78	**33.** 5938 × 19	**34.** 61724 × 73	**35.** 53806 × 46

Solve.

36. How many ounces are there in 85 pounds?

37. How many eggs are there in 144 dozen?

38. Jan can average 18 kilometers per hour on her bicycle. At that rate, how far could she ride in 13 hours?

39. A pair of tennis shoes costs \$15.60 and a pair of socks costs \$1.59. What is the total price of a pair of tennis shoes and 3 pairs of socks?

Copy and give the missing digits.

40. ■5■ × 4 = 636

41. ■■3 × 7 = 8■■

42.
```
   537
  ×2■
  ■■■8
 ■■■4■
 ■■■■■
```

Multiplying by larger numbers

Study this example.

421 × 345 2105 ← 5×421 16840 ← 40×421 126300 ← 300×421 145,245	421 × 345 2105 16840 126300 145,245

Other examples.

742 ×526 4452 1484 3710 390,292	3821 ×295 19105 34389 7642 1,127,195	53674 ×1539 483066 161022 268370 53674 82,604,286

EXERCISES

Multiply.

1. 48 × 26	**2.** 37 × 41	**3.** 58 × 63	**4.** 218 × 43	**5.** 516 × 38
6. 472 × 218	**7.** 523 × 412	**8.** 398 × 136	**9.** 473 × 247	**10.** 5143 × 326
11. 438 × 324	**12.** 695 × 118	**13.** 634 × 176	**14.** \$5.82 × 245	**15.** \$6.25 × 335
16. 1389 × 256	**17.** 1256 × 389	**18.** 2756 × 395	**19.** 5812 × 426	**20.** 6805 × 258
21. 1625 × 1245	**22.** 3782 × 1242	**23.** 5916 × 3251	**24.** 3428 × 8162	**25.** 7596 × 5349

To solve a problem:

1. Read the problem and find the question.
2. What are the facts?
3. Decide what to do.
4. Answer the question.
5. Estimate! Does your answer seem right?

Solve these problems.

26. Dave bought 2 records for $5.98 each and 3 comic books for $.50 each. How much did he spend?

27. Mrs. Helder bought 200 shares of stock for $37.25 per share. Later she sold them for $44.50 each. How much more was the total selling price than the total purchase price?

28. 18 busloads of children went to Disneyland. There were 48 students in each of 16 buses and 54 in each of the other two buses. How many students went on the trip?

29. A truck was loaded with 28 crates that weighed 35 kilograms each. At his first stop, the driver unloaded 11 crates. How much weight was left on the truck?

Keeping Skills Sharp

1. $\begin{array}{r} 742 \\ -\ 158 \\ \hline \end{array}$

2. $\begin{array}{r} 296 \\ -\ 159 \\ \hline \end{array}$

3. $\begin{array}{r} 711 \\ -\ 325 \\ \hline \end{array}$

4. $\begin{array}{r} 803 \\ -\ 184 \\ \hline \end{array}$

5. $\begin{array}{r} 600 \\ -\ 375 \\ \hline \end{array}$

6. $\begin{array}{r} 3261 \\ -\ 1589 \\ \hline \end{array}$

7. $\begin{array}{r} 8621 \\ -\ 2759 \\ \hline \end{array}$

8. $\begin{array}{r} 4205 \\ -\ 1759 \\ \hline \end{array}$

9. $\begin{array}{r} 8003 \\ -\ 2749 \\ \hline \end{array}$

10. $\begin{array}{r} 9000 \\ -\ 4661 \\ \hline \end{array}$

Practice and estimating

It is easy to make a careless mistake when multiplying. For example, John forgot to line up the product under the 4 when he multiplied. If he had first estimated the product, he would have known that his answer was wrong.

Here is how he could have estimated the product.

Step 1. Round each factor.

$$\begin{array}{rcr} 327 & \longrightarrow & 300 \\ \times\,403 & \longrightarrow & \times\,400 \\ \hline \end{array}$$

Step 2. Multiply.

$$\begin{array}{rcr} 327 & \longrightarrow & 300 \\ \times\,403 & \longrightarrow & \times\,400 \\ \hline & & 120{,}000 \end{array}$$

From the estimate we know that John's answer is much too small.

EXERCISES

First estimate the product. Then multiply.

1. 19×18	**2.** 22×18	**3.** 32×29
4. 53×29	**5.** 76×31	**6.** 58×49
7. 74×38	**8.** 83×61	**9.** 92×27
10. 69×52	**11.** 48×76	**12.** 95×87

13. 168×12	**14.** 292×29	**15.** 507×48	**16.** 436×92	**17.** 645×34
18. 212×83	**19.** 586×51	**20.** 633×48	**21.** 782×57	**22.** 983×43
23. 1877×643	**24.** 2927×231	**25.** 3826×931	**26.** 2956×998	**27.** 8934×362
28. 392×306	**29.** 468×502	**30.** 534×403	**31.** 674×206	**32.** 921×305
33. 5638×508	**34.** 2614×604	**35.** 9375×4107	**36.** 7618×3004	**37.** 4936×2001

Solve. [To solve these problems you have to estimate the height of the bars.]

38. How many pounds of beef are eaten by each person per year?

39. A family of 4 would eat how much pork in a year? How much chicken?

40. Suppose that the average price of potatoes is 22¢ a pound. What would a family of 3 spend for potatoes in a year?

41. If fresh fruit averages 59¢ a pound, how much would a family of 5 spend for fresh fruit in a year?

42. A family of 6 would eat how many more pounds of fresh vegetables in a year than fresh fruit?

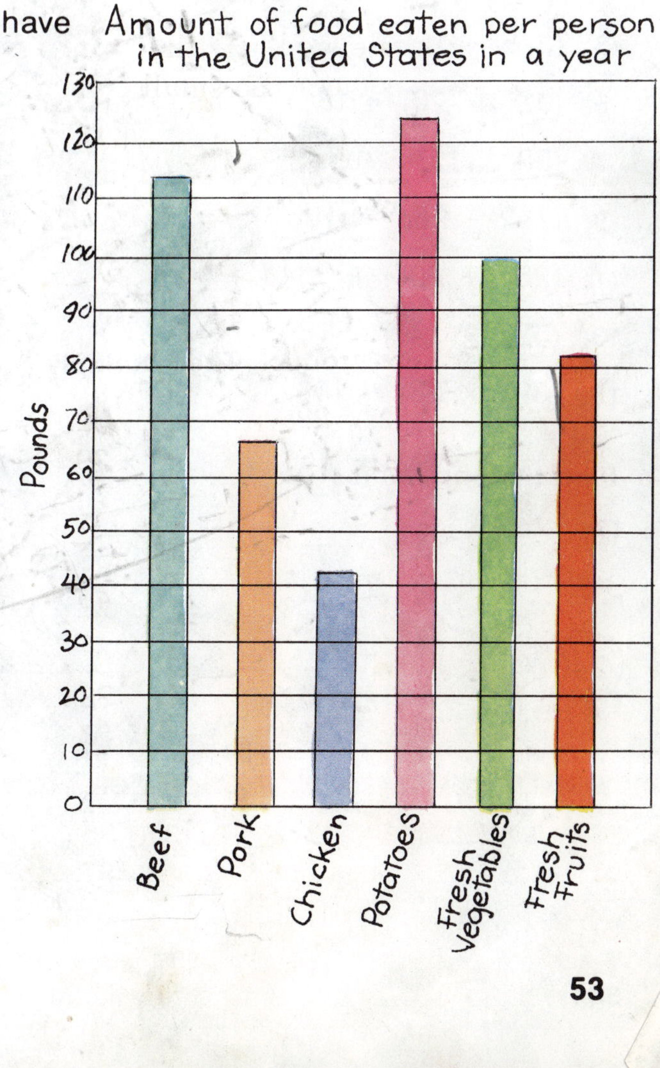

Practice exercises

Always work inside the grouping symbols first.

$726 + (36 \times 15) = 1266$

EXERCISES

Compute.

1. $(96 + 38) \times 24$
2. $96 + (38 \times 24)$
3. $(375 - 12) \times 10$
4. $375 - (12 \times 10)$
5. $(926 \times 20) + 18$
6. $926 \times (20 + 18)$
7. $(38 \times 15) \times 10$
8. $38 \times (15 \times 10)$
9. $(752 - 284) - 106$
10. $752 - (284 - 106)$
11. $(253 + 206) + 184$
12. $253 + (206 + 184)$
13. $(527 \times 142) - 68$
14. $527 \times (142 - 68)$
15. $(397 \times 23) + 156$
16. $397 \times (23 + 156)$

Find the end number.

17.

Start

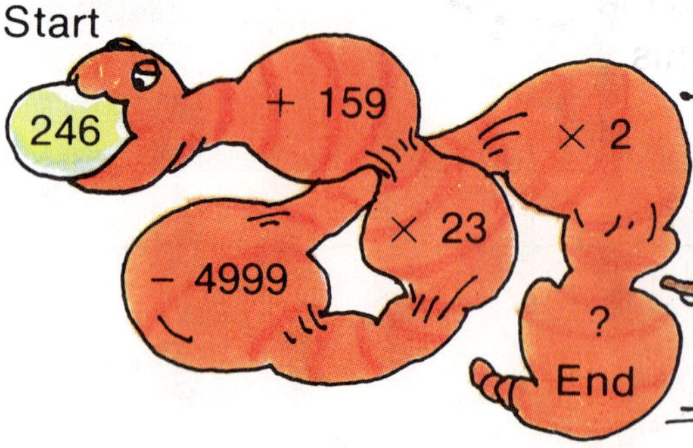

18.

Start

Skill Game

Juan was 105 over the target number, 3000.

Find how close these players came to the target number.

1. Susan (56 × 41) + 183 → 3000
2. Bill (81 × 35) + 614 → 3000
3. Seth (38 × 65) + 411 → 3000
4. Allen (53 × 61) + 148 → 3000
5. Who came closest to the target number?

Play the game.

1. Make a digit card for each of the digits 0 through 9.
2. Choose a game leader.
3. Draw a table like this on your paper. (☐☐ × ☐☐) + ☐☐☐ → 3000
4. The leader picks a card. Each player writes the digit that is on the card in any square of his table.
5. The leader replaces the card.
6. Steps 4 and 5 are repeated until all the squares are filled in.
7. The player who gets closest to the target number wins the game.

Problem solving

Solve.

1. The Smith family and the Wong family drove in two cars from their homes to the amusement park. They drove for 2 hours, averaging 75 kilometers per hour. How far did they drive in the trip to the amusement park and back home?

2. There were 5 in the Wong family and 7 in the Smith family. The admission charge was $6.50 per person. What was the total admission price?

3. A large ferris wheel had a total of 24 cars. Each car could hold 3 people. How many people could ride the ferris wheel at one time?

4. At lunchtime each of the Smiths and Wongs ate 2 hot dogs costing $.80 each and drank a lemonade costing $.45. How much did they spend in all?

5. One year, the average daily attendance at the amusement park was 8520. The park was open 120 days that year. How many people attended the park that year?

6. All the rides were free except the River Boat, which cost an extra $1.50. All but two of the group went on that ride. Mr. Smith bought the tickets with a $20 bill. How much change did he get?

★7. One ride was a boat trip through a haunted cave. There were 16 boats in all. Each boat held 24 people. Each boat made 50 trips a day. How many people could take the haunted-cave ride in a day?

8. On the way home, both cars stopped for gas. One car got 30 liters of gas and the other got 35 liters. A liter cost 31¢. How much did they spend for gas?

CHAPTER CHECKUP

Give the least common multiple. [pages 38–39]

1. 3, 4
2. 3, 6
3. 4, 6
4. 8, 10
5. 6, 9

Multiply. [pages 34–37, 40–42]

6. 24×2
7. 43×3
8. 57×5
9. 28×6
10. 41×8
11. 253×3
12. 375×8
13. 298×7
14. 1586×6
15. 6853×9

Multiply. [pages 34–37, 44–55]

16. 426×12
17. 592×32
18. 627×28
19. 418×36
20. 906×42
21. 3215×152
22. 5261×326
23. 3072×402
24. 8712×801
25. 5436×500

Solve. [pages 43, 56–57]

26. On a coast-to-coast flight an airliner averaged 594 miles per hour. At that rate, how far did the airliner fly in 3 hours?

27. O'Hare International Airport at Chicago is the busiest airport in the world. During one year there were an average of 1757 takeoffs and landings each day. How many takeoffs and landings were there during the year?

Project

How far can your class reach?

1. Measure the arm span in centimeters of each person in your class.
2. Record the arm spans in a table like this one.

Arm span in Centimeters	Number of Students	Total
130	///	390
131	/	131
132	/	132
133	//	266
134	//	268
135	///	405
136	//	272
137	////	548
138		0
139		0
140	~~////~~	700
141		0
142	//	284
143		0
144	/	144
145		0

CLASS TOTAL: ☐

CHAPTER REVIEW

Give the next 3 multiples of the number on the card.

1. [4] 0, 4, 8, ?, ?, ? 2. [5] 0, 5, 10, ?, ?, ? 3. [6] 0, 6, 12, ?, ?, ?

Give 3 common multiples.

4. [3] 0, 3, 6, 9, 12, 15, 18, 21, 24,
   [4] 0, 4, 8, 12, 16, 20, 24

5. [6] 0, 6, 12, 18, 24, 30, 36
   [9] 0, 9, 18, 27, 36

Give the least common multiple.

6. 3, 4 7. 6, 9 8. 6, 8 9. 4, 6 10. 5, 6

$$\begin{array}{r} \scriptstyle 12 \\ 438 \\ \times 3 \\ \hline 1314 \end{array}$$

Multiply.

11. 521×4 12. 683×5 13. 605×8 14. 4213×7

$$\begin{array}{r} 357 \\ \times 100 \\ \hline 35{,}700 \end{array}$$

Multiply.

15. 67×10 16. 180×10 17. 523×100 18. 246×1000

$$\begin{array}{rl} 48 & \\ \times 23 & \\ \hline 144 & \leftarrow 3 \times 48 \\ 960 & \leftarrow 20 \times 48 \\ \hline 1104 & \end{array}$$

Multiply.

19. 73×16 20. 85×34 21. 253×26 22. 738×49

23. 756×138 24. 593×421 25. 637×209 26. 491×306

CHAPTER CHALLENGE

Sometimes we have to use the same factor several times.

Exponent

$$6 \times 6 \times 6 \times 6 = 6^4$$

The exponent 4 tells us how many times the number 6 is used as a factor.

Other examples.

$5 \times 5 \times 5 = 5^3$ $\quad$ $3 \times 3 \times 3 \times 3 \times 3 = 3^5$

Give the exponent.

1. $2 \times 2 \times 2 = 2^?$
2. $3 \times 3 = 3^?$
3. $5 \times 5 \times 5 \times 5 = 5^?$
4. $8 \times 8 \times 8 \times 8 \times 8 = 8^?$
5. $10 \times 10 \times 10 = 10^?$
6. $4 \times 4 \times 4 \times 4 \times 4 \times 4 = 4^?$

Write, using an exponent.

7. $3 \times 3 \times 3 \times 3$
8. 5×5
9. $4 \times 4 \times 4 \times 4 \times 4$
10. $8 \times 8 \times 8$
11. $2 \times 2 \times 2 \times 2 \times 2 \times 2$
12. 6×6

Write the standard numeral.

$2 \times 2 \times 2 = 8$

13. 2^3
14. 2^2
15. 2^4
16. 3^2
17. 1^5
18. 5^2
19. 4^3
20. 5^3
21. 3^3
22. 4^2
23. 2^6
24. 6^2
25. 10^1
26. 10^2
27. 10^3
28. 10^4
29. 10^5
30. 10^6

MAJOR CHECKUP

Standardized Format

Choose the correct letter.

1. The standard numeral for six hundred million seventy-two is

- **a.** 600,720,000
- **b.** 600,000,072
- **c.** 600,072
- **d.** none of these

2. 59,431 rounded to the nearest thousand is

- **a.** 50,000
- **b.** 59,000
- **c.** 60,000
- **d.** none of these

3. Which number is greater than 499,999?

- **a.** 50,000
- **b.** 499,990
- **c.** 409,999
- **d.** none of these

4. Add.

$$\begin{array}{r} 673 \\ 421 \\ +846 \\ \hline \end{array}$$

- **a.** 1840
- **b.** 1830
- **c.** 1940
- **d.** none of these

5. Add.

$$\begin{array}{r} 3574 \\ +8763 \\ \hline \end{array}$$

- **a.** 12,337
- **b.** 11,337
- **c.** 12,347
- **d.** none of these

6. Subtract.

$$\begin{array}{r} 3751 \\ -1884 \\ \hline \end{array}$$

- **a.** 2767
- **b.** 1567
- **c.** 1767
- **d.** none of these

7. Subtract.

$$\begin{array}{r} 5003 \\ -1794 \\ \hline \end{array}$$

- **a.** 3219
- **b.** 3209
- **c.** 4209
- **d.** none of these

8. Which number is a multiple of 3?

- **a.** 32
- **b.** 23
- **c.** 27
- **d.** none of these

9. Multiply.

$$\begin{array}{r} 38 \\ \times 6 \\ \hline \end{array}$$

- **a.** 228
- **b.** 208
- **c.** 188
- **d.** none of these

10. Multiply.

$$\begin{array}{r} 673 \\ \times 10 \\ \hline \end{array}$$

- **a.** 6703
- **b.** 67,300
- **c.** 6730
- **d.** none of these

11. Multiply.

$$\begin{array}{r} 802 \\ \times 42 \\ \hline \end{array}$$

- **a.** 4812
- **b.** 33,684
- **c.** 34,104
- **d.** none of these

12. Multiply.

$$\begin{array}{r} 521 \\ \times 304 \\ \hline \end{array}$$

- **a.** 17,714
- **b.** 3647
- **c.** 158,384
- **d.** none of these

3
Division

Basic division facts

If you have a division problem such as 15 ÷ 3 you can solve it in several ways.

You can lay out 15 things in *groups of 3* and count the number of groups

$15 \div 3 = 5$

3 ↑ number in each group

5 ↑ number of groups

or

you can lay out 15 things in *3 equal groups* and count the number in each group.

$15 \div 3 = 5$

3 ↑ number of groups

5 ↑ number in each group

The best way is to *memorize* the basic division facts. Learning the basic division facts is easy if you know the basic multiplication facts.

$15 \div 3 = \mathbf{5}$ because $\mathbf{5} \times 3 = 15$

Remember: Here is another way to write a division problem.

$$3\overline{)15}$$

5 ← quotient

15 ← dividend

divisor → 3

Remember: Sometimes a division does not come out even.

$$\begin{array}{r} 5R1 \\ 3\overline{)16} \\ -15 \\ \hline 1 \end{array}$$

5R1 ← quotient and remainder

EXERCISES

Have someone time you as you write the quotients.

1. $4\overline{)36}$	**2.** $5\overline{)15}$	**3.** $9\overline{)54}$	**4.** $3\overline{)18}$	**5.** $6\overline{)30}$
6. $7\overline{)49}$	**7.** $9\overline{)72}$	**8.** $8\overline{)40}$	**9.** $7\overline{)28}$	**10.** $5\overline{)40}$
11. $4\overline{)20}$	**12.** $8\overline{)64}$	**13.** $7\overline{)42}$	**14.** $9\overline{)27}$	**15.** $9\overline{)45}$
16. $7\overline{)21}$	**17.** $4\overline{)28}$	**18.** $8\overline{)72}$	**19.** $7\overline{)35}$	**20.** $7\overline{)56}$
21. $6\overline{)42}$	**22.** $8\overline{)24}$	**23.** $3\overline{)12}$	**24.** $9\overline{)63}$	**25.** $4\overline{)24}$
26. $5\overline{)35}$	**27.** $5\overline{)45}$	**28.** $5\overline{)30}$	**29.** $9\overline{)81}$	**30.** $2\overline{)18}$
31. $3\overline{)24}$	**32.** $4\overline{)16}$	**33.** $9\overline{)36}$	**34.** $8\overline{)32}$	**35.** $8\overline{)48}$
36. $8\overline{)16}$	**37.** $8\overline{)56}$	**38.** $1\overline{)9}$	**39.** $3\overline{)27}$	**40.** $3\overline{)21}$
41. $5\overline{)25}$	**42.** $7\overline{)63}$	**43.** $6\overline{)18}$	**44.** $6\overline{)48}$	**45.** $5\overline{)20}$
46. $6\overline{)54}$	**47.** $6\overline{)24}$	**48.** $3\overline{)15}$	**49.** $4\overline{)32}$	**50.** $6\overline{)36}$

51. Check your answers below.

52. Make graphs as you did on page 36.

Divide.

53. $5\overline{)31}$	**54.** $7\overline{)26}$	**55.** $8\overline{)19}$	**56.** $5\overline{)43}$	**57.** $6\overline{)51}$
58. $7\overline{)51}$	**59.** $8\overline{)36}$	**60.** $9\overline{)37}$	**61.** $4\overline{)32}$	**62.** $5\overline{)17}$
63. $6\overline{)50}$	**64.** $8\overline{)48}$	**65.** $9\overline{)44}$	**66.** $8\overline{)44}$	**67.** $7\overline{)44}$
68. $6\overline{)44}$	**69.** $5\overline{)44}$	**70.** $9\overline{)72}$	**71.** $8\overline{)51}$	**72.** $7\overline{)39}$
73. $8\overline{)67}$	**74.** $8\overline{)72}$	**75.** $9\overline{)54}$	**76.** $6\overline{)30}$	

Answers.

1. 9	**2.** 3	**3.** 6	**4.** 6	**5.** 5	**6.** 7	**7.** 8	**8.** 5	**9.** 4	**10.** 8
11. 5	**12.** 8	**13.** 6	**14.** 3	**15.** 5	**16.** 3	**17.** 7	**18.** 9	**19.** 5	**20.** 8
21. 7	**22.** 3	**23.** 4	**24.** 7	**25.** 6	**26.** 7	**27.** 9	**28.** 6	**29.** 9	**30.** 9
31. 8	**32.** 4	**33.** 4	**34.** 4	**35.** 6	**36.** 2	**37.** 7	**38.** 9	**39.** 9	**40.** 7
41. 5	**42.** 9	**43.** 3	**44.** 8	**45.** 4	**46.** 9	**47.** 4	**48.** 5	**49.** 8	**50.** 6

Factors

Since 3 divides 24 evenly, 3 is a **factor,** or **divisor,** of 24.

$$\begin{array}{r} 8 \\ 3\overline{)24} \end{array}$$

5 does not divide 24 evenly. Therefore, 5 is not a factor of 24.

$$\begin{array}{r} 4\ R4 \\ 5\overline{)24} \\ -20 \\ \hline 4 \end{array}$$

Here are all the factors of 24: 1, 2, 3, 4, 6, 8, 12, 24

Here are all the factors of 18: 1, 2, 3, 6, 9, 18

1, 2, 3, and 6 are factors of *both* 18 and 24. They are called **common factors** of 18 and 24. 6 is the **greatest common factor** (greatest common divisor) of 24 and 18.

EXERCISES

Give all factors.

1. 2 **2.** 3 **3.** 4 **4.** 5 **5.** 6 **6.** 7 **7.** 8 **8.** 1

9. 15 **10.** 12 **11.** 35 **12.** 36 **13.** 40 **14.** 50 **15.** 39 **16.** 46

Give all common factors.

17. 6, 8 **18.** 9, 12 **19.** 2, 3 **20.** 5, 10 **21.** 10, 21

Give the greatest common factor.

22. 6, 9 **23.** 15, 24 **24.** 27, 25 **25.** 15, 16 **26.** 35, 42

True or false?

27. 5 is a factor of 15.

28. 15 is a multiple of 5.

29. 6 is a factor of 9.

30. 9 is a multiple of 6.

31. 3 is the greatest common factor of 12 and 18.

32. 36 is the least common multiple of 12 and 18.

33. 0 is a factor of 3. (*Remember:* We don't divide by 0.)

How many factors does each of these numbers have?

1. 8
2. 6
3. 4
4. 9
5. 11
6. 12
7. 14
8. 20
9. 18
10. 15
11. 16
12. 24

Play the game.

1. Divide the class into two teams, team A and team B.
2. Team A gives team B a number from 1 through 99 (for example, 12).
3. Team B gets 1 point for each factor of the number that it gives. (For 12, team B could earn a total of 6 points, 1 point for each factor of 12.)
4. Now team B gives team A a number. (12 cannot be used again.) Team A earns 1 point for each factor it gives.
5. The first team to earn a total of 30 points wins!

Prime numbers

12 can be **factored** into a product of two smaller numbers.

$12 = 2 \times 6 \qquad 12 = 3 \times 4$

These numbers cannot be factored into products of two smaller numbers.

3 5 17

Each number above has only two factors, 1 and the number itself. Such numbers are called **prime numbers**, or **primes.**

2 is a prime number.

7 is a prime number.

8 is *not* a prime number.

Numbers greater than 1 that are not prime numbers are called **composite numbers.**

8 is a composite number.

6 is a composite number.

Each composite number can be factored into a product of prime numbers. This is called the **prime factorization** of the composite number. Study these "factor trees."

```
     12              18
     / \             / \
    4 × 3           3 × 6   ← 6 is not prime
   / \   \         /   / \
  2 × 2 × 3       3 × 2 × 3   ← prime
                                factorization
```

EXERCISES

Copy and complete this table.

	Number	Factors	Prime or composite?
1.	6	1, 2, 3, 6	composite
2.	7		
3.	11		
4.	12		
5.	15		
6.	16		
7.	21		
8.	23		
9.	27		
10.	29		

Copy and complete each factor tree.

11.

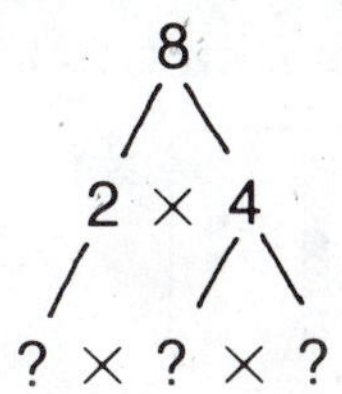

12.

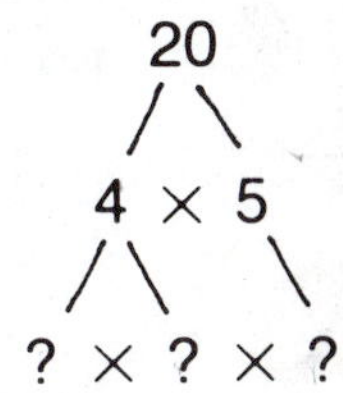

13.

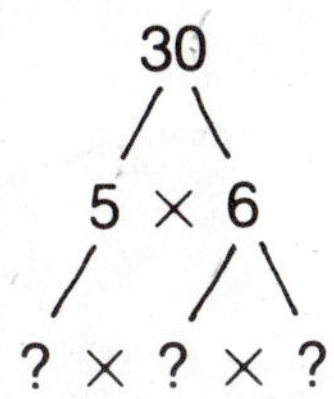

★**14.**

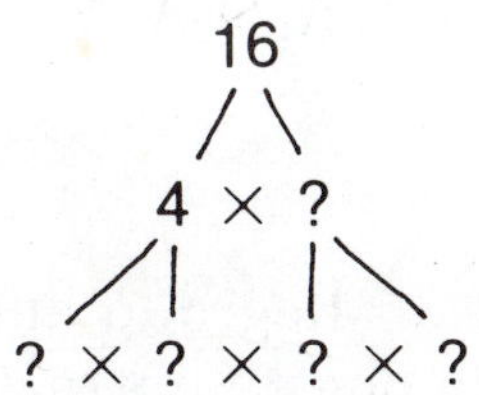

★**15.**

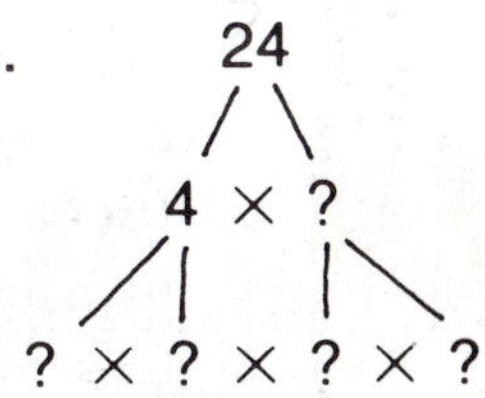

★**16.**

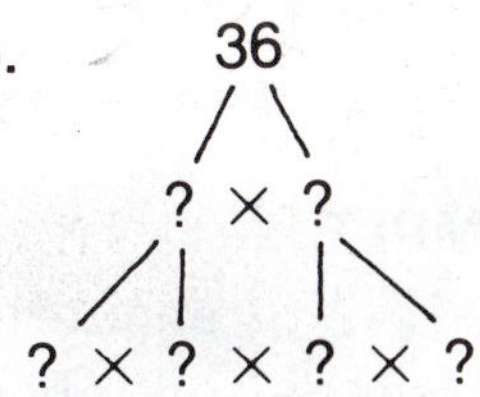

Give the prime factorization.

17. 10 *Answer: 2×5* **18.** 15 **19.** 18 **20.** 20 **21.** 24 **22.** 25

23. 30 **24.** 27 **25.** 28 **26.** 49 **27.** 45 **28.** 42

Dividing by a 1-digit number

Remember that you divide from left to right one place at a time.

EXAMPLE 1.

Step 1. Divide tens.

```
   2
2 )46
  -4
   6
```

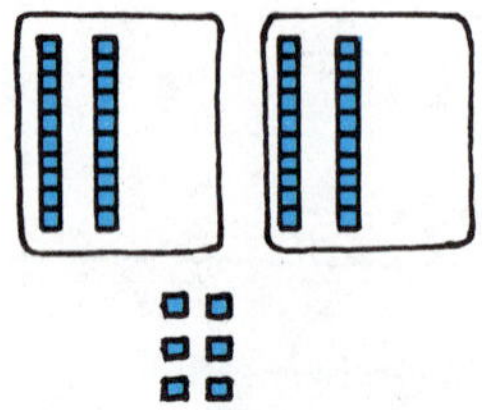

Step 2. Divide ones.

```
   23
2 )46
  -4
   6
  -6
   0
```

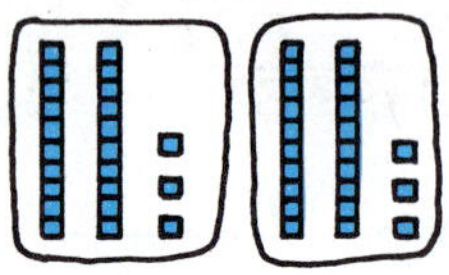

EXAMPLE 2.

Step 1. Divide tens. Subtract.

```
   2
2 )57
  -4
   1
```

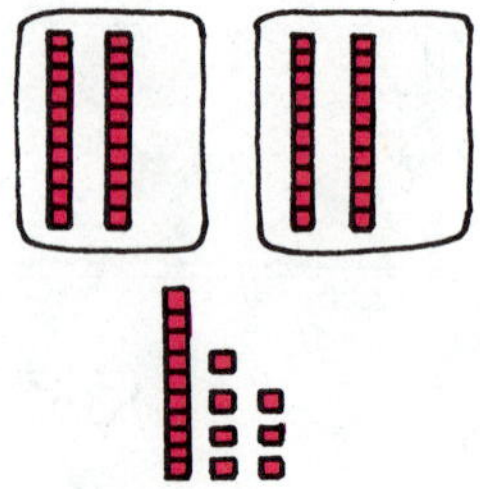

Step 2. Regroup 1 ten for 10 ones.

```
   2
2 )57
  -4
   17
```

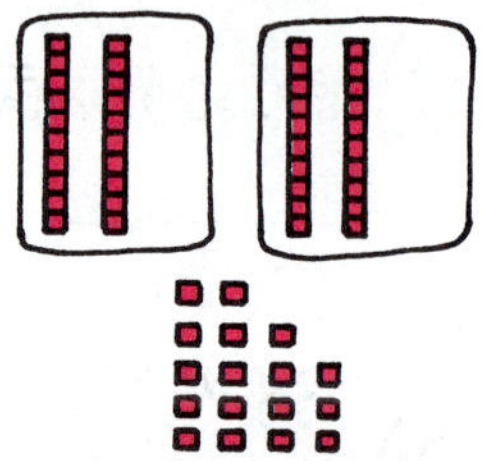

Step 3. Divide ones. Subtract.

```
   28 R1
2 )57
  -4
   17
  -16
    1
```

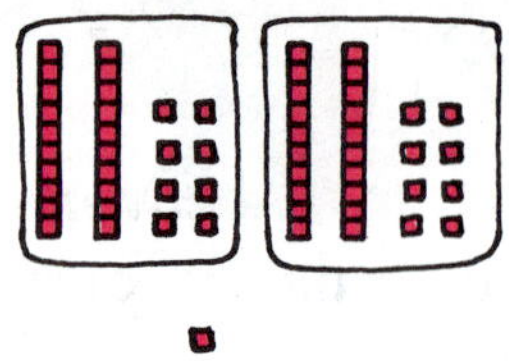

EXAMPLE 3.

Step 1. Not enough hundreds. Think about regrouping 1 hundred for 10 tens.

```
5 )137
```

Step 2. Divide tens. Subtract.

```
    2
5 )137
  -10
    3
```

Step 3. Regroup and divide ones. Subtract.

```
    27R2
5 )137
  -10
    37
   -35
     2
```

EXERCISES

Divide.

1. $2\overline{)64}$	**2.** $3\overline{)63}$	**3.** $2\overline{)48}$	**4.** $4\overline{)48}$	**5.** $3\overline{)39}$
6. $4\overline{)84}$	**7.** $2\overline{)66}$	**8.** $3\overline{)69}$	**9.** $4\overline{)80}$	**10.** $2\overline{)86}$
11. $2\overline{)37}$	**12.** $4\overline{)93}$	**13.** $3\overline{)57}$	**14.** $5\overline{)78}$	**15.** $7\overline{)94}$
16. $8\overline{)93}$	**17.** $2\overline{)39}$	**18.** $3\overline{)42}$	**19.** $3\overline{)81}$	**20.** $2\overline{)80}$
21. $3\overline{)158}$	**22.** $5\overline{)236}$	**23.** $4\overline{)395}$	**24.** $7\overline{)654}$	**25.** $6\overline{)503}$
26. $2\overline{)175}$	**27.** $5\overline{)362}$	**28.** $2\overline{)129}$	**29.** $7\overline{)506}$	**30.** $7\overline{)438}$
31. $8\overline{)652}$	**32.** $3\overline{)253}$	**33.** $3\overline{)148}$	**34.** $5\overline{)438}$	**35.** $9\overline{)621}$
36. $4\overline{)\$5.92}$	**37.** $9\overline{)\$7.56}$	**38.** $4\overline{)\$9.44}$	**39.** $6\overline{)\$7.32}$	**40.** $8\overline{)\$3.92}$

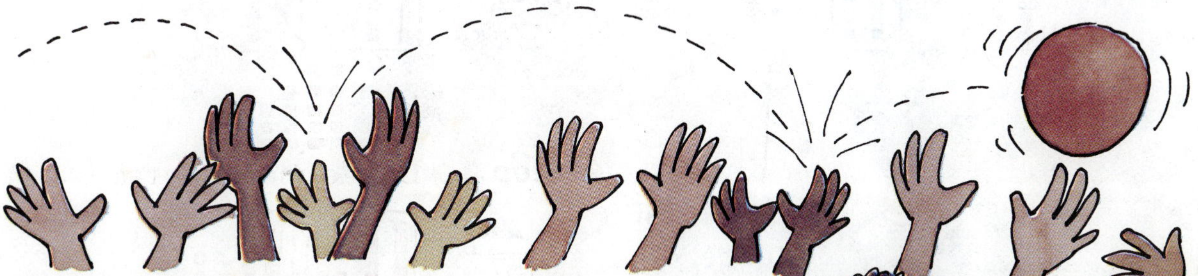

The two numbers are covered. Tell whether you would add, subtract, multiply, or divide the two numbers to answer the question.

41. There are ■ players on ● teams. Each team has the same number of players. How many players are on each team?

42. There are ■ girls playing basketball and ● boys playing baseball. How many more girls are playing basketball?

43. There are only ■ players on a team. There should be ● players on the team. How many more players are needed?

44. There are ■ students on one team and ● students on the other team. How many students are on the two teams?

45. There are ■ teams with ● players on each team. How many players are there?

46. There are ■ players in all. ● players are on each team. How many teams are there?

Dividing larger numbers

There are 3184 red beads to be used in 4 belts. We divide to find the number of red beads for each belt.

```
   796
4)3184
 -28
   38
  -36
    24
   -24
     0
```

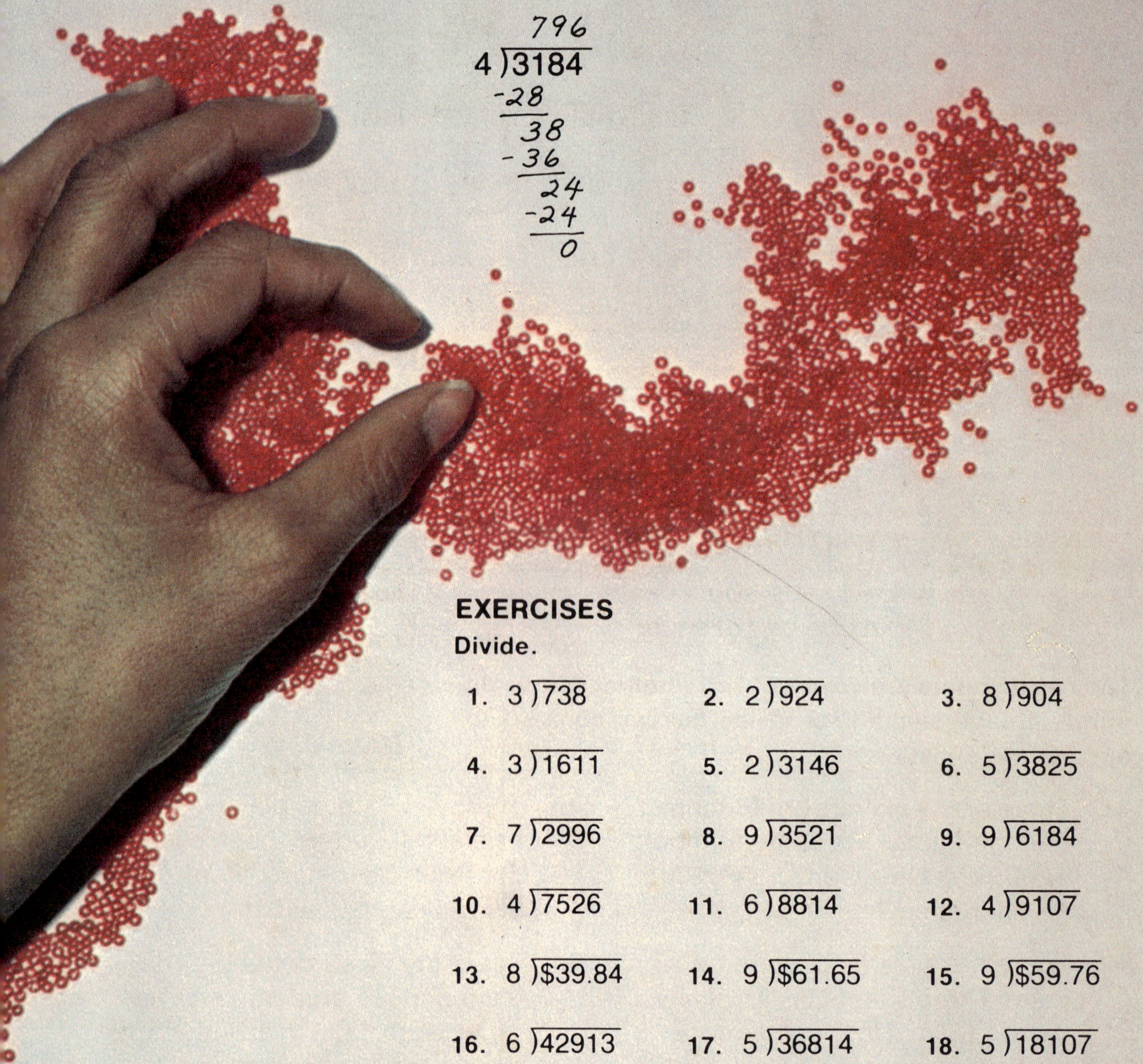

EXERCISES

Divide.

1. 3)738	**2.** 2)924	**3.** 8)904
4. 3)1611	**5.** 2)3146	**6.** 5)3825
7. 7)2996	**8.** 9)3521	**9.** 9)6184
10. 4)7526	**11.** 6)8814	**12.** 4)9107
13. 8)$39.84	**14.** 9)$61.65	**15.** 9)$59.76
16. 6)42913	**17.** 5)36814	**18.** 5)18107
19. 6)67386	**20.** 8)52714	**21.** 7)80375
22. 7)$628.32	**23.** 3)$529.17	**24.** 4)$129.68

Solve.

	Number of beads			
Belt Design	*Red*	*White*	*Blue*	*Yellow*
Thunderbird	632	504	1072	848
Arrow	416	824	792	256
Lightning	326	600	356	528
Star	500	336	500	648

25. How many beads are needed for a Thunderbird belt?

26. How many white beads would be needed to make one belt of each design?

27. Terry has 238 blue beads. How many more are needed to make an Arrow belt?

28. Beth wants to make three lightning belts. She has 852 white beads. How many more does she need?

29. Together Terry, Beth, and Anglice bought a box of yellow beads. There were 4840 beads in the box. If they divided the beads evenly, how many beads should each get? How many beads would be left over?

30. John has 3294 blue beads. Does he have enough beads for two Thunderbird belts and an Arrow belt? Will he have any blue beads left? If so, how many?

31. Mary bought an assortment of 24,000 beads. She wants to use the same number of beads in each of 9 belts. What is the greatest number of beads that she can use in each belt?

Practice and estimating

Karen forgot to write a 0 in the quotient. To find the quotient she should have done this:

Step 1. Divide, subtract, and write the 1 below.

```
   2
3)615
 -6
  1
```

Step 2. Divide.

```
   20
3)615
 -6
  1
```

Step 3. Write the 5 below.

```
   20
3)615
 -6
  15
```

Step 4. Divide.

```
   205
3)615
 -6
  15
 -15
   0
```

Karen would have known that she had made a mistake if she had first estimated the quotient. She could have estimated the quotient like this:

Step 1. Find the first digit of the quotient.

```
3)615   ...   2
            3)615
```

Step 2. Write 0s in the remaining places.

```
3)615   ...   200
            3)615
```

The quotient is near 200.

Study these examples of estimating.

```
5)2617   ...   500
             5)2617
```

The quotient is near 500.

```
4)32297   ...   8000
              4)32297
```

The quotient is near 8000.

EXERCISES

1. Here is a student's paper. Estimate each quotient. Which answers are incorrect according to your estimates?

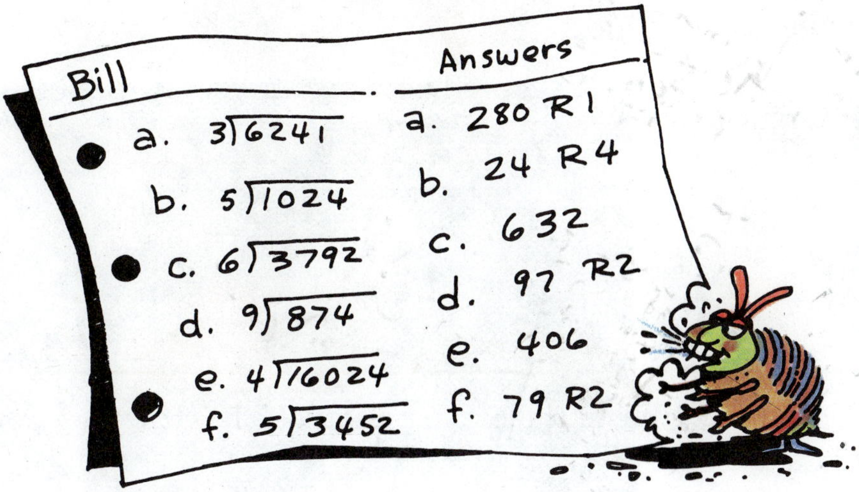

2. Exercise d is incorrect. Did estimating help you find that error? Why or why not?

First estimate the quotient. Then divide.

3. 6)1830	4. 7)2154	5. 8)1754
6. 6)2931	7. 5)1574	8. 9)4563
9. 4)2873	10. 7)5624	11. 3)5937
12. 2)4132	13. 6)3674	14. 8)5218
15. 4)7315	16. 8)6214	17. 5)5938
18. 7)7529	19. 9)63452	20. 4)59386
21. 3)74291	22. 5)76538	23. 5)29614
24. 6)58391	25. 7)42653	26. 6)36748
27. 3)52135	28. 9)27814	29. 7)59634

Keeping Skills Sharp

Multiply.

1. 26×31
2. 47×53
3. 82×28
4. 45×45
5. 45×54
6. 348×27
7. 596×31
8. 842×68
9. 305×48
10. 6012×73

Averages

In 5 games Gerry scored 15, 12, 18, 14, and 16 points. Here is how to find the *average* number of points scored by Gerry in each game.

Step 1. Find the total number of points scored.

$$\begin{array}{r} 15 \\ 12 \\ 18 \\ 14 \\ +16 \\ \hline 75 \end{array}$$

Step 2. Divide the total by the number of games.

$$\begin{array}{r} 15 \\ 5\overline{)75} \\ -5 \\ \hline 25 \\ -25 \\ \hline 0 \end{array}$$

Gerry's average was 15 points per game. If her points had been evenly divided among the games, she would have scored 15 points in each game.

EXERCISES

This table gives some facts for Gerry's team. Use the table to answer the questions below.

Player	Position	Height in cm	Weight in kg	Points scored Game 1	Game 2	Game 3	Game 4	Game 5
Gerry	center	154	39	15	12	18	14	16
Kate	center	156	40	4	6	5	3	2
Alix	forward	149	37	6	5	7	2	6
Mary	forward	148	36	3	2	1	7	0
Karen	forward	146	35	4	5	3	2	4
Ann	guard	138	33	5	3	4	1	5
Laura	guard	138	32	6	2	8	3	5
Amy	guard	136	32	0	0	2	1	0
Nancy	guard	140	31	2	0	4	1	2

1. What is the average height of the team?
2. What is the average weight of the team?
3. How many points did Kate average per game?
4. How many points did Alix average during the first 3 games?
5. What is the average height of the guards?
6. What is the average weight of the forwards?

Solve.

7. The attendance for the first five games was 142, 136, 121, 108, and 122. What was the average attendance? Round your answer to the nearest whole number.
8. A pep club sold popcorn at the first 4 games. At these games they collected $20.50, $18.50, $19.00, and $16.00. What was the average amount collected?

Dividing by 2-digit numbers

The only thing that makes division by a 2-digit divisor difficult is that we usually don't know the multiplication facts for the divisor. If the multiplication facts are given, the division is easy.

35 ×0	35 ×1	35 ×2	35 ×3	35 ×4
0	35	70	105	140

35 ×5	35 ×6	35 ×7	35 ×8	35 ×9
175	210	245	280	315

Step 1. Not enough thousands.

$35\overline{)1194}$

Step 2. Not enough hundreds.

$35\overline{)1194}$

Step 3. Divide tens. Subtract.

```
       3
35 )1194
   -105
     14
```

35 ×3 = 105

Step 4. Regroup and divide ones. Subtract.

```
     34 R4
35 )1194
   -105
     144
    -140
       4
```

35 ×4 = 140

EXERCISES

Divide. Use the multiplication facts given above.

1. $35\overline{)980}$
2. $35\overline{)1190}$
3. $35\overline{)1435}$
4. $35\overline{)1820}$
5. $35\overline{)2385}$
6. $35\overline{)2835}$
7. $35\overline{)7175}$
8. $35\overline{)7455}$
9. $35\overline{)11655}$
10. $35\overline{)12936}$
11. $35\overline{)\$113.40}$
12. $35\overline{)\$203.00}$

Solve.

13. 35 students in each class
455 students in all
How many classes?

14. Jim scored 35 points on each of his spelling tests. He has a total of 385 points. How many tests has he taken?

Divide. Use these multiplication facts.

67	67	67	67	67	67	67	67	67	67
×0	×1	×2	×3	×4	×5	×6	×7	×8	×9
0	67	134	201	268	335	402	469	536	603

15. $67\overline{)1407}$ **16.** $67\overline{)4154}$ **17.** $67\overline{)5427}$ **18.** $67\overline{)21507}$

19. $67\overline{)15477}$ **20.** $67\overline{)8844}$ **21.** $67\overline{)\$209.04}$ **22.** $67\overline{)\$63.65}$

23. $67\overline{)38755}$ **24.** $67\overline{)41783}$ **25.** $67\overline{)33768}$ **26.** $67\overline{)53801}$

Solve.

27. A trucker drove 3484 km in 67 hours. How many km did he average each hour?

28. An orchard owner bought 8500 apple trees. She decided to plant 67 trees in each row. How many full rows could she plant? How many trees would be left over?

Divide. You should know the multiplication facts for these divisions.

29. $40\overline{)673}$ **30.** $40\overline{)1865}$ **31.** $50\overline{)3416}$ **32.** $80\overline{)2051}$

33. $30\overline{)6735}$ **34.** $90\overline{)3041}$ **35.** $60\overline{)2935}$ **36.** $20\overline{)8142}$

★**37.** What is the greatest remainder you can have at any step when you divide by 20? By 70? By 95?

More about dividing

It would be helpful if someone would give us the multiplication facts for any divisor. But it isn't possible. Of course, we could compute the multiplication facts each time we have a division problem. But here is an easier way. Study these examples.

EXAMPLE 1.

Step 1. Round the divisor. Think about dividing by 40. The first digit would be 6.

(40) 42)2753

Step 2. Try 6.

42)2753 — quotient: 6
-252
23

$42 \times 6 = 252$

It works!

Step 3. Regroup. Think about dividing by 40. The next digit would be 5.

(40) 42)2753 — quotient: 6
-252
233

Step 4. Try 5.

42)2753 — quotient: 65 R23
-252
233
-210
23

$42 \times 5 = 210$

It works!

EXAMPLE 2.

Step 1. Round the divisor. Think about dividing by 70. The first digit would be 6.

(70) 66)4639

Step 2. Try 6.

66)4639 — quotient: 6
-396
67

$66 \times 6 = 396$

Too big ↗ 67

Each remainder must be less than the divisor.

Step 3. Try 7.

66)4639 — quotient: 7
-462
1

$66 \times 7 = 462$

It works!

Step 4. Regroup and divide.

66)4639 — quotient: 70 R19
-462
19

EXERCISES

What divisor will you think about?

1. 71)5324 (?)
2. 86)2359 (?)
3. 93)8774 (?)
4. 28)11352 (?)

What will be the first digit that you will try?

5. 31)2381 (30)
6. 47)2988 (50)
7. 59)4235 (60)
8. 62)1433 (60)
9. 48)15523
10. 68)31592
11. 53)28871
12. 85)62351

Divide.

13. 43)258
14. 82)328
15. 59)413
16. 38)304
17. 27)165
18. 48)305
19. 54)400
20. 56)510
21. 41)301
22. 18)103
23. 19)152
24. 49)450
25. 21)315
26. 38)1596
27. 82)2624
28. 92)4148
29. 56)1381
30. 53)2204
31. 58)1550
32. 63)1575
33. 78)2808
34. 92)4073
35. 12)5400
36. 19)1103

Who am I?

37. If you divide me by 36, you get a quotient of 19 and no remainder.
38. If you divide me by 45, you get a quotient of 37 and a remainder of 25.
39. If you multiply me by 56, you get 7952.
40. If you multiply me by 58, you get 12 less than 2100.

Practice and estimating

During 14 years Wilt Chamberlain scored 31,419 points. We can divide to find how many points he averaged each year. But before we divide, let's look at two ways to estimate the quotient.

$14\overline{)31419}$

$$10\overline{)31000} = 3000$$

$$14\overline{)31000} = 2000$$

The less we round the divisor the closer the estimate will be to the quotient. Which estimate do you think is closer?

$$\begin{array}{r} 2244 \\ 14\overline{)31419} \\ -28 \\ \hline 34 \\ -28 \\ \hline 61 \\ -56 \\ \hline 59 \\ -56 \\ \hline 3 \end{array}$$

Rounded to the nearest whole number, his yearly average was 2244 points.

EXERCISES

First estimate the quotient. Then divide.

1. $57\overline{)4640}$	2. $63\overline{)1323}$	3. $21\overline{)8509}$
4. $87\overline{)3654}$	5. $39\overline{)9638}$	6. $53\overline{)5974}$
7. $92\overline{)1926}$	8. $60\overline{)5784}$	9. $81\overline{)31956}$
10. $79\overline{)74261}$	11. $57\overline{)35780}$	12. $83\overline{)92165}$
13. $56\overline{)89763}$	14. $38\overline{)52917}$	15. $62\overline{)58359}$
16. $58\overline{)74693}$	17. $78\overline{)15825}$	18. $94\overline{)63941}$
19. $42\overline{)75061}$	20. $44\overline{)23561}$	21. $51\overline{)42968}$
22. $63\overline{)83961}$	23. $72\overline{)50000}$	24. $93\overline{)63819}$
25. $82\overline{)39216}$	26. $49\overline{)74281}$	27. $69\overline{)39654}$

Solve.

28. During the 9 years that Jim Brown played professional football, he scored 126 touchdowns. How many touchdowns did he average each year? How many points did he score during his career? (A touchdown is worth 6 points.)

29. Y.A. Tittle gained 33,070 yards passing during his 17 years of professional football. How many yards did he average a year? Give your answer to the nearest yard.

30. Ray Berry played professional football for 13 years. During his professional career he caught 631 passes. What was the average number he caught each year? Give your answer to the nearest whole number.

Keeping Skills Sharp

Add.

1.
$$\begin{array}{r} 57 \\ 38 \\ 26 \\ 43 \\ +51 \\ \hline \end{array}$$

2.
$$\begin{array}{r} 83 \\ 59 \\ 64 \\ 28 \\ +16 \\ \hline \end{array}$$

3.
$$\begin{array}{r} 93 \\ 57 \\ 18 \\ 52 \\ +14 \\ \hline \end{array}$$

4.
$$\begin{array}{r} 297 \\ 358 \\ 553 \\ 462 \\ +116 \\ \hline \end{array}$$

5.
$$\begin{array}{r} 8732 \\ 4167 \\ 3218 \\ 5174 \\ +2674 \\ \hline \end{array}$$

Dividing by a 3-digit number

Step 1. Try 6.

$$\begin{array}{r} 6 \\ \boxed{3}21\overline{)\boxed{19}9362} \\ -1926 \\ \hline 67 \end{array}$$

Step 2. Try 2.

$$\begin{array}{r} 62 \\ \boxed{3}21\overline{)199362} \\ -1926 \\ \hline \boxed{6}76 \\ -642 \\ \hline 34 \end{array}$$

Step 3. Try 1.

$$\begin{array}{r} 621 \text{ R21} \\ \boxed{3}21\overline{)199362} \\ -1926 \\ \hline 676 \\ -642 \\ \hline \boxed{3}42 \\ -321 \\ \hline 21 \end{array}$$

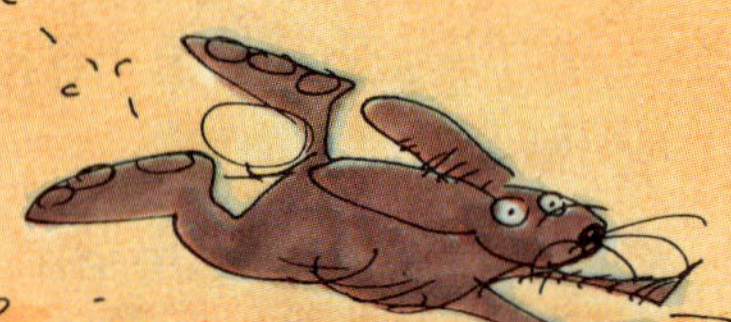

EXERCISES

Divide.

1. $\boxed{2}58\overline{)\boxed{3}1734}$
2. $\boxed{3}22\overline{)\boxed{14}1036}$
3. $\boxed{5}18\overline{)\boxed{12}4838}$
4. $\boxed{3}41\overline{)\boxed{8}7296}$
5. $\boxed{5}56\overline{)\boxed{21}9064}$
6. $\boxed{4}32\overline{)\boxed{26}2656}$
7. $\boxed{5}19\overline{)\boxed{19}2549}$
8. $\boxed{6}23\overline{)\boxed{14}2667}$
9. $\boxed{7}05\overline{)\boxed{26}9310}$
10. $\boxed{6}18\overline{)\boxed{33}3102}$
11. $\boxed{9}22\overline{)\boxed{66}1996}$
12. $\boxed{8}34\overline{)\boxed{60}4650}$

Divide.

13. $518\overline{)592875}$

14. $356\overline{)628374}$

15. $628\overline{)956138}$

16. $477\overline{)756382}$

17. $611\overline{)591765}$

18. $346\overline{)378295}$

19. $578\overline{)469273}$

20. $397\overline{)359683}$

21. $702\overline{)400239}$

22. $835\overline{)657542}$

23. $623\overline{)839617}$

24. $948\overline{)528391}$

25. $495\overline{)742398}$

26. $258\overline{)658371}$

27. $492\overline{)295683}$

28. $748\overline{)496933}$

29. $918\overline{)342608}$

30. $806\overline{)739524}$

31. $759\overline{)520314}$

32. $682\overline{)829613}$

33. $719\overline{)516784}$

34. $942\overline{)296351}$

35. $634\overline{)742978}$

36. $724\overline{)396542}$

Solve.

37. Ms. Allen's car used 146 liters of gasoline on a 1752-kilometer trip. How many kilometers per liter was that?

38. One month Mr. Adams worked 172 hours. He earned a total of $1100.80. How much did he earn per hour?

Find the starting number.

★39.

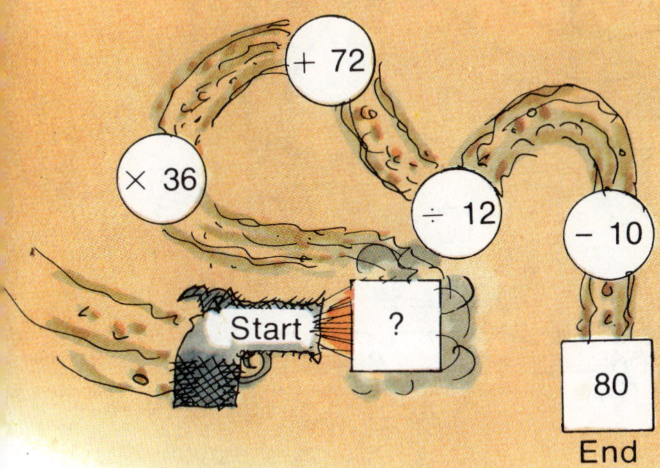

★40.

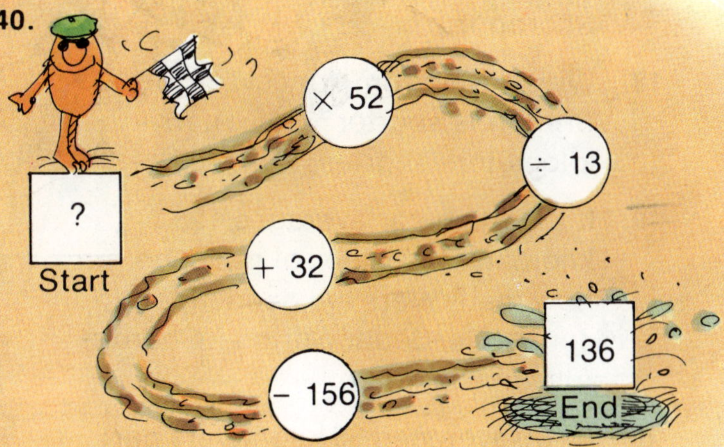

Addition, subtraction, multiplication, and division

Remember to work inside the grouping symbols first.

EXERCISES

Compute.

1. (743 − 258) + 139
2. 743 − (258 + 139)
3. (925 + 607) + 274
4. 925 + (607 + 274)
5. (564 + 42) × 18
6. 564 + (42 × 18)
7. (8192 ÷ 128) ÷ 64
8. 8192 ÷ (128 ÷ 64)
9. (6624 ÷ 138) × 16
10. 6624 ÷ (138 × 16)
11. (853 − 256) − 197
12. 853 − (256 − 197)
13. (567 + 285) × 17
14. 567 + (285 × 17)
15. (1944 ÷ 54) − 27
16. 1944 ÷ (54 − 27)

Solve.

17. How many kilometers is a Jonesville–Cedar Springs–Sharon trip?
18. Which way from Jonesville to Sharon is shorter? How many kilometers shorter?
19. If a certain car averages 12 kilometers per liter of gasoline, how many liters are needed for a Jonesville–Cedar Springs–Sharon–Allison trip?
20. If you averaged 68 kilometers per hour, how long would it take you to drive the "loop"?

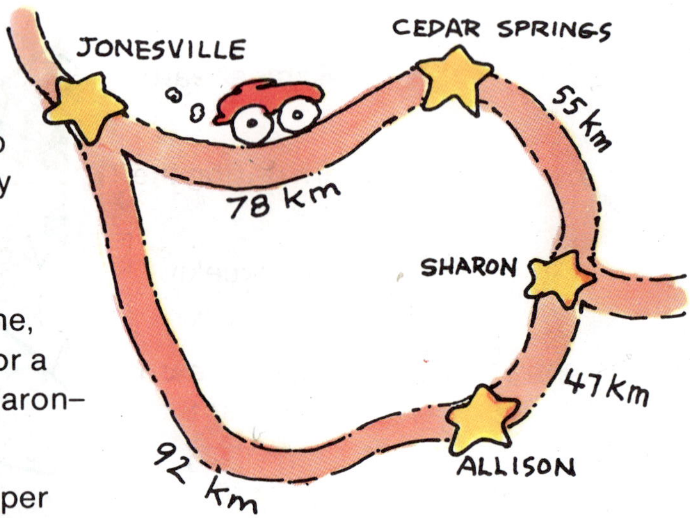

Samuel kept a graph of the money he earned each week on his paper route.

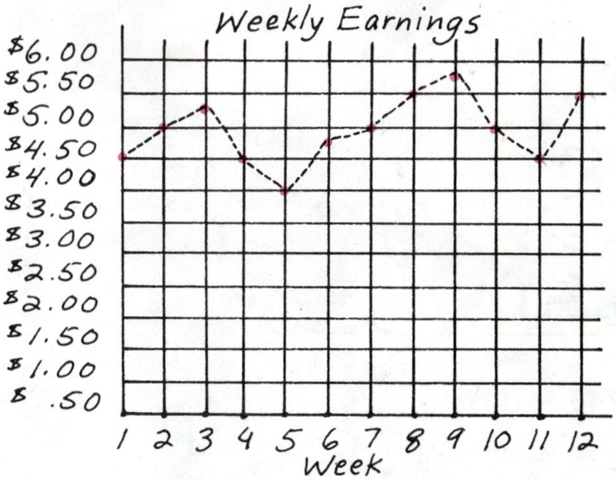

21. In how many weeks did Samuel earn more than $4.50?

22. In how many weeks did Samuel earn less than $5.00?

23. What was his greatest weekly earning?

24. What was his least weekly earning?

25. In how many weeks did he earn $5.25?

26. In which week did Samuel earn $4.75?

27. How much did he earn during the 12 weeks?

28. What were his average weekly earnings?

29. In how many weeks did he earn more than his average? Less than his average?

30. At his average rate, how much would Samuel earn in a year?

Problem solving

Big problems are usually made up of small problems.

1. **a.** A drill team was made up of 5 rows with 5 girls in each row. How many girls were there in all?

   b. The drill team also had 4 alternates who performed if any of the regular marchers were ill. How many members did the drill team have?

c. When the drill team was organized, 56 girls tried out. How many of the girls who tried out did not make the team?

2. One drill team had 5 rows with 12 girls in each row. They also had 4 majorettes and a drum major. How many on the team?

3. To raise money for a trip, a drill team decided to have a cookie sale. They baked 552 cookies and sold them for $.90 a dozen. How much money did they raise?

4. To raise more money they decided to have a car wash. They washed 23 cars for $1.75 each. If their supplies cost $3.46, how much profit did they make?

5. All 65 team members went to a contest. Forty-two traveled on a bus and the rest by car. How many cars were needed if no more than 5 team members rode in one car?

6. They had to travel 256 km to the contest. They started at 7:00 A.M. and averaged 64 km per hour. What time did they arrive?

7. For lunch, each member had a hamburger for $.89, a soft drink for $.35, and french fries for $.60. What was the total price for the 65-member team?

8. For the opening march, all the drill teams were combined into two large teams. One large team had 15 rows with 24 girls in each row. The other large team had 12 rows with 26 girls in each row. How many in all?

CHAPTER CHECKUP

List all common factors of the two numbers.
Underline the greatest common factor. [pages 66–67]

1. 8, 12 **2.** 12, 15 **3.** 7, 9 **4.** 12, 24 **5.** 18, 30

Prime or composite? [pages 68–69]

6. 9 **7.** 13 **8.** 18 **9.** 27 **10.** 31 **11.** 36

Give the prime factorization. [pages 68–69]

12. 6 **13.** 8 **14.** 12 **15.** 18 **16.** 24 **17.** 42

Divide. [pages 64–65, 70–75]

18. $3\overline{)225}$ **19.** $6\overline{)1812}$ **20.** $9\overline{)3715}$ **21.** $8\overline{)51432}$

Divide. [pages 78–86]

22. $24\overline{)1344}$ **23.** $56\overline{)11592}$ **24.** $39\overline{)18563}$ **25.** $82\overline{)75141}$

26. $168\overline{)378219}$ **27.** $321\overline{)592746}$ **28.** $405\overline{)893426}$ **29.** $529\overline{)493826}$

Solve. [pages 76–77, 86–89]

30. How much change should you get from a $20 bill if you buy 2 gallons of paint?

31. If a gallon of paint covers 425 square feet, how many gallons of paint should you buy to paint 2800 square feet?

32. A shipment of 12,000 tires is received at an auto plant. If 5 tires go to each car, how many cars will the shipment supply?

33. A small factory had a weekly payroll of $4716 for its 23 employees. The manager earned $360 per week. What was the average pay of the other employees?

Project

Do you watch more television than your classmates?

1. Take a survey to find out how many minutes of television each of your classmates watched yesterday. Have them round the time to the nearest 30 minutes.
2. Show your results on a bar graph like the one shown here.
3. How many classmates watched more than you did? Less than you did?
4. What was the average number of minutes spent watching television?
5. How many classmates watched more than the average? Watched less than the average?

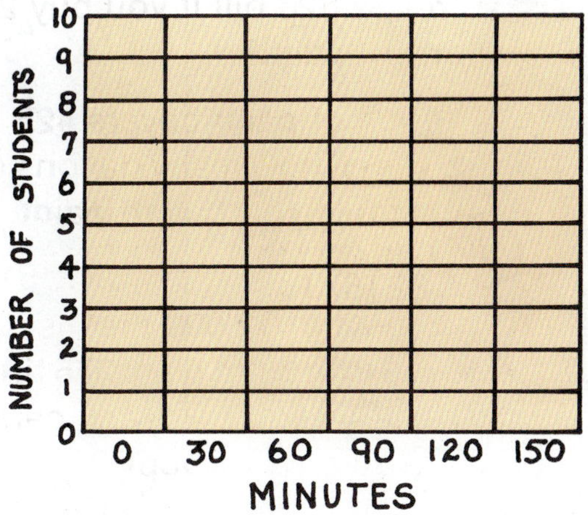

CHAPTER REVIEW

Example:

$$\begin{array}{r} 8\text{ R}2 \\ 3\overline{)26} \\ -24 \\ \hline 2 \end{array}$$

Divide.

1. $8\overline{)43}$ 2. $9\overline{)29}$ 3. $5\overline{)41}$ 4. $2\overline{)17}$

5. $6\overline{)48}$ 6. $7\overline{)35}$ 7. $3\overline{)20}$ 8. $5\overline{)34}$

Example:

$$\begin{array}{r} 229\text{ R}1 \\ 6\overline{)1375} \\ -12 \\ \hline 17 \\ -12 \\ \hline 55 \\ -54 \\ \hline 1 \end{array}$$

Divide.

9. $7\overline{)413}$ 10. $8\overline{)685}$ 11. $4\overline{)916}$

12. $8\overline{)936}$ 13. $4\overline{)816}$ 14. $5\overline{)3295}$

15. $4\overline{)7731}$ 16. $6\overline{)1224}$ 17. $8\overline{)65163}$

Divide.

48	48	48	48	48	48	48	48	48	48
×0	×1	×2	×3	×4	×5	×6	×7	×8	×9
0	48	96	144	192	240	288	336	384	432

Example: (48 × 6 = 288)

$$\begin{array}{r} 6\text{ R}42 \\ 48\overline{)330} \\ -288 \\ \hline 42 \end{array}$$

18. $48\overline{)381}$ 19. $48\overline{)1008}$ 20. $48\overline{)2980}$

21. $48\overline{)9744}$ 22. $48\overline{)2600}$ 23. $48\overline{)28992}$

Divide.

Example: (30)

$$\begin{array}{r} 27\text{ R}9 \\ 32\overline{)873} \\ -64 \\ \hline 233 \\ -224 \\ \hline 9 \end{array}$$

24. $51\overline{)268}$ 25. $62\overline{)5371}$ 26. $88\overline{)7063}$

27. $21\overline{)5314}$ 28. $39\overline{)16761}$ 29. $56\overline{)28743}$

CHAPTER CHALLENGE

Eratosthenes, an ancient Greek mathematician, invented a way to find prime numbers. It is called the Sieve of Eratosthenes.

Follow these steps to make your own sieve.

1. Make a number table as shown.
2. Cross off all multiples of 2 except 2.
3. 3 is the next number that is not crossed off. Cross off all multiples of 3 except 3.
4. What is the next number that is not crossed off? Cross off all multiples of that number except the number itself.
5. Repeat step 4 until no more numbers can be crossed off. The numbers not crossed off are prime numbers.

	2	3	4	5	6	7	8	9	10
11	12	13	14	15	16	17	18	19	20
21	22	23	24	25	26	27	28	29	30
31	32	33	34	35	36	37	38	39	40
41	42	43	44	45	46	47	48	49	50
51	52	53	54	55	56	57	58	59	60
61	62	63	64	65	66	67	68	69	70
71	72	73	74	75	76	77	78	79	80
81	82	83	84	85	86	87	88	89	90
91	92	93	94	95	96	97	98	99	100

Use your sieve to answer the following questions.

6. What is the greatest prime number that is less than 100?
7. How many prime numbers are less than 100?
8. **Twin primes** are prime numbers that have a difference of 2. List the twin primes that are less than 100.

MAJOR CHECKUP

Standardized Format

Choose the correct letter.

1. The standard numeral for eight million eight thousand is
   - a. 8,008,000
   - b. 8,000,008,000
   - c. 8,080,000
   - d. none of these

2. Add.
   $$\begin{array}{r} 6834 \\ +5979 \\ \hline \end{array}$$
   - a. 12,713
   - b. 11,813
   - c. 12,803
   - d. none of these

3. Add.
   $$\begin{array}{r} 58 \\ 27 \\ 34 \\ +16 \\ \hline \end{array}$$
   - a. 125
   - b. 115
   - c. 135
   - d. none of these

4. Subtract.
   $$\begin{array}{r} 4836 \\ -1914 \\ \hline \end{array}$$
   - a. 2122
   - b. 3922
   - c. 2922
   - d. none of these

5. Subtract.
   $$\begin{array}{r} 6005 \\ -1297 \\ \hline \end{array}$$
   - a. 4818
   - b. 5292
   - c. 4708
   - d. none of these

6. The least common multiple of 20 and 30 is
   - a. 1
   - b. 60
   - c. 10
   - d. none of these

7. The greatest common factor of 20 and 30 is
   - a. 1
   - b. 60
   - c. 10
   - d. none of these

8. Multiply.
   $$\begin{array}{r} 1503 \\ \times 6 \\ \hline \end{array}$$
   - a. 9018
   - b. 918
   - c. 90,018
   - d. none of these

9. Multiply.
   $$\begin{array}{r} 461 \\ \times 205 \\ \hline \end{array}$$
   - a. 11,525
   - b. 94,505
   - c. 95,505
   - d. none of these

10. Divide.
    $4\overline{)39}$
    - a. 6 R3
    - b. 10 R1
    - c. 9 R3
    - d. none of these

11. Divide.
    $7\overline{)4163}$
    - a. 594 R5
    - b. 596 R3
    - c. 584 R5
    - d. none of these

12. Divide.
    $58\overline{)2378}$
    - a. 31
    - b. 401
    - c. 41
    - d. none of these

4 Geometry

READY OR NOT ?

Match.

1. point
2. line
3. plane
4. ray
5. segment

a. 

b. 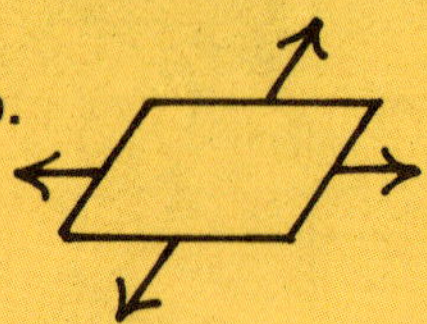

c. •

d. 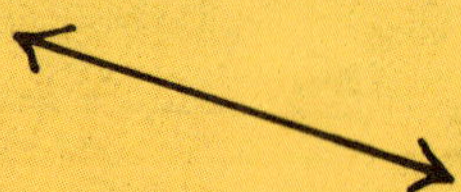

e. 

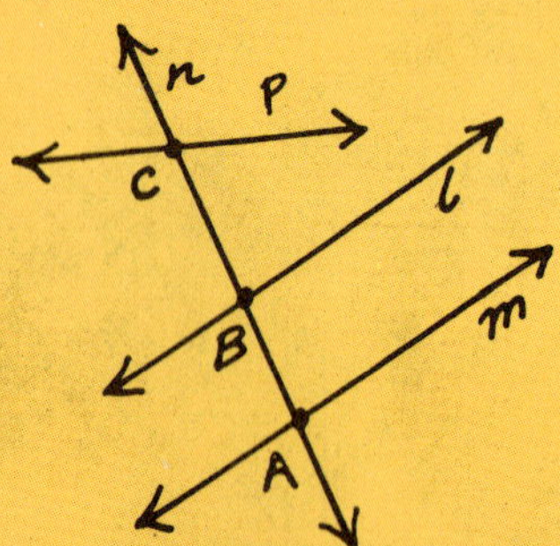

6. Which line intersects line *n* at point *A*?
7. Which line intersects line *n* at point *B*?

Congruent figures

Here are three cookie cutters. They make cookies in the shapes of some geometric figures.

All cookies made by the first cutter are the same size and shape.

Figures that are the same size and shape are called **congruent figures.** Triangle 1 is congruent to triangle 2. A tracing of triangle 1 fits on triangle 2.

Triangle 1 is not congruent to triangle 3.

EXERCISES

These three cookies were made with cookie cutters.

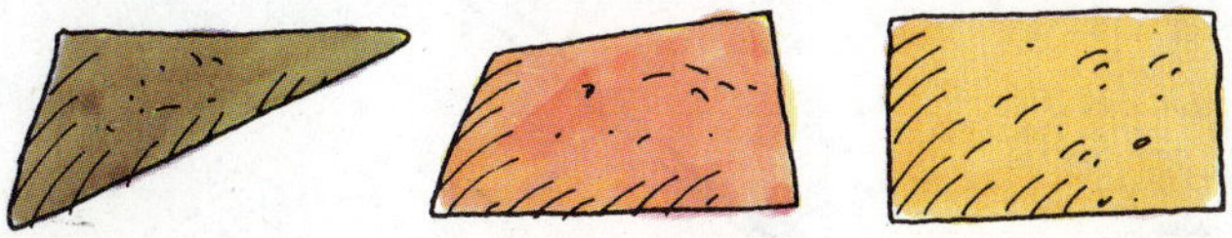

Use tracings to find which cookies below were cut by the same cutters.

1. 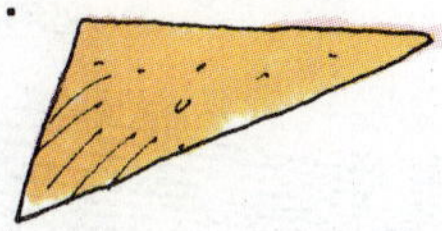

2. 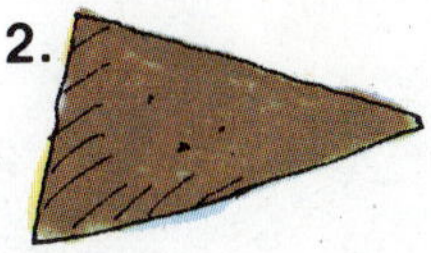

3. 

4. 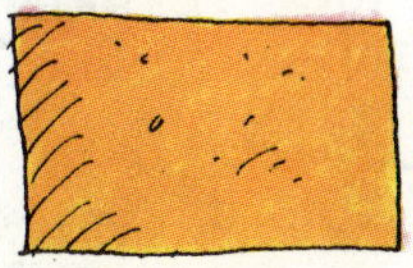

5. 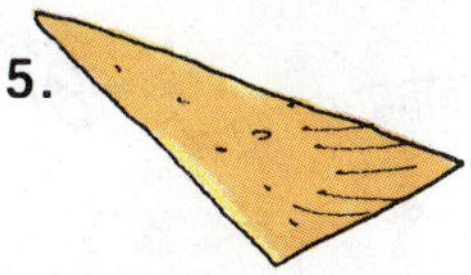

6. 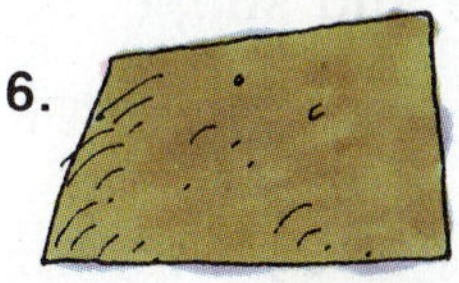

7. 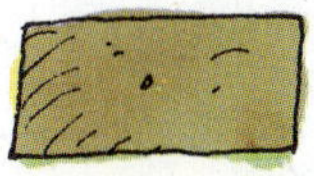

8. 

Are the two figures congruent? Use tracings.

9. 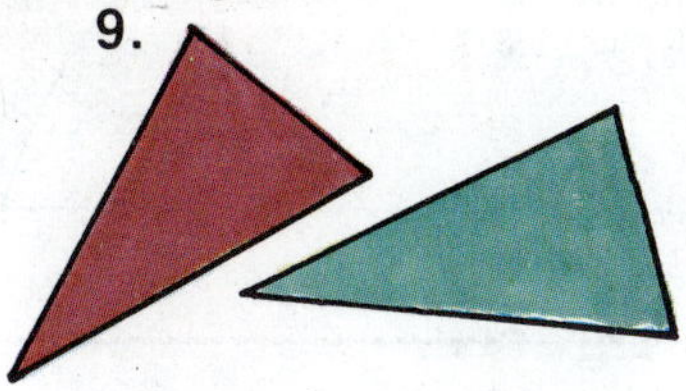

10. 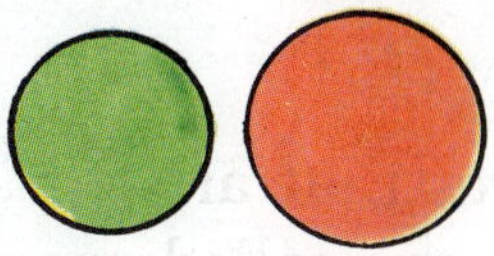

11. 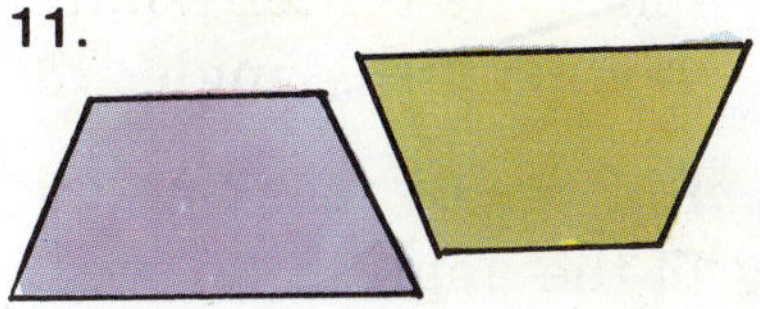

12. 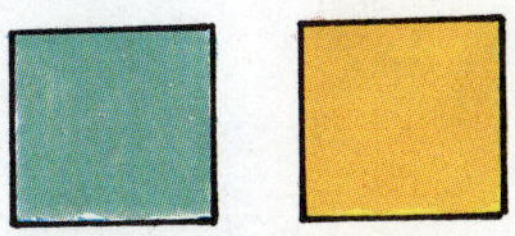

13. 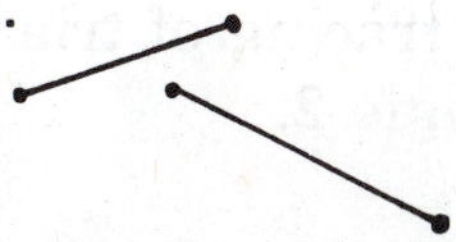

14. 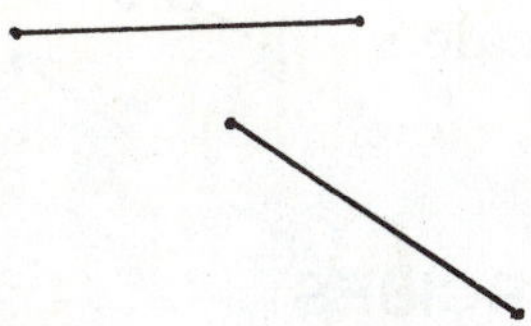

15. 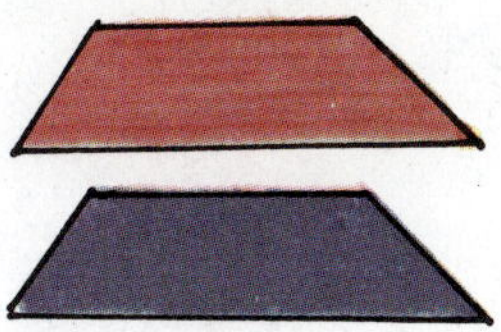

16. Is segment *AB* congruent to segment *CD*?

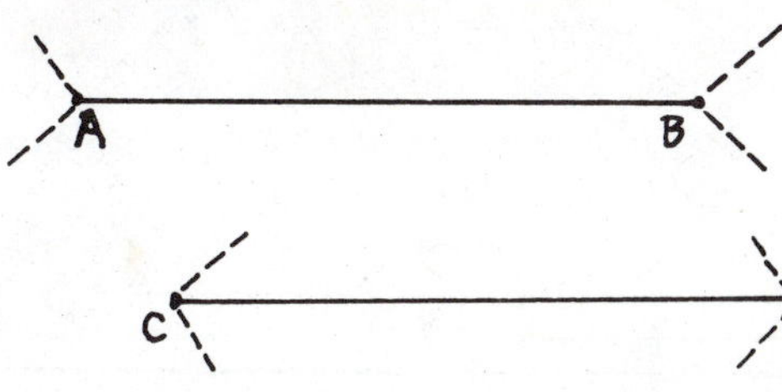

Measuring angles

Two congruent angles have the same measure. In this lesson you will learn to measure angles.
The unit for measuring angles is the **degree.**

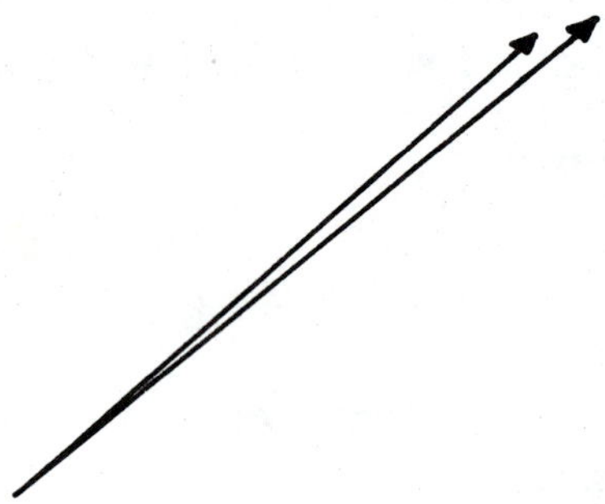

This is a 1° (1 degree) angle. To measure an angle, we find out how many 1° angles it takes to "fill up" the angle.

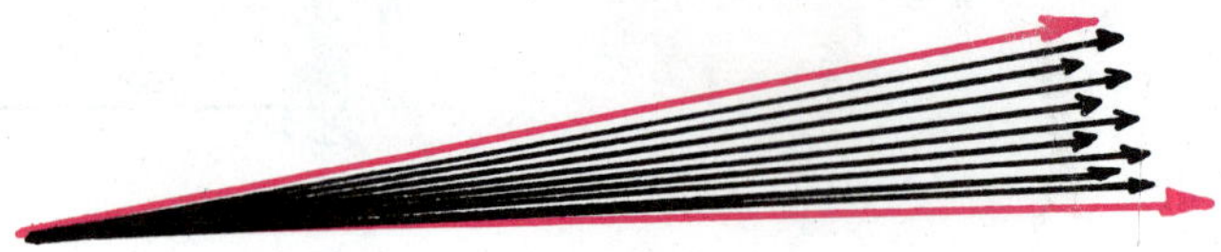

The angle shown in red is a 10° angle.

We usually use a **protractor** for measuring angles.

Here is how a protractor is used.

1. Place the center point on the **vertex** of the angle.
2. Put the 0° mark on one **side** of the angle.
3. Read the measure of the angle where the other side crosses the scale.

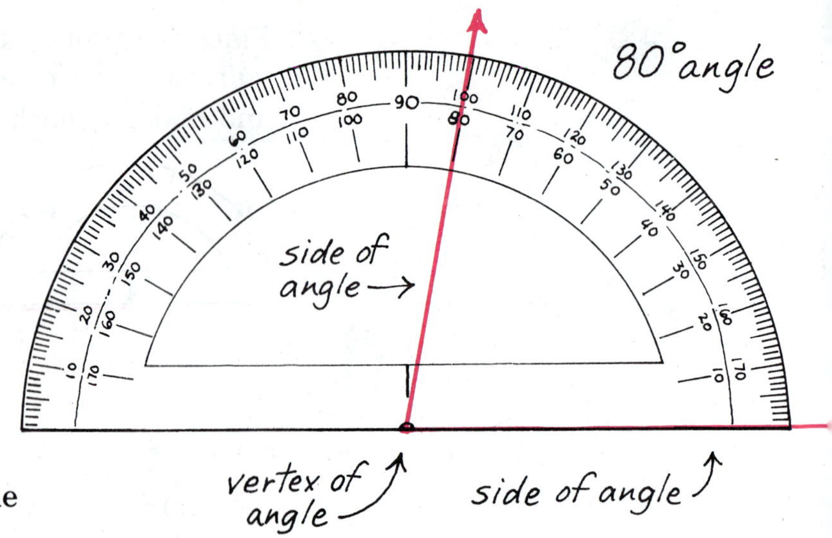

EXERCISES

What is the measure?

1.

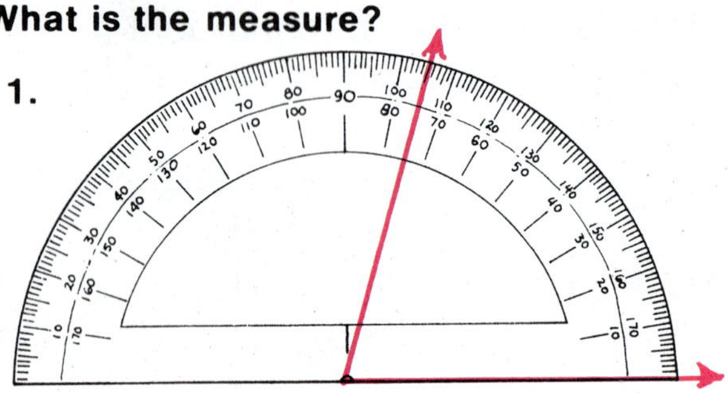

2.

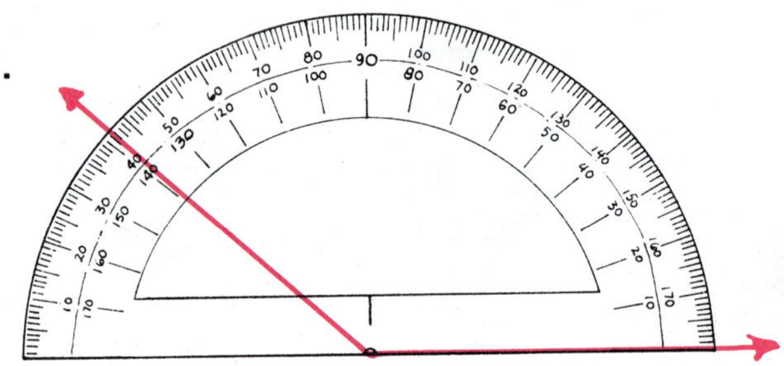

Measure each angle.

3. 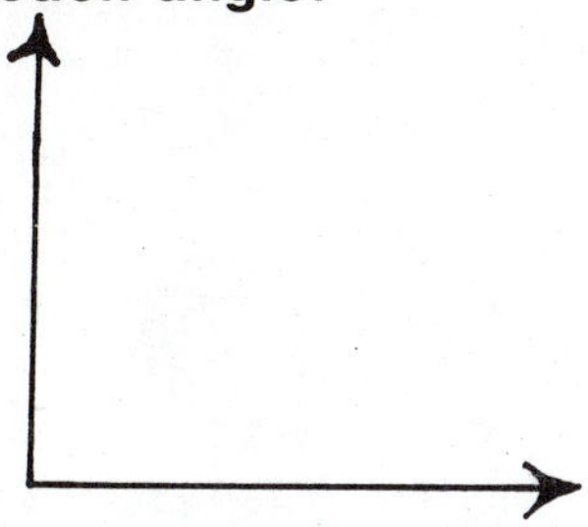

4. 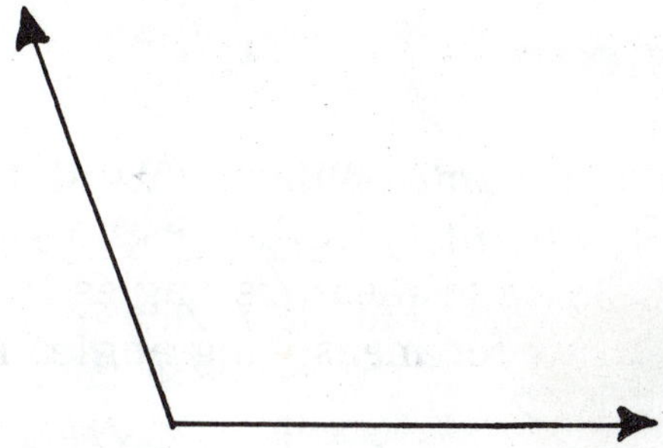

5. 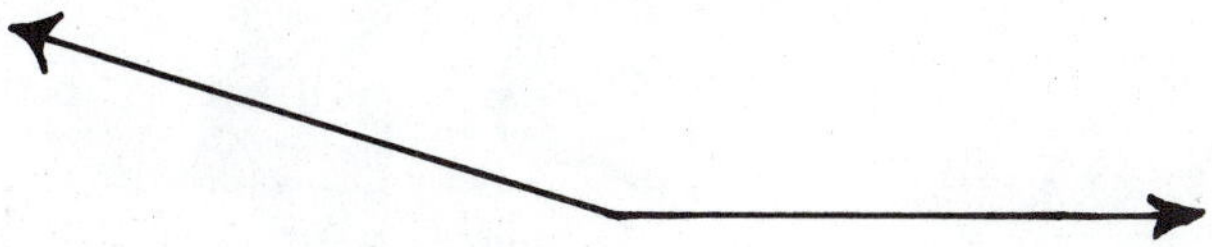

6. 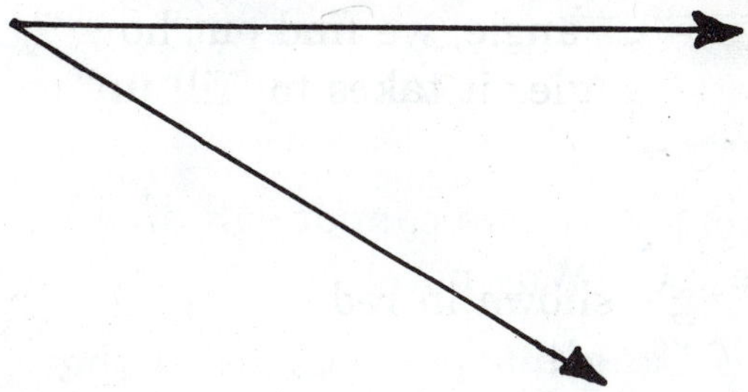

Here is how to draw a 50° angle.

1. Draw one side.

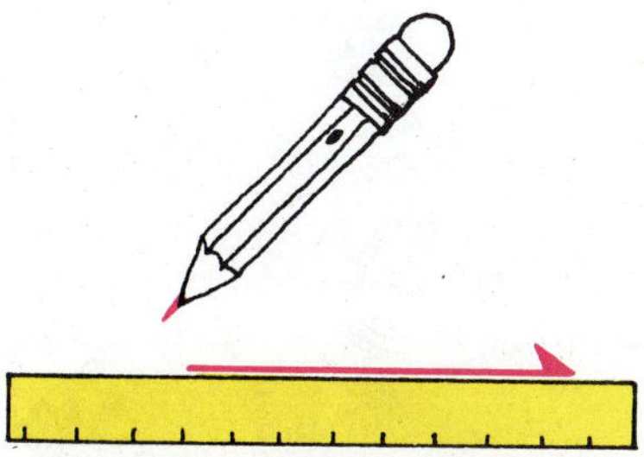

2. Place the protractor as you would for measuring and make a mark at 50°.

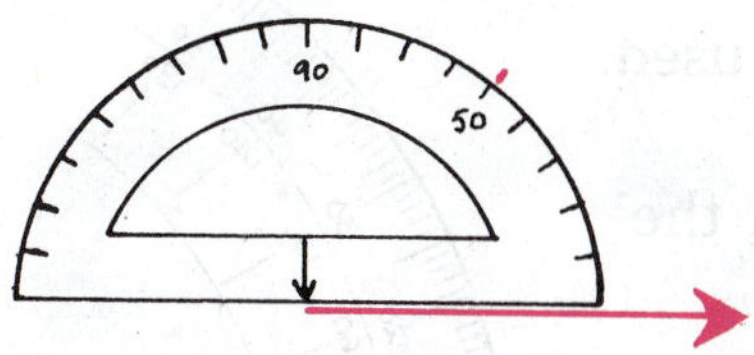

3. Draw the other side.

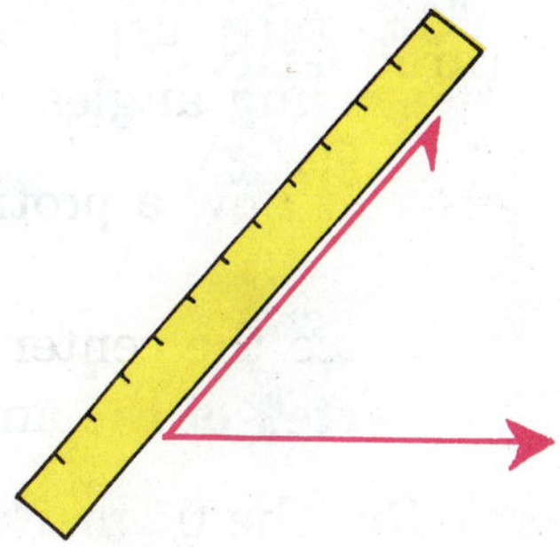

Draw angles having these measures.

7. 30° **8.** 60° **9.** 90° **10.** 120° **11.** 150° **12.** 45°

Estimate the measures of the angles shown in red. Then measure the angles.

13.

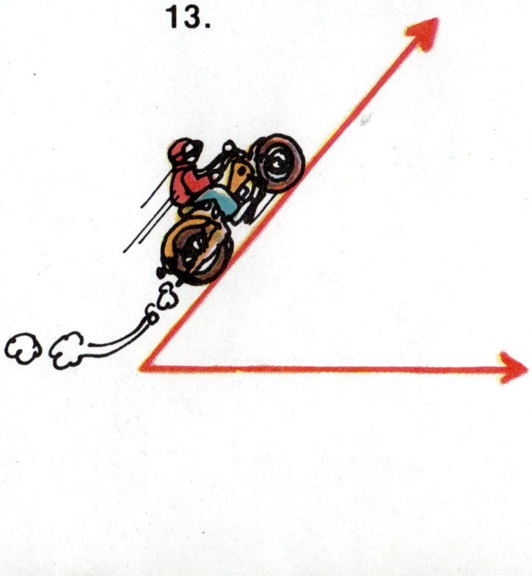

14.

15.

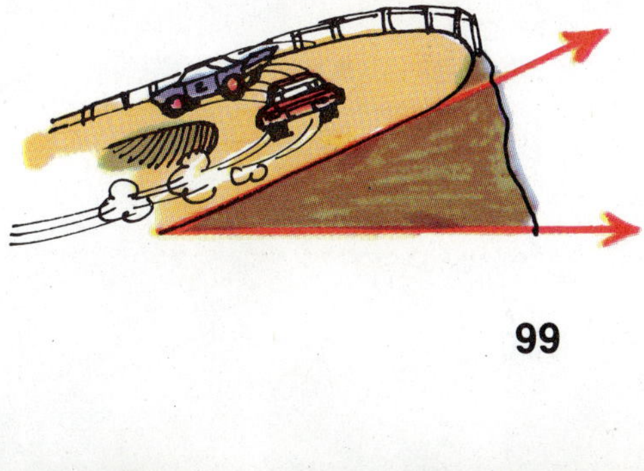

Angles

We can name angles by using the name of the vertex point.

In this picture we cannot call any angle ∠C. (Why not?)

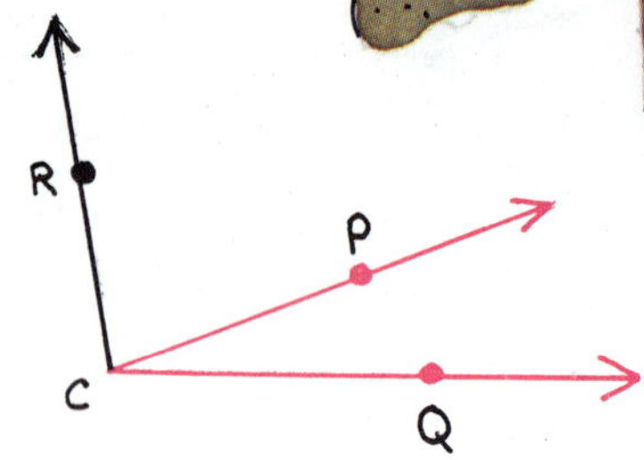

So we use points on the sides of the angle to help name the angle. Here is how to name the red angle using three letters:

∠M is a **right angle.** Its measure is 90°.

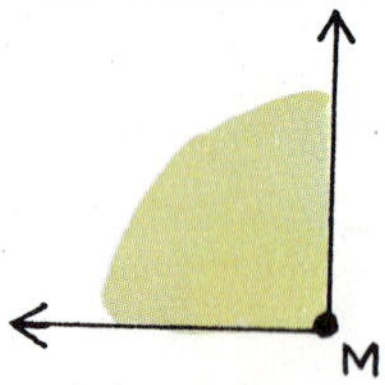

∠RST is an **acute angle.** It is smaller than a right angle.

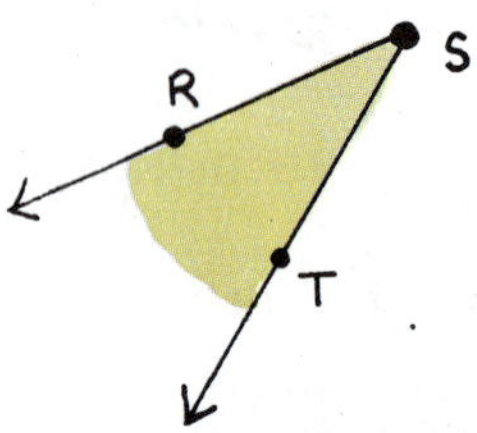

∠MGR is an **obtuse angle.** It is greater than a right angle.

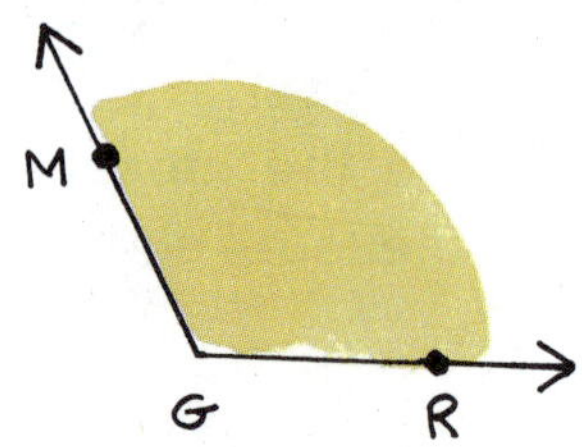

EXERCISES

Name the red angle.

1.

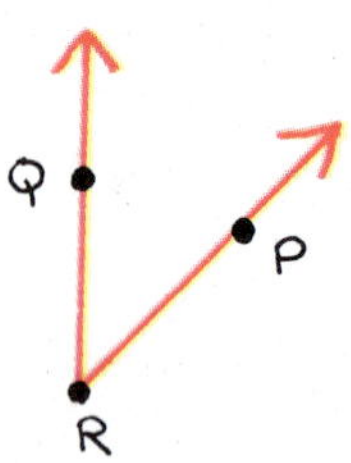

2.

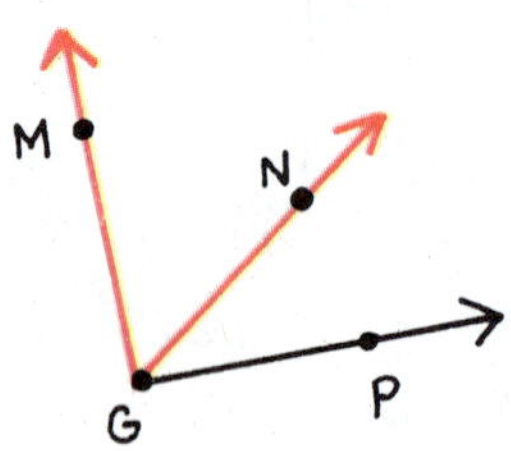

3.

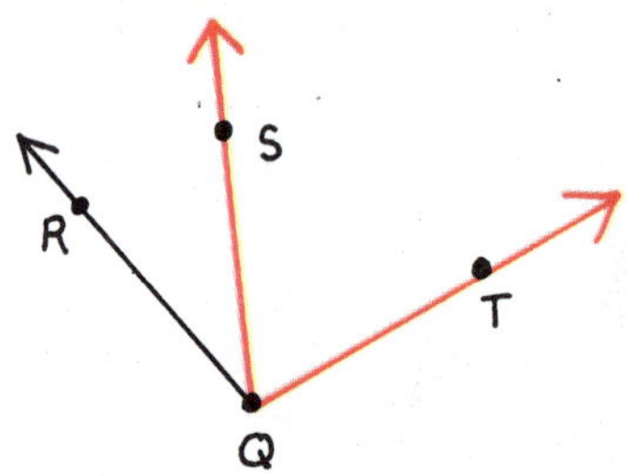

4.

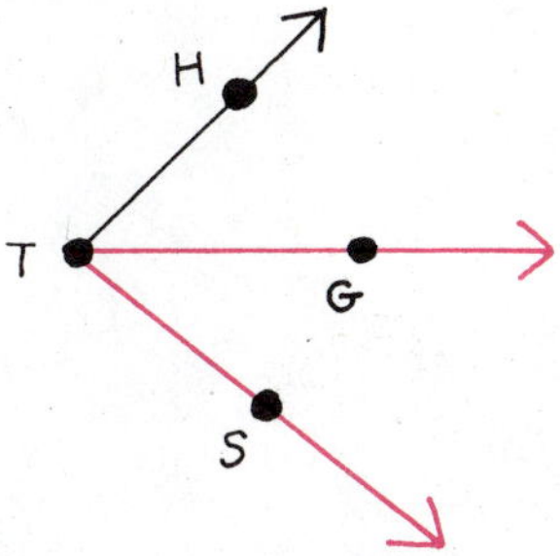

5.

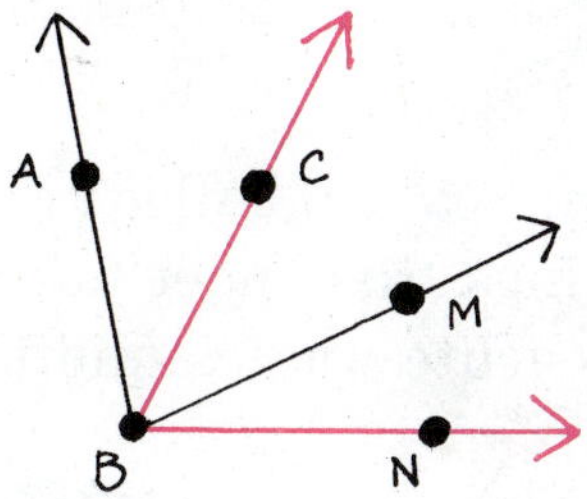

★ 6.

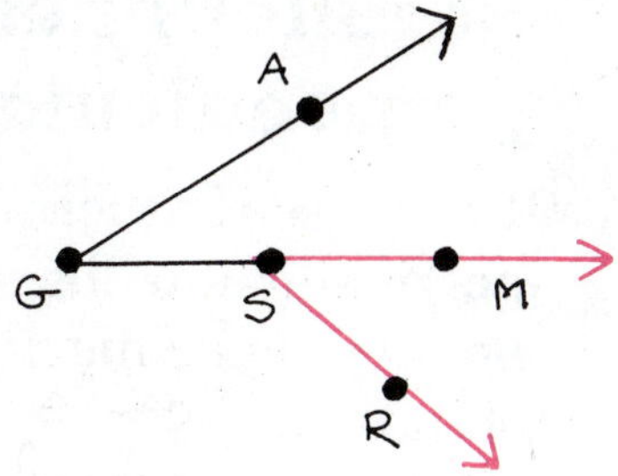

Acute, right, or obtuse?

7.

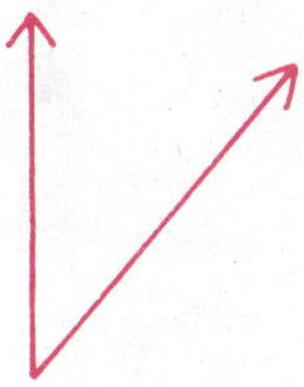

8.

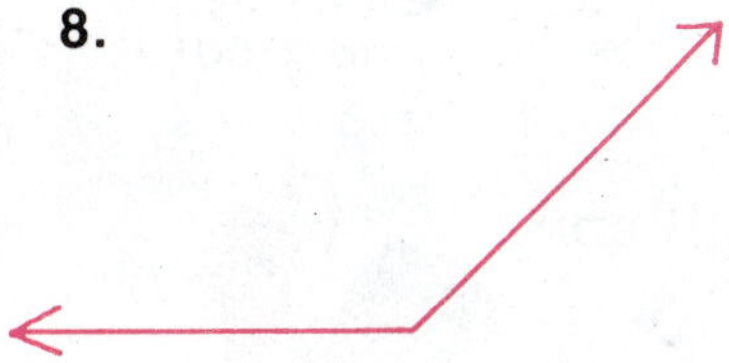

9.

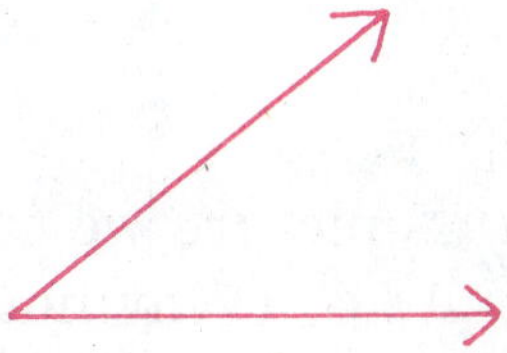

10.

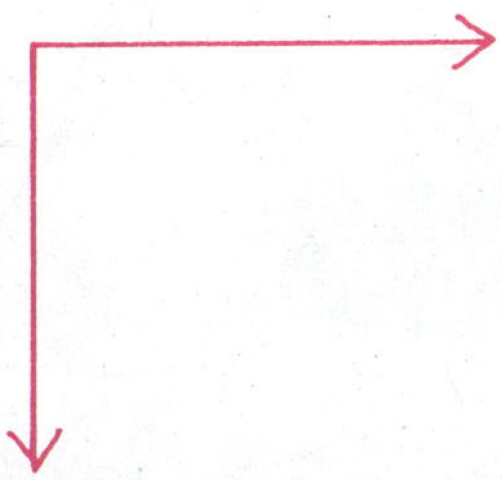

11.

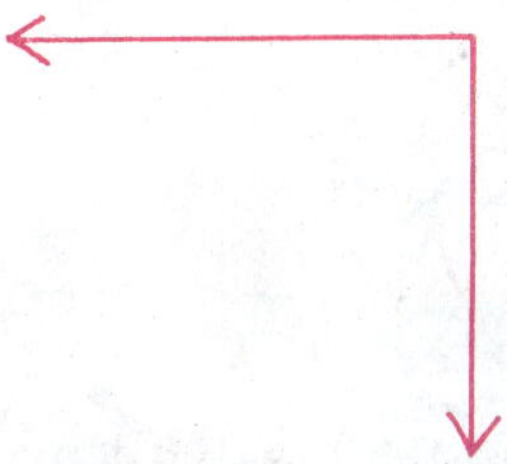

12.

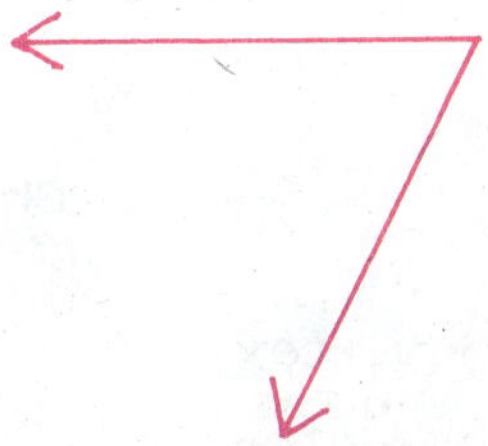

13. a. Draw a line and a ray as shown here.
    b. Measure ∠*AMB*.
    c. Measure ∠*BMC*.
    d. What is the sum of the two measures?
    e. Repeat with other pictures like the one shown.

14. Draw two lines that form a right angle. What is the measure of each of the other three angles?

15. Draw two lines that cross but do not form a right angle. What is the measure of each of the four angles?

Draw a triangle with

16. 3 acute angles

17. 2 obtuse angles

18. 1 right angle

Parallel lines and perpendicular lines

Two lines that form right angles are called **perpendicular lines.** Two lines that meet but are not perpendicular form two acute angles and two obtuse angles.

Perpendicular lines

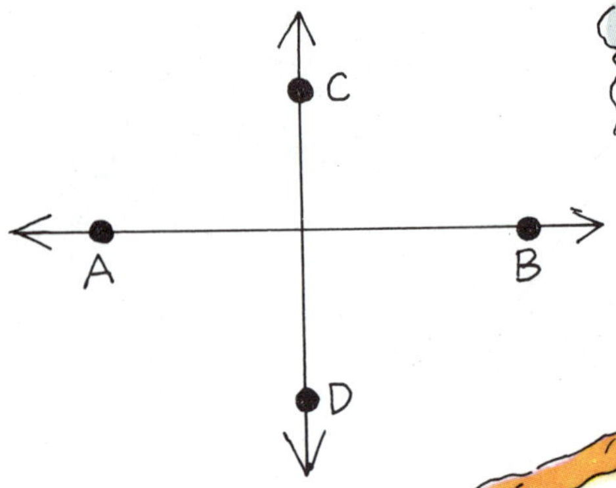

$\overleftrightarrow{AB} \perp \overleftrightarrow{CD}$

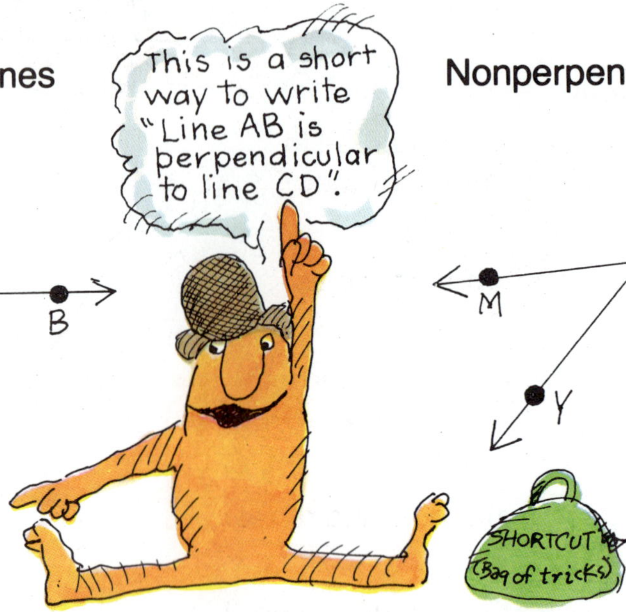

Nonperpendicular lines

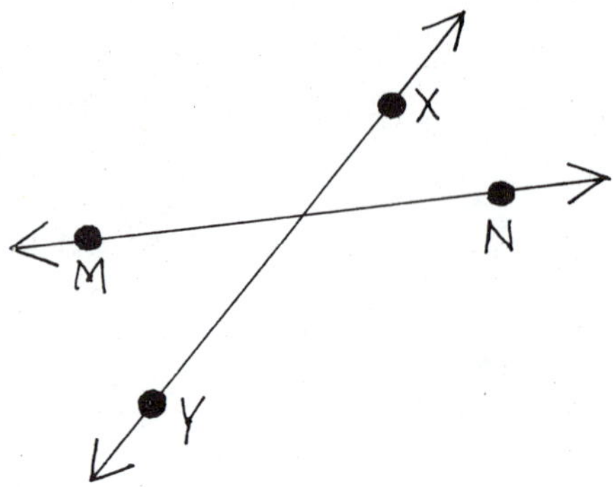

Two lines in a plane that don't meet at all are called **parallel lines.**

Parallel lines

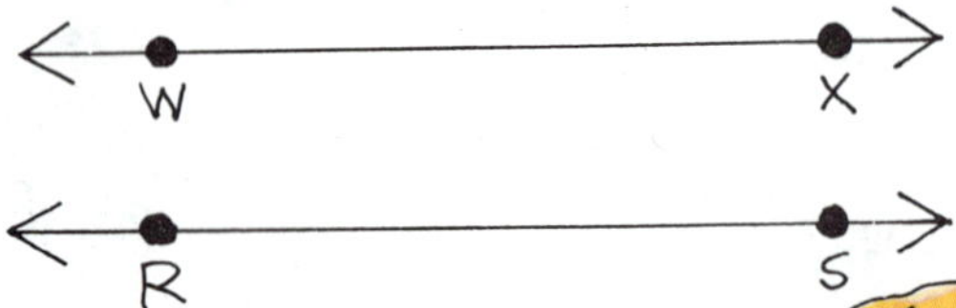

$\overleftrightarrow{WX} \parallel \overleftrightarrow{RS}$

Segments also can be parallel or perpendicular.

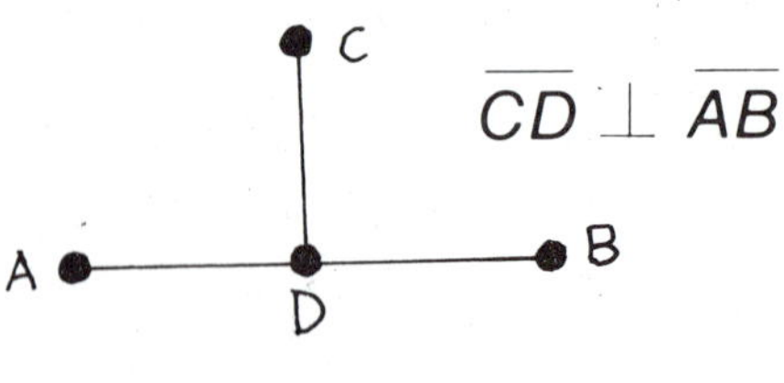

$\overline{CD} \perp \overline{AB}$

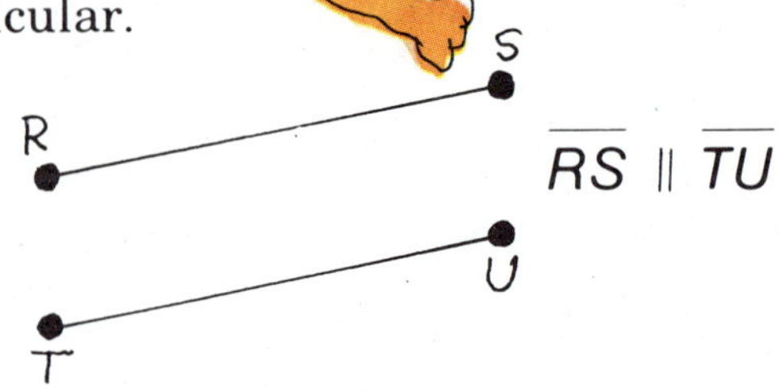

$\overline{RS} \parallel \overline{TU}$

EXERCISES

Parallel or perpendicular?

1. Main St. is ? to First St.
2. Second St. is ? to Broadway.
3. Oak St. is ? to Elm St.
4. Maple St. is ? to Third St.

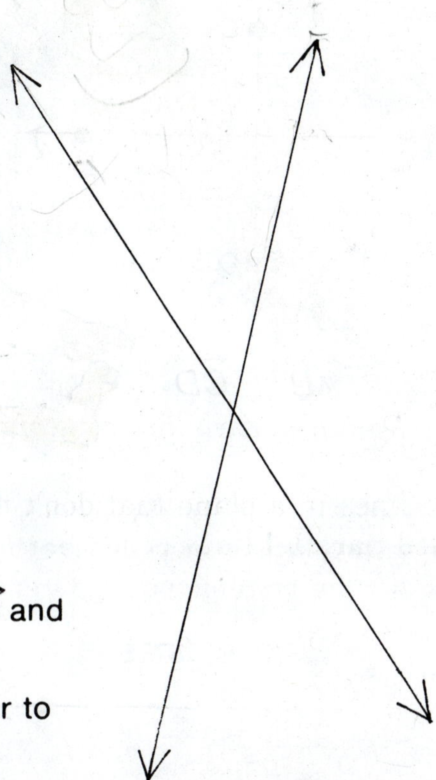

5. Measure the two acute angles in the figure at the right. Are they congruent?
6. Measure the two obtuse angles. Are they congruent?
7. Add the measure of one of the acute angles to the measure of one of the obtuse angles.
8. Draw two lines that cross, and repeat exercises 5–7.
9. Use your protractor and draw two perpendicular lines.
10. a. Draw a line and label it $\overleftrightarrow{AB}$.

    b. Draw a line that is perpendicular to $\overleftrightarrow{AB}$ and label it $\overleftrightarrow{CD}$.

    c. Draw another line that is perpendicular to $\overleftrightarrow{AB}$ and label it $\overleftrightarrow{EF}$.

    d. How are $\overleftrightarrow{CD}$ and $\overleftrightarrow{EF}$ related to each other?

11. 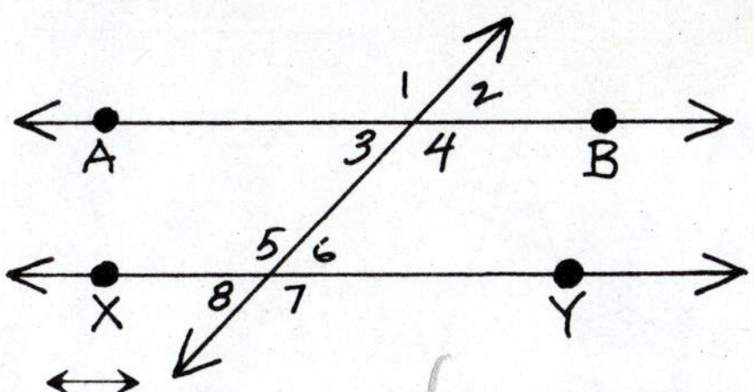

$\overleftrightarrow{AB} \parallel \overleftrightarrow{XY}$. Eight angles have been numbered. List all angles that are congruent to $\angle 1$. To $\angle 2$.

12. 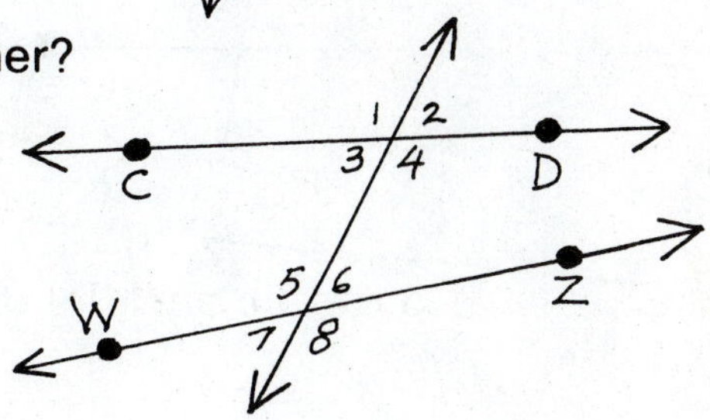

Lines CD and WZ are not parallel. List all angles that are congruent to $\angle 1$. To $\angle 2$.

Fitting congruent figures together

These two triangles are congruent. A tracing of one fits on the other. Parts that fit together are called **corresponding parts.**

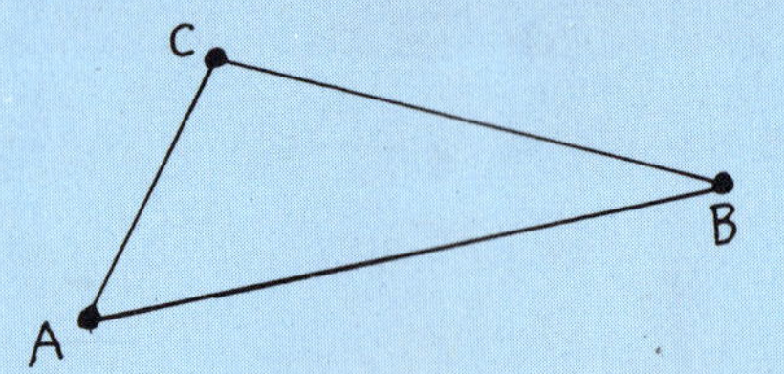

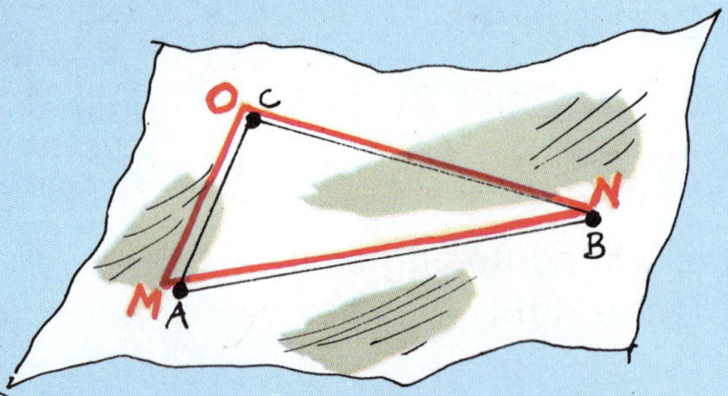

Corresponding Sides	Corresponding Angles
$\overline{AB} \longleftrightarrow \overline{MN}$	$\angle A \longleftrightarrow \angle M$
$\overline{BC} \longleftrightarrow \overline{NO}$	$\angle B \longleftrightarrow \angle N$
$\overline{CA} \longleftrightarrow \overline{OM}$	$\angle C \longleftrightarrow \angle O$

Corresponding parts are congruent to each other.

EXERCISES

Find a way to fit one congruent figure on the other. Then list the corresponding parts.

1.

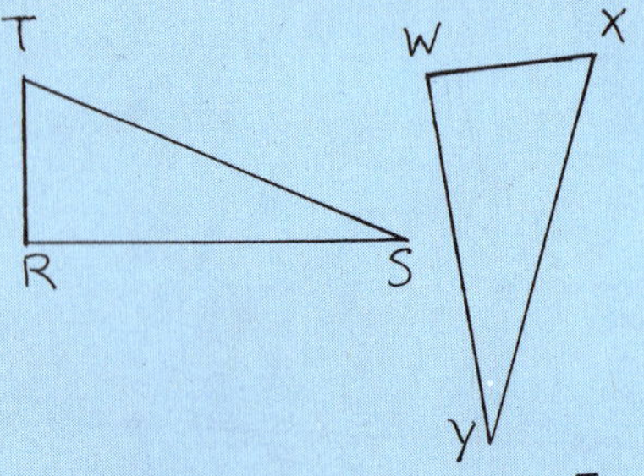

2.

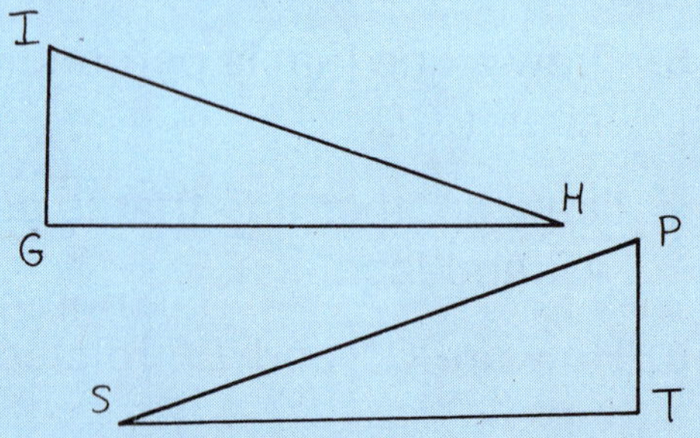

3.

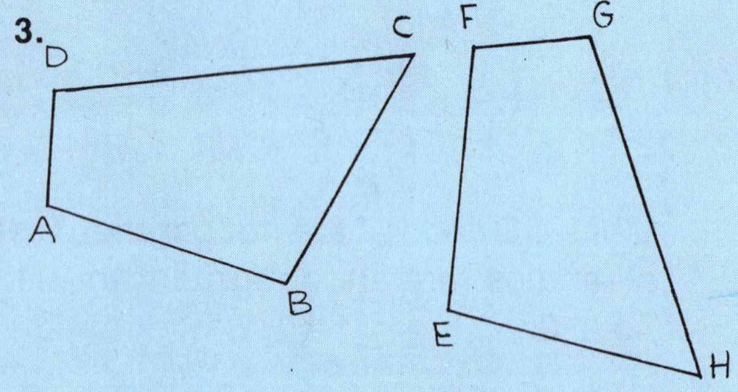

4.

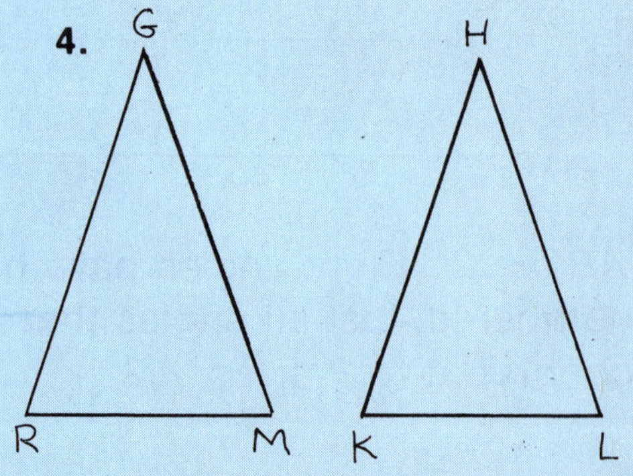

5. There are two congruent triangles in this figure. Find them and list the corresponding parts.

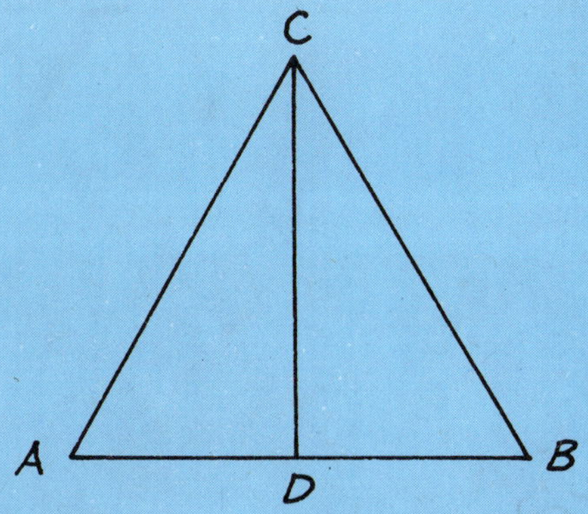

6. Here is a triangle with angles of 50°, 60°, and 70°. Can you draw a triangle that has the same size angles and is not congruent to this one?

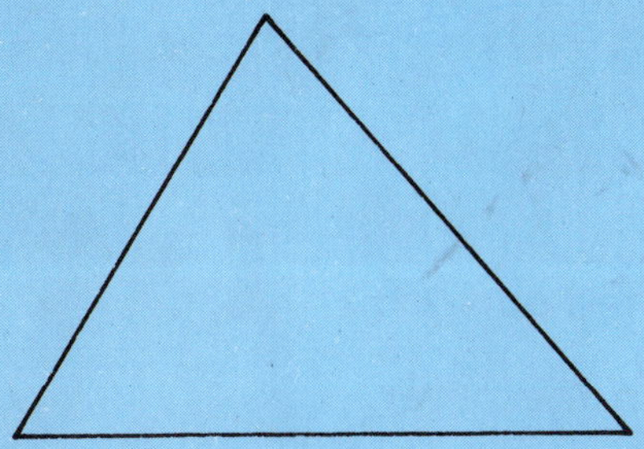

7. Here is a triangle with sides of 5 cm, 6 cm, and 8 cm. Can you draw a triangle that has sides the same length and is not congruent to this one?

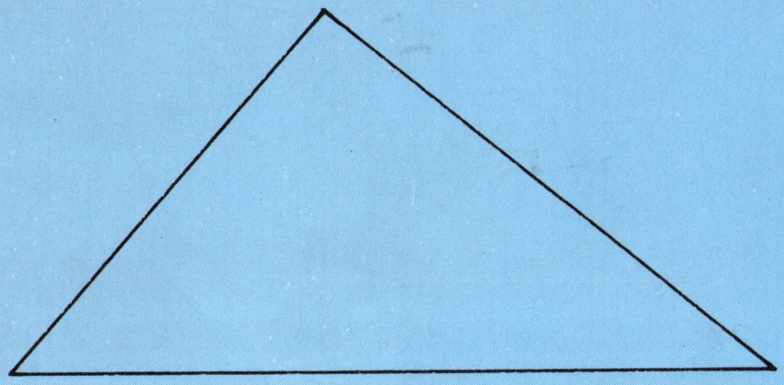

Keeping Skills Sharp

Divide.

1. $35\overline{)945}$
2. $48\overline{)302}$
3. $19\overline{)108}$
4. $39\overline{)218}$
5. $17\overline{)1903}$
6. $32\overline{)4965}$
7. $58\overline{)4073}$
8. $47\overline{)3196}$
9. $75\overline{)6805}$
10. $39\overline{)1536}$
11. $43\overline{)17458}$
12. $59\overline{)29677}$
13. $85\overline{)41310}$

Lines of symmetry

When this heart is folded on the dashed line, the two parts fit together exactly.

Because the two parts fit together, the dashed line is called a **line of symmetry**.

EXERCISES

Is the dashed line a line of symmetry of the figure? If you need to, you can trace the figure and fold it.

1. 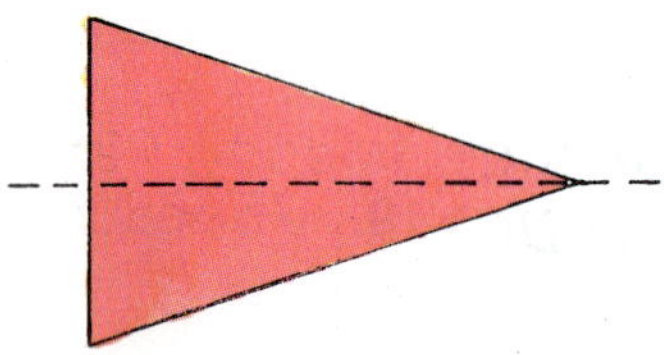

2. 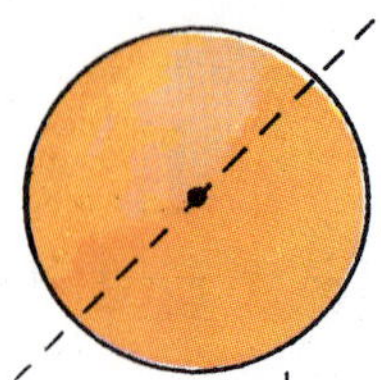

3. 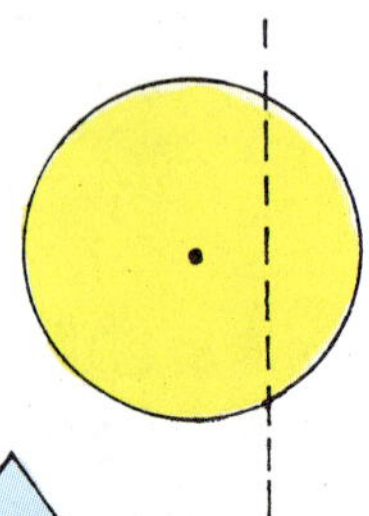

4. 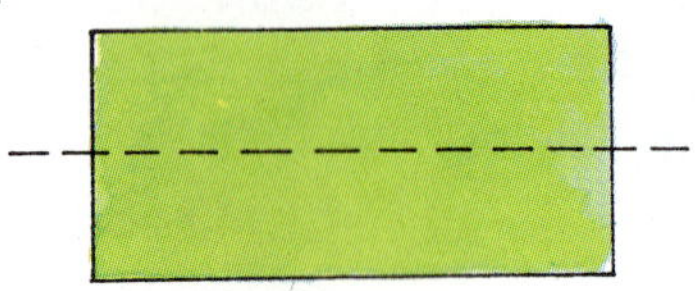

5. 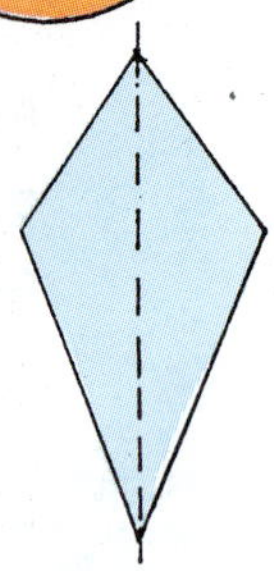

6. 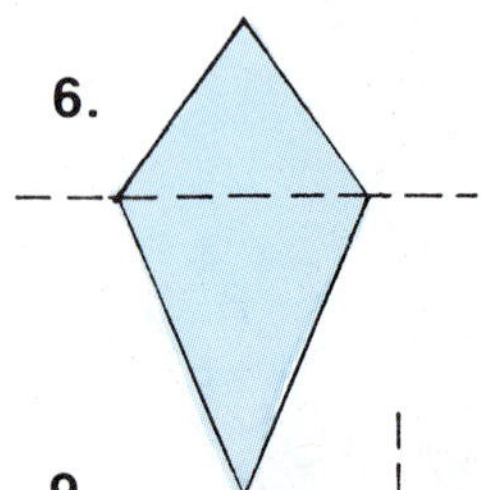

7. 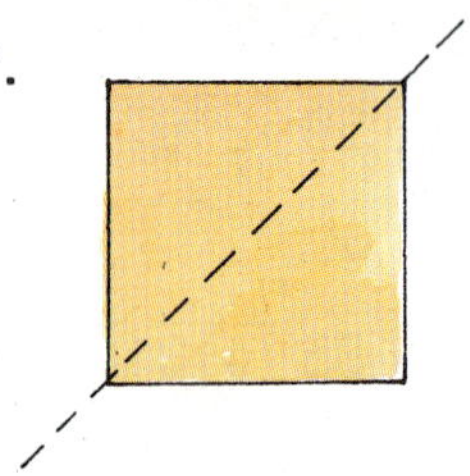

8. 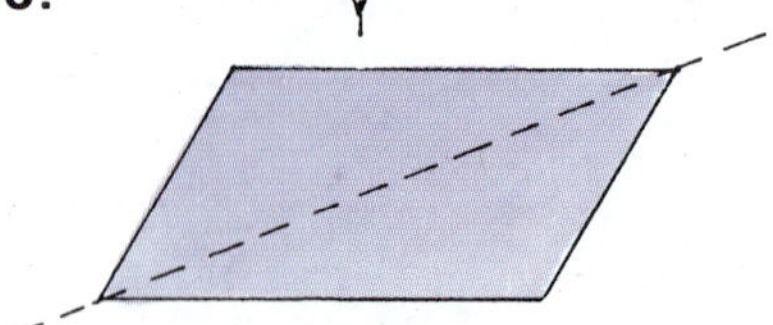

9. 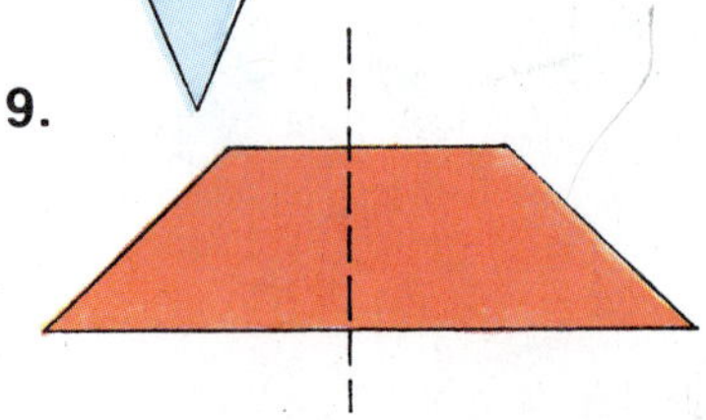

10.

11.

Trace each figure and draw *all* lines of symmetry.

12.

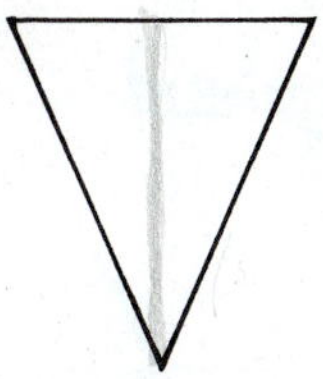

13.

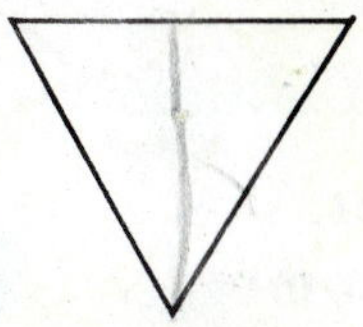

14.

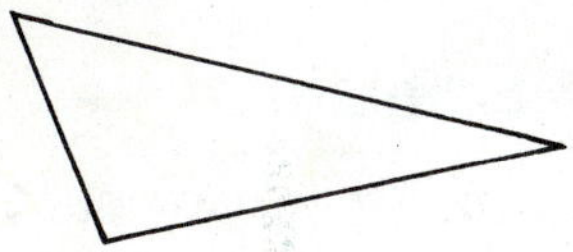

15.

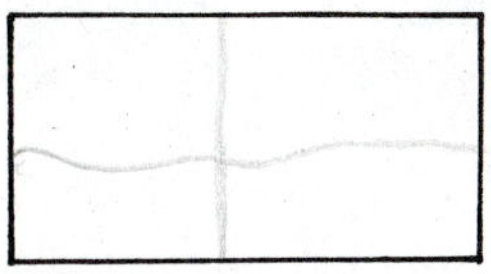

16.

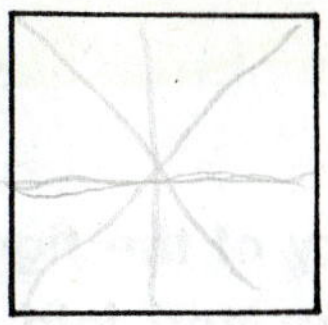

17.

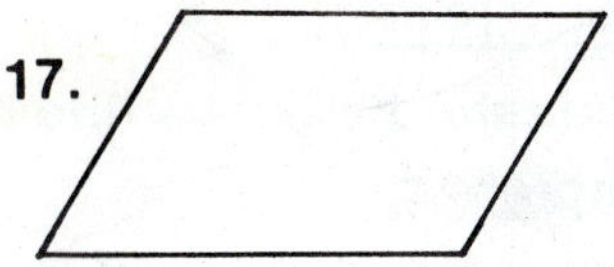

The dashed line is a line of symmetry of triangle *ABC*.

18. Which triangle is congruent to triangle *ADC*?

For the two congruent triangles of exercise 18, complete the corresponding parts.

19. $\overline{AC} \leftrightarrow$?

20. $\angle B \leftrightarrow$?

21. $\overline{AD} \leftrightarrow$?

22. $\overline{CD} \leftrightarrow$?

23. $\angle ACD \leftrightarrow$?

Triangles

A triangle with 3 lines of symmetry is called an **equilateral triangle.**

A triangle with 1 line of symmetry is called an **isosceles triangle.**

A triangle with 0 lines of symmetry is called a **scalene triangle.**

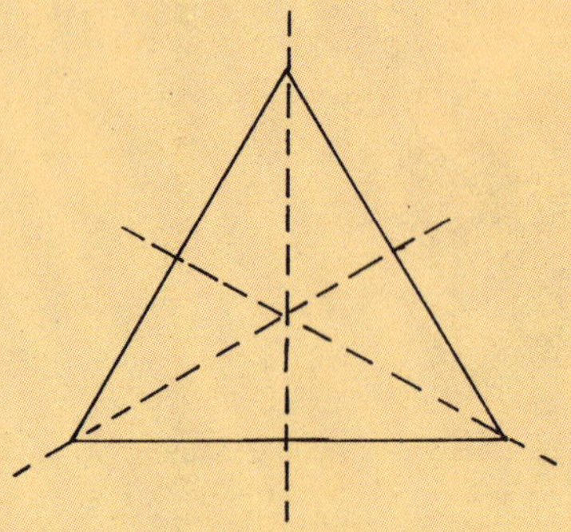

Equilateral

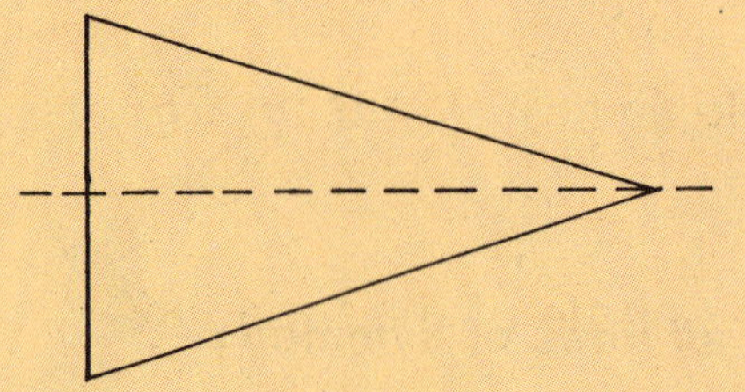

Isosceles

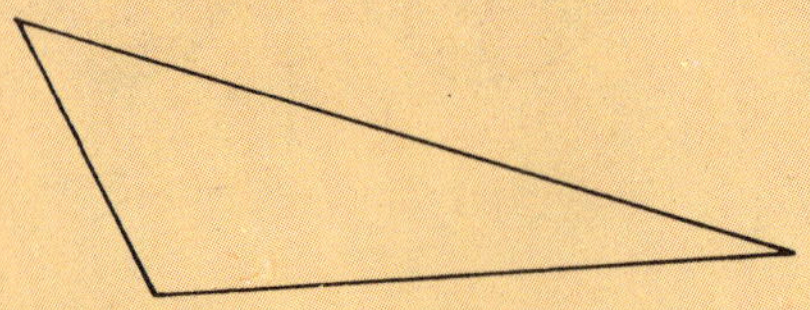

Scalene

EXERCISES

The dashed line is a line of symmetry of triangle *MNO*. Complete.

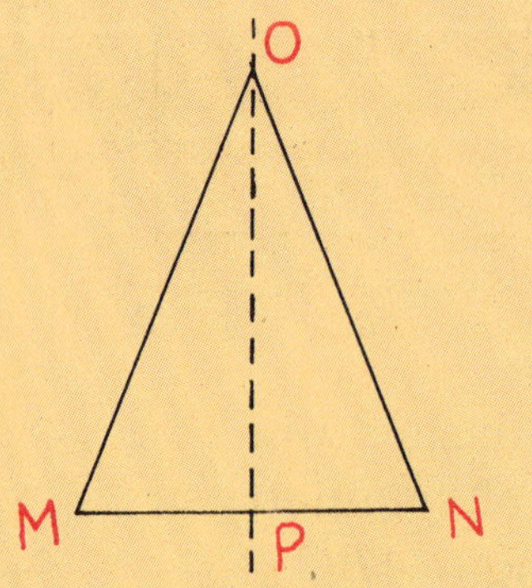

1. Triangle *MPO* is congruent to triangle ?.
2. $\overline{MO}$ corresponds to ?.
3. $\overline{MO}$ is congruent to ?.
4. $\angle M$ is congruent to ?.
5. An isosceles triangle has ? congruent sides and ? congruent angles.

Trace this equilateral triangle and the three lines of symmetry.

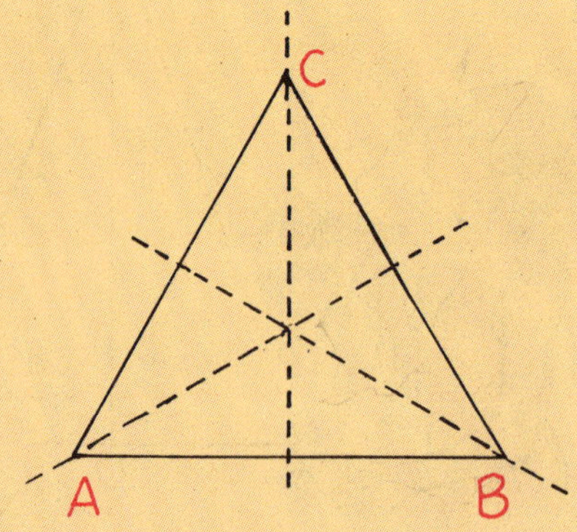

6. Can you fold the triangle along a line of symmetry so that $\overline{AB}$ and $\overline{AC}$ fit?
7. Exercise 6 shows that $\overline{AB}$ is congruent to ?.
8. Can you fold the triangle so that $\overline{AB}$ and $\overline{BC}$ fit?
9. Exercise 8 shows that $\overline{AB}$ is congruent to ?.

10. Is $\overline{AC}$ congruent to $\overline{BC}$? How do you know?

11. $\angle A$ is congruent to _?_. How do you know?

12. $\angle C$ is congruent to _?_. How do you know?

13. An equilateral triangle has _?_ congruent sides and _?_ congruent angles.

Trace scalene triangle *RST*.

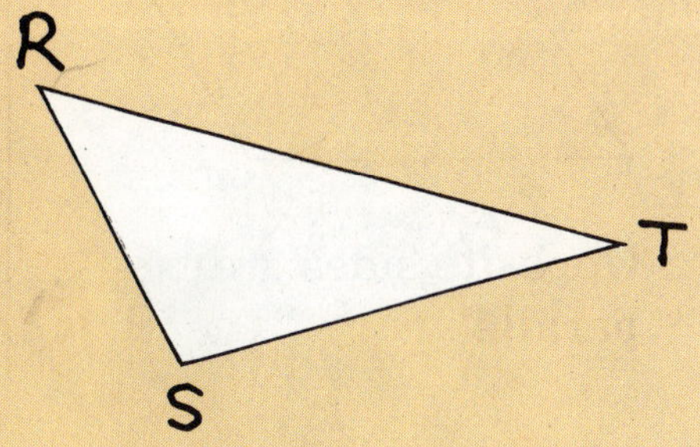

14. Can you fold the triangle so that $\overline{RS}$ fits on $\overline{ST}$?

15. Will $\overline{RS}$ fit on $\overline{RT}$?

16. Will $\angle S$ fit on any other angle of the triangle?

17. A scalene triangle has _?_ congruent sides and _?_ congruent angles.

You should have found out that

An equilateral triangle has 3 congruent sides and 3 congruent angles.

An isosceles triangle has 2 congruent sides and 2 congruent angles.

A scalene triangle has 0 congruent sides and 0 congruent angles.

Keeping Skills Sharp

Multiply.

1. 631 × 528
2. 467 × 599
3. 859 × 468
4. 293 × 675
5. 206 × 321
6. 321 × 206
7. 5834 × 305
8. 2973 × 764

Quadrilaterals

Quadrilaterals are 4-sided figures. Here are some special quadrilaterals.

Parallelogram

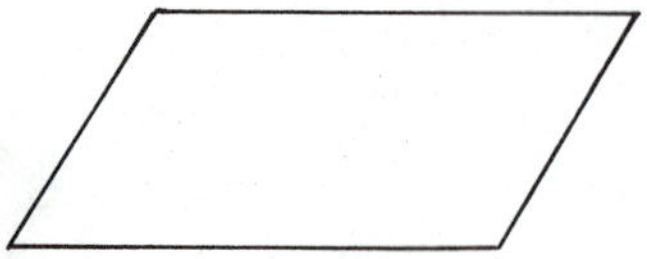

Opposite sides are parallel.

Rectangle

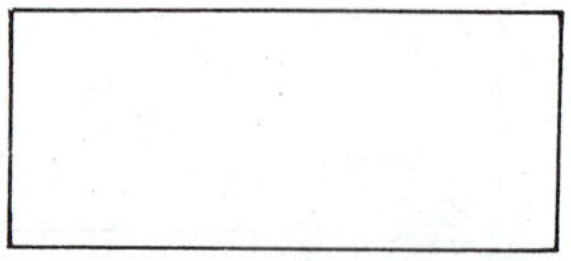

This is a special parallelogram with 4 right angles.

Rhombus

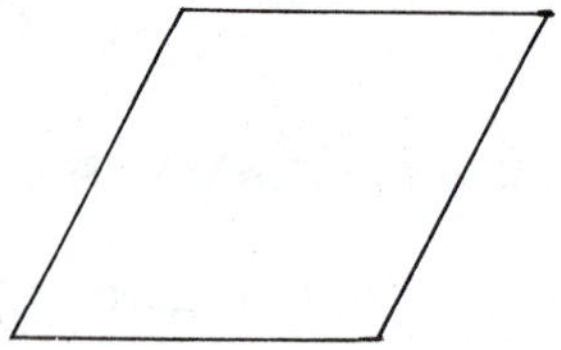

This is a special parallelogram with 4 congruent sides.

Square

This is both a rectangle and a rhombus. It has 4 right angles and 4 congruent sides.

Trapezoid

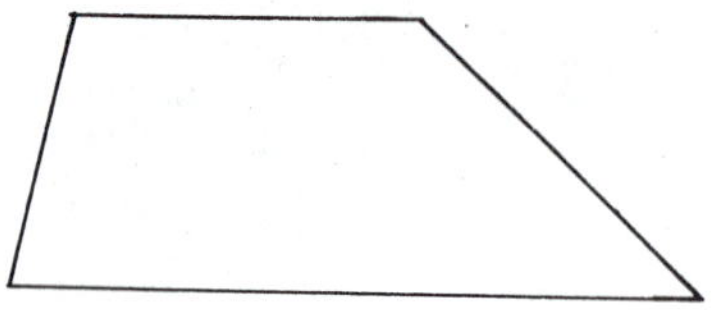

Two sides are parallel.

Isosceles Trapezoid

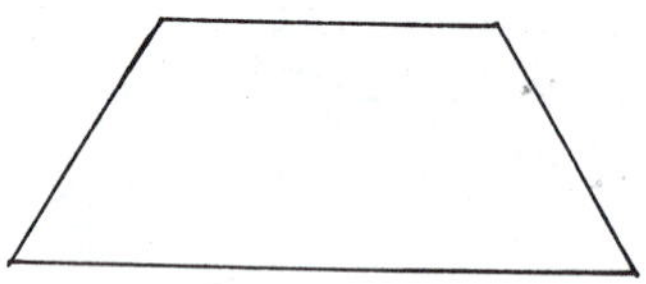

This is a trapezoid with 2 congruent sides.

Kite

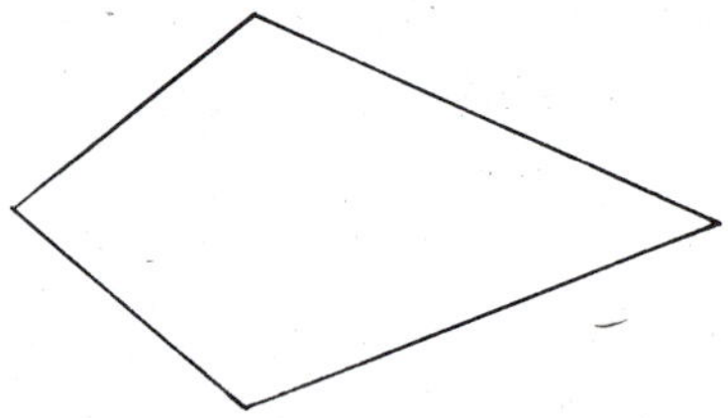

This figure has 2 pairs of congruent sides.

EXERCISES

1. Does this parallelogram have any lines of symmetry? Trace and fold if necessary.

2. Is $\overline{AB}$ congruent to $\overline{DC}$?
3. Is $\angle A$ congruent to $\angle B$?
4. Is $\angle A$ congruent to $\angle C$?
5. Is $\angle B$ congruent to $\angle D$?
6. Is $\overline{DA}$ congruent to $\overline{CB}$?
7. Is $\overline{DA}$ congruent to $\overline{AB}$?
8. Does a rectangle have lines of symmetry? Does a rhombus?
9. How are the lines of symmetry of a rhombus different from those of a rectangle?
10. How many lines of symmetry does a square have? How can you tell just from lines of symmetry that a square is both a rectangle and a rhombus?

11. Does an isosceles trapezoid have any lines of symmetry?
12. Does this isosceles trapezoid have any congruent angles? How do you know?

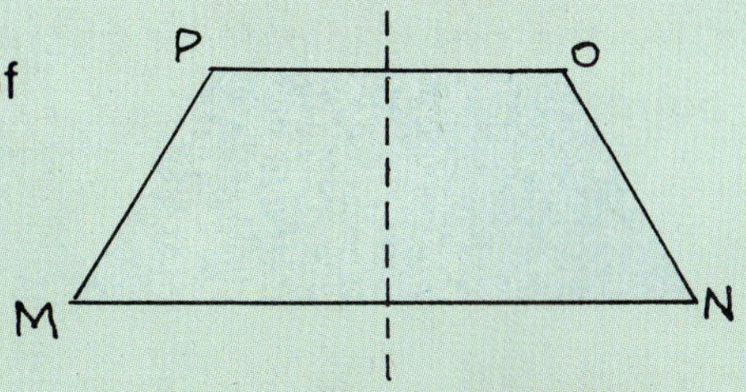

13. Does a kite have any lines of symmetry?
14. How are the lines of symmetry of a kite different from the lines of symmetry of an isosceles trapezoid?
15. Does this kite have any congruent angles? How do you know?

Points on a grid

We can identify points on this grid by using pairs of numbers. For example, the point *G* goes with the pair (5, 2). The first number tells how far to the right of 0, and the second number tells how far up from 0.

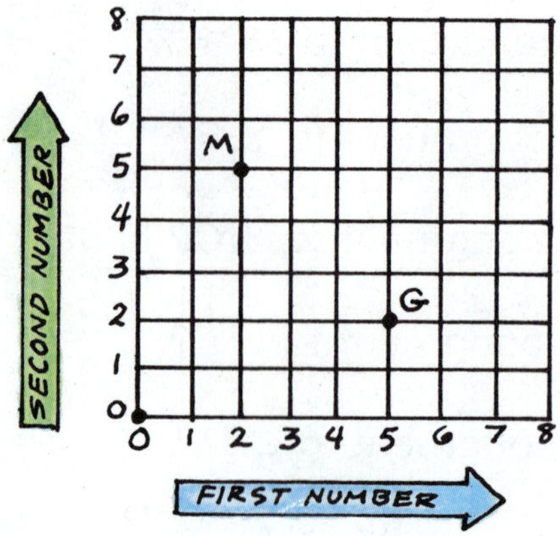

The order of the two numbers is important. The point for (5, 2) is *G* and the point for (2, 5) is *M*. This is why the number pair is called an **ordered pair.**

EXERCISES

Give the ordered pair for each of these points.

1. *A* **2.** *B* **3.** *C*

4. *D* **5.** *E* **6.** *F*

7. *G* **8.** *H* **9.** *I*

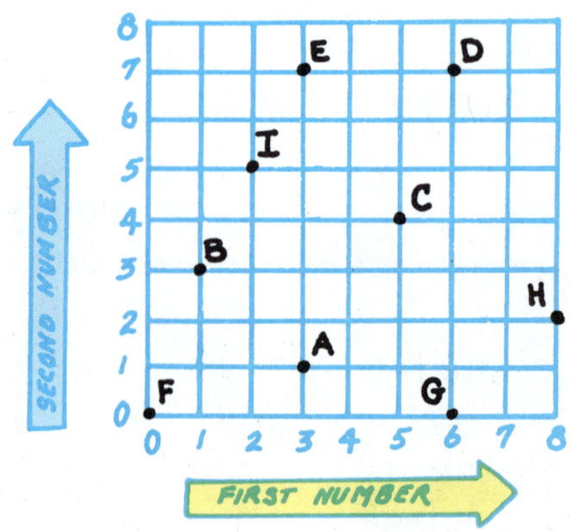

Draw a grid like the one shown above. Label the points with the letters.

10. (7, 3) *Z* **11.** (0, 4) *Y* **12.** (4, 0) *X*

13. (2, 1) *W* **14.** (6, 6) *V* **15.** (3, 5) *U*

16. (0, 0) *T* **17.** (5, 3) *S* **18.** (7, 8) *R*

Make a grid like this one.

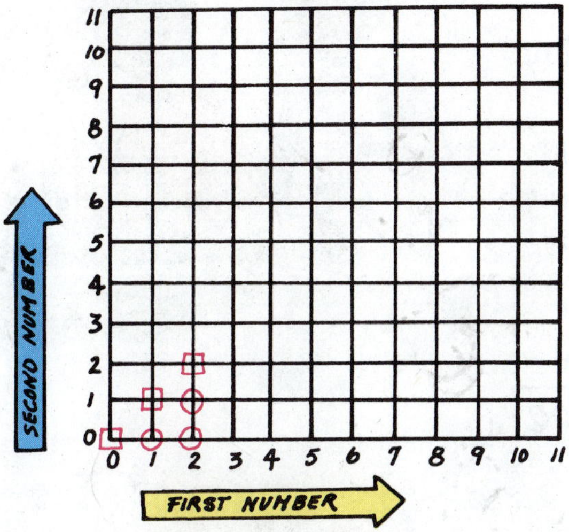

19. Put a □ on each point that has an ordered pair of whole numbers in which the first and second numbers are equal.

20. Put a ○ on each point for which the ordered pair of whole numbers has a first number greater than the second number.

Make another grid like the one shown above.

21. Put a □ on each point for which the sum of the first and second numbers is 5.

22. Put a ○ on each point for which the second number is 2 times the first number.

Draw a square grid and number the lines 0 – 6.

23. Draw segments from (1, 5) to (1, 4) to (1, 3) to (1, 2) to (1, 1).

24. Draw segments from (1, 3) to (2, 3) to (3, 3) to (3, 4) to (3, 5).

25. Draw segments from (3, 3) to (3, 2) to (3, 1).

26. Draw segments from (5, 1) to (5, 2) to (5, 3).

27. Put a dot at (5, 5).

Keeping Skills Sharp

Divide.

1. $6\overline{)216}$
2. $7\overline{)348}$
3. $8\overline{)576}$
4. $9\overline{)243}$
5. $8\overline{)321}$
6. $7\overline{)4235}$
7. $5\overline{)4235}$
8. $6\overline{)3604}$
9. $5\overline{)31026}$
10. $9\overline{)54103}$
11. $7\overline{)63113}$
12. $6\overline{)61095}$
13. $8\overline{)15029}$

Line graphs

You can use what you learned in the preceding lesson to make line graphs.

Suppose that you put a container of very cold water over a burner and heated it. If you kept a record of the temperature every minute, your record might look like this:

Time in minutes	0	1	2	3	4	5	6
Temperature in degrees Celsius	5	20	30	50	80	100	100

Here is how to make a line graph.

1. Make a grid and label the lines so that all information will fit.
2. Mark the dots for the pairs of numbers in the table above.
3. Join the dots with segments.

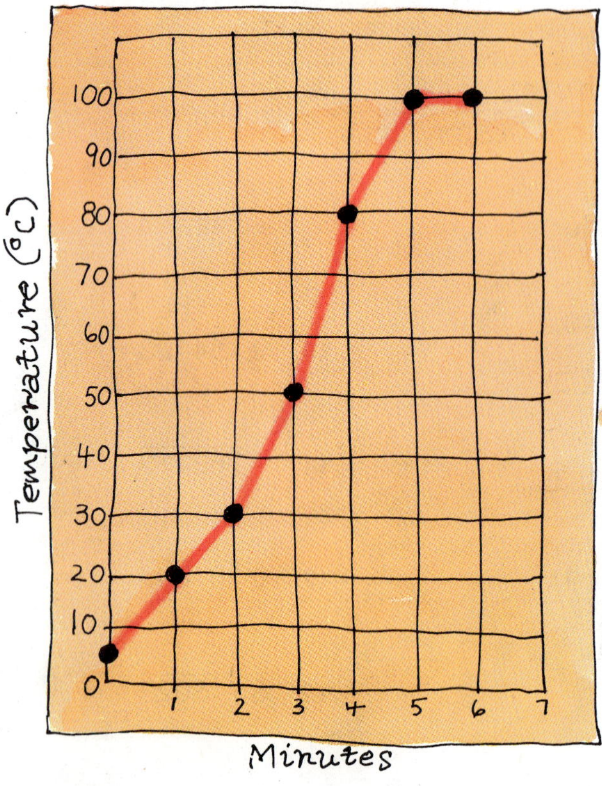

EXERCISES

Here is a line graph showing the cost of a share of stock in Kirby Car Company for a period of 10 weeks.

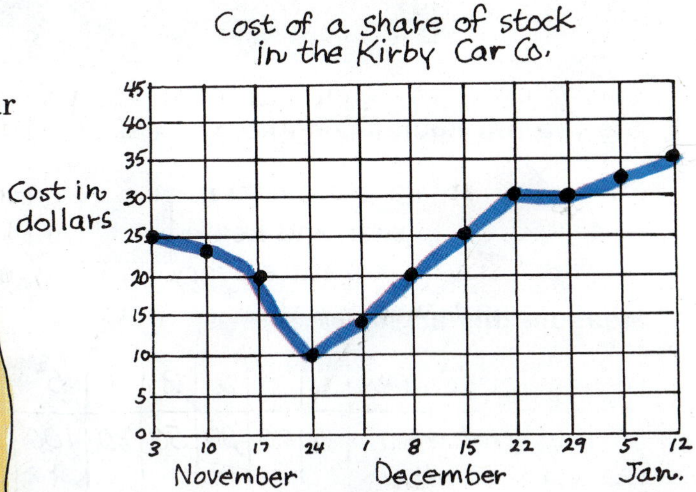

Answer these questions.

1. What was the cost at the beginning of the 10-week period (Nov. 3)?
2. What was the cost at the end of the tenth week (Jan. 12)?

3. What was the total increase in price?
4. What was the lowest cost?
5. What was the highest cost?
6. What was the difference between the lowest and highest costs?
7. What was the cost on December 22?
8. When was the cost $25?

9. Here is a set of information. Make a line graph to show the information.

	A. M.				P. M.					
Time	9:00	10:00	11:00	12:00 noon	1:00	2:00	3:00	4:00	5:00	6:00
Temperature	4°C	6°C	8°C	9°C	12°C	13°C	14°C	12°C	7°C	5°C

CHAPTER CHECKUP

Are the two figures congruent? [pages 96–97, 104–105]

1. 

2. 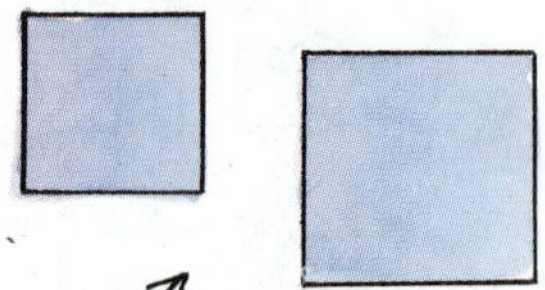

3. 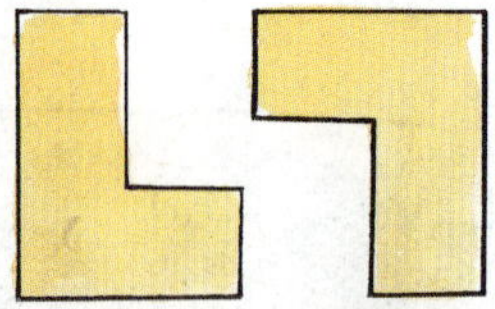

Measure this angle. [pages 98–99]

4. 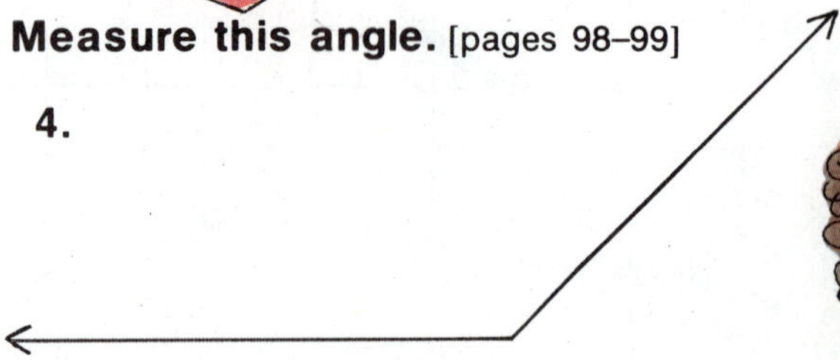

Acute, obtuse, or right? [pages 100–101]

5. 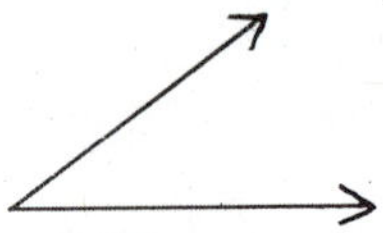

6. 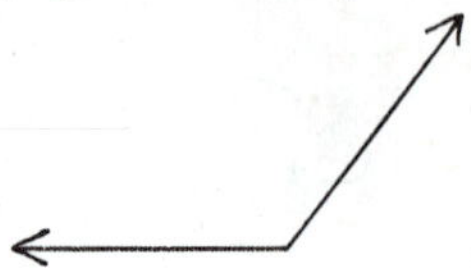

7. 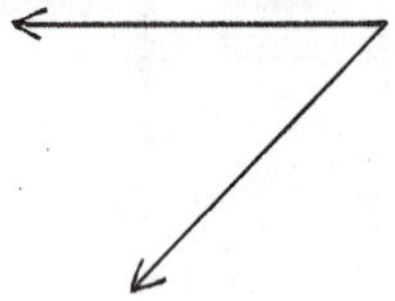

8. 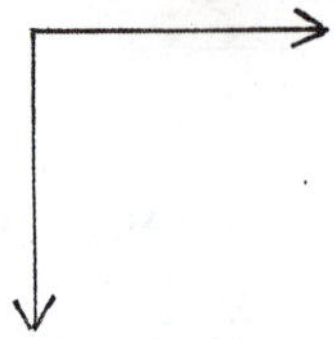

Is the dashed line a line of symmetry? [pages 106–107]

9. 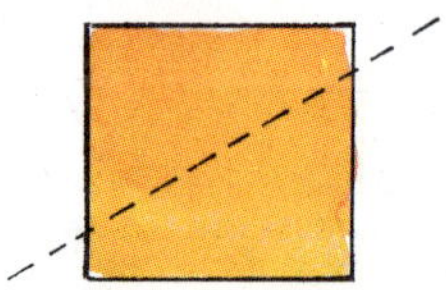

10. 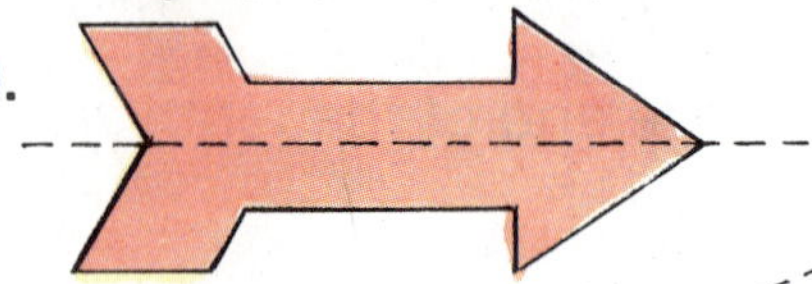

11. 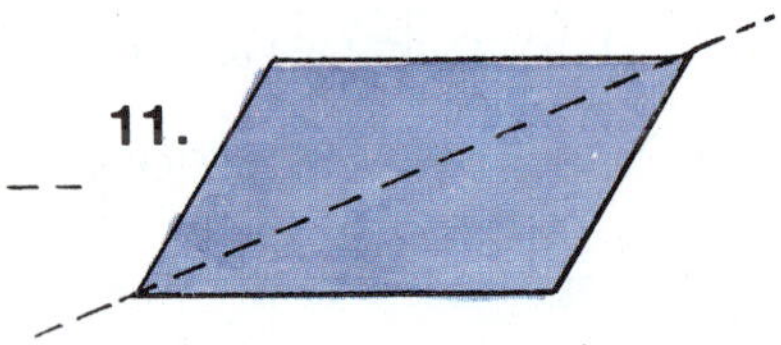

Match. [pages 102–103, 108–111]

12. square
13. rectangle
14. equilateral triangle
15. isosceles triangle
16. parallelogram

a. 3 sides, 2 sides are congruent
b. parallelogram, 4 right angles
c. rectangle, 4 congruent sides
d. 3 sides, 3 congruent angles
e. 4 sides, 2 pairs of parallel sides

Give the letter for each ordered pair. [pages 112–115]

17. (2, 4)
18. (0, 3)
19. (4, 2)
20. (1, 1)

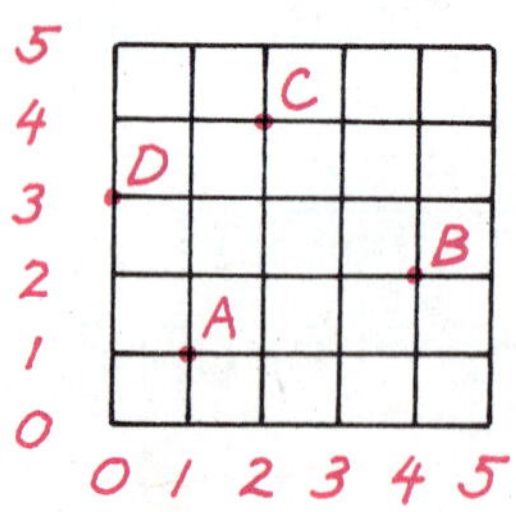

Project

Make a geometry mobile.
Cut figures like these from thin cardboard.

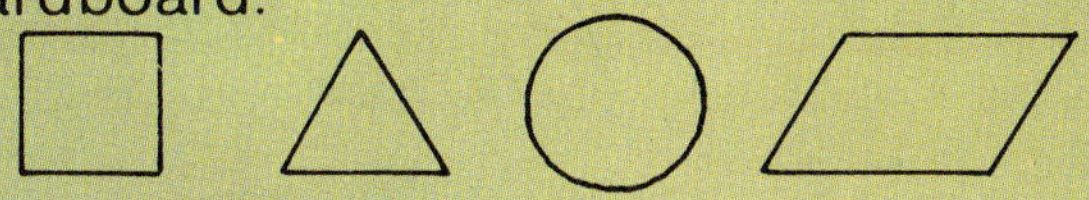

Cut out other figures if you wish.

Get some stiff wire, some thin wooden sticks or drinking straws, and some thread.

Make a mobile. The hard thing is to make the parts balance each other.

CHAPTER REVIEW

Trace one figure. Try to fit the tracing on the other figure. You may have to flip the tracing face down. Are the two figures congruent?

1. 

2. 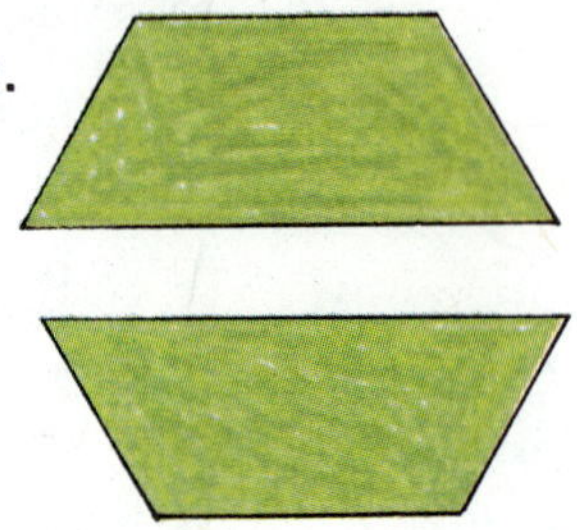

3. 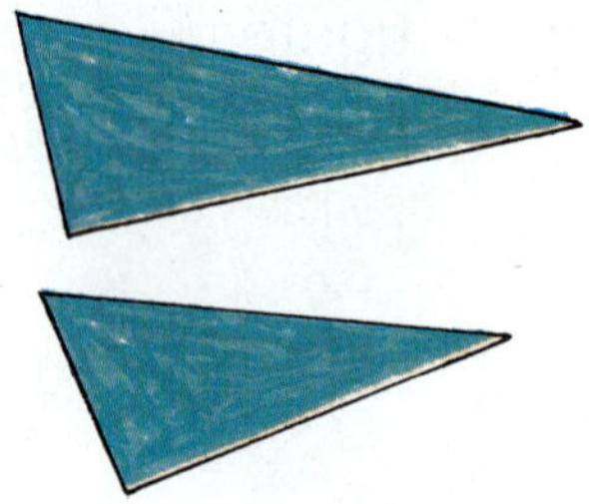

What is the measure of each angle?

4. 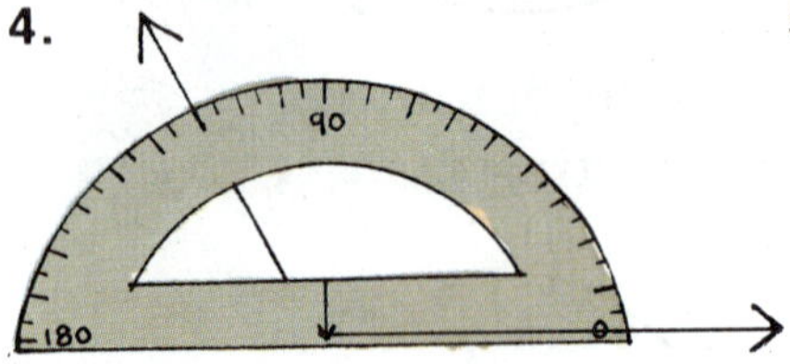

5. 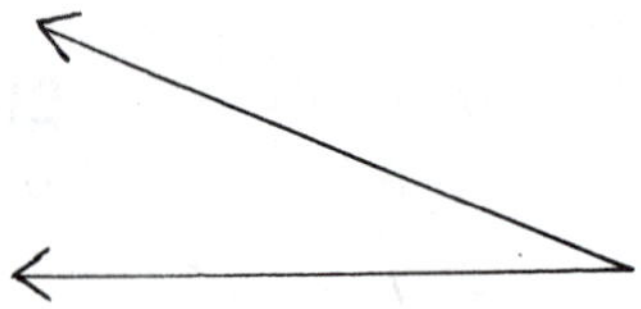

6. 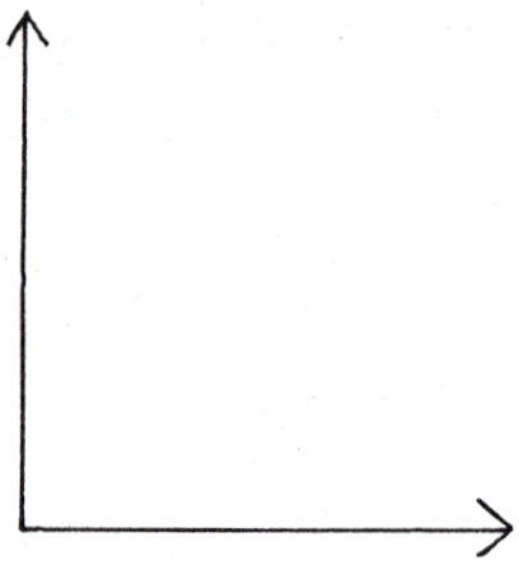

7–9. Is each angle in exercises 4–6 acute, right, or obtuse?

Trace the figures below. Fold the tracings on the dashed lines. Do the two parts fit? Is the dashed line a line of symmetry?

10. 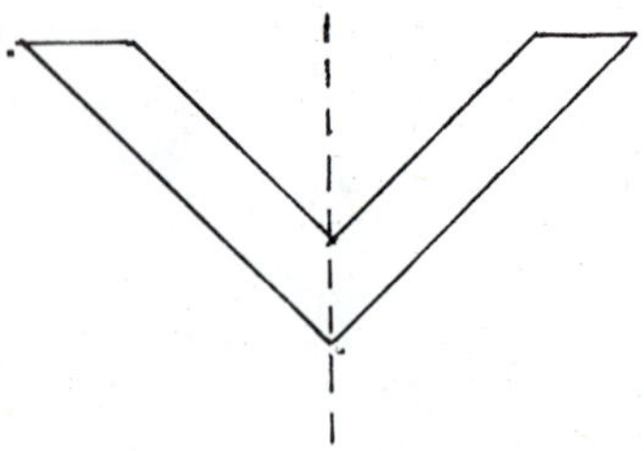

11. 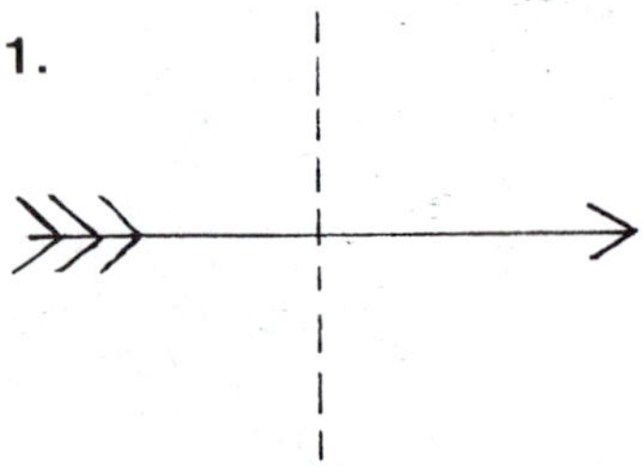

12. 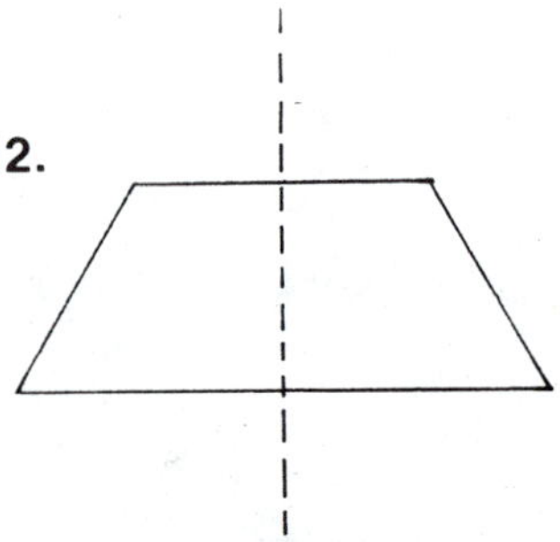

13. 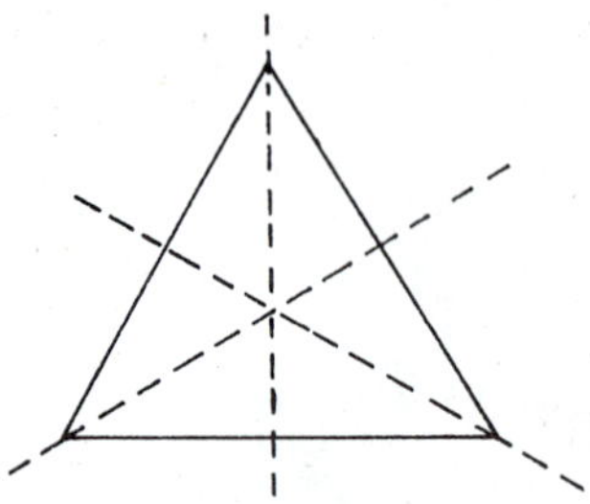

14. 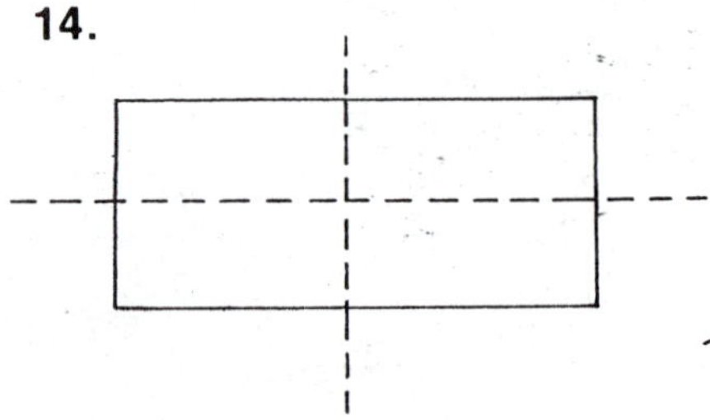

15. 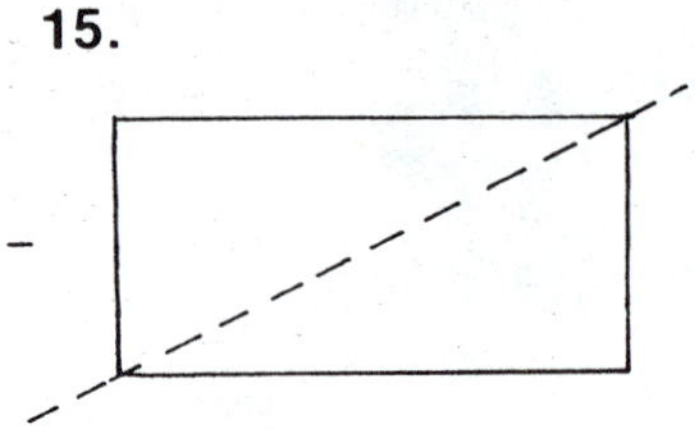

CHAPTER CHALLENGE

Here is how to use a compass and a straightedge to make a hexagon (6-sided figure).

Step 1. Draw a circle.

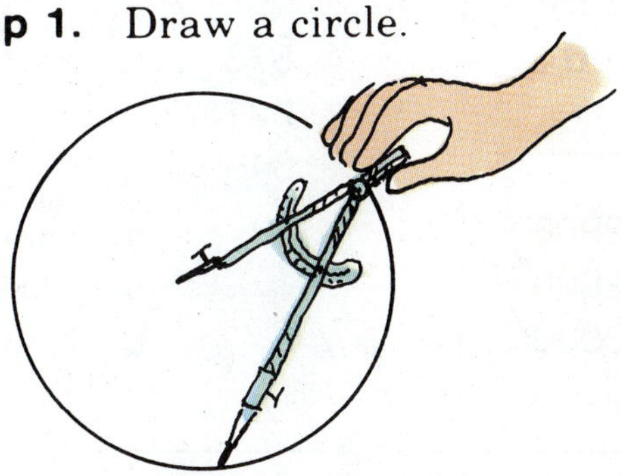

Step 2. Do not change the compass setting. Put the compass point on the circle and make a mark on the circle.

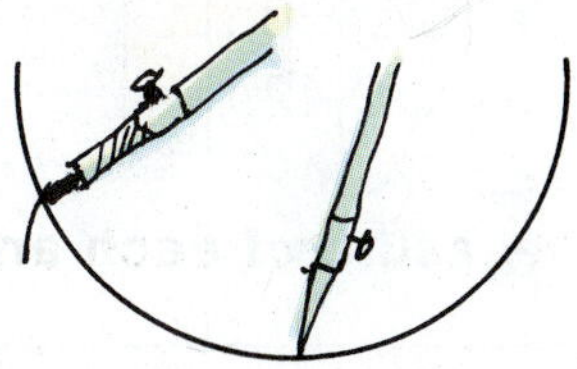

Step 3. Move your compass point to the new mark and repeat.

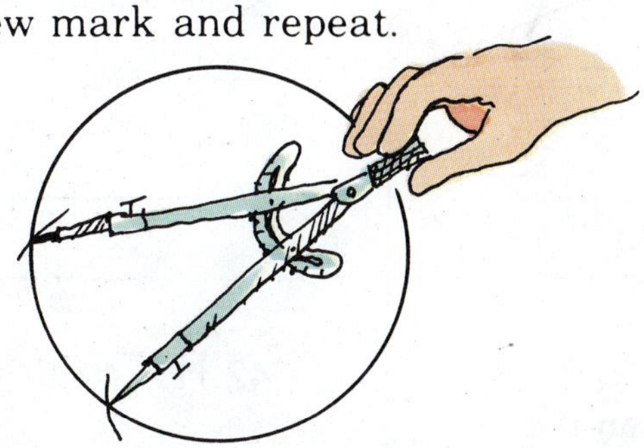

Step 4. Repeat until you get back to your starting point. Connect the points.

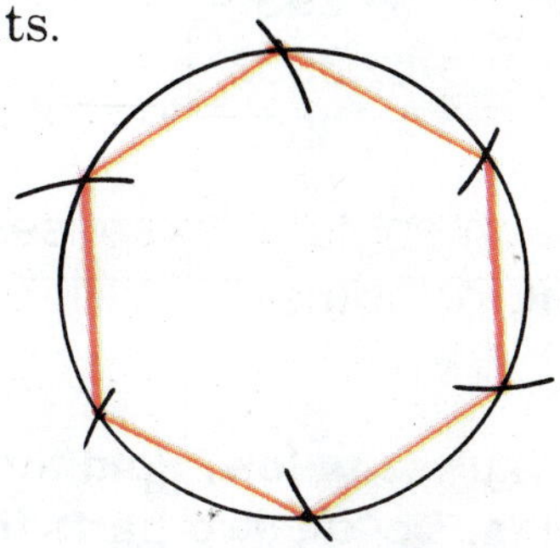

Are the angles all congruent? Are the sides all congruent?

These designs were made using the hexagon as a basis. Can you make these designs?

Make your own designs.

MAJOR CHECKUP

Standardized Format

Choose the correct letter.

1. The standard numeral for fifty-seven thousand forty-two is
   - a. 57,420
   - b. 570,420
   - c. 57,042
   - d. none of these

2. Round 2850 to the nearest hundred.
   - a. 2900
   - b. 2800
   - c. 3000
   - d. none of these

3. Add.

$$\begin{array}{r} 673 \\ 221 \\ +346 \\ \hline \end{array}$$

   - a. 12,040
   - b. 1231
   - c. 1240
   - d. none of these

4. Subtract.

$$\begin{array}{r} 4834 \\ -1897 \\ \hline \end{array}$$

   - a. 2937
   - b. 3063
   - c. 3037
   - d. none of these

5. Subtract.

$$\begin{array}{r} \$6.05 \\ -\$1.26 \\ \hline \end{array}$$

   - a. $4.89
   - b. $4.79
   - c. $5.89
   - d. none of these

6. Multiply.

$$\begin{array}{r} 631 \\ \times 4 \\ \hline \end{array}$$

   - a. 1924
   - b. 2524
   - c. 2520
   - d. none of these

7. Multiply.

$$\begin{array}{r} 281 \\ \times 35 \\ \hline \end{array}$$

   - a. 9835
   - b. 8575
   - c. 2248
   - d. none of these

8. Divide.

$$6\overline{)5734}$$

   - a. 955 R4
   - b. 955
   - c. 954
   - d. none of these

9. Divide.

$$42\overline{)1134}$$

   - a. 32
   - b. 28
   - c. 27
   - d. none of these

10. Which figure is an isosceles triangle?
   - a. 
   - b. 
   - c. 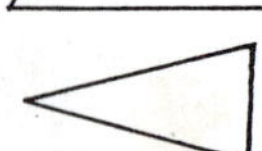
   - d. none of these

11. Which line is a line of symmetry?

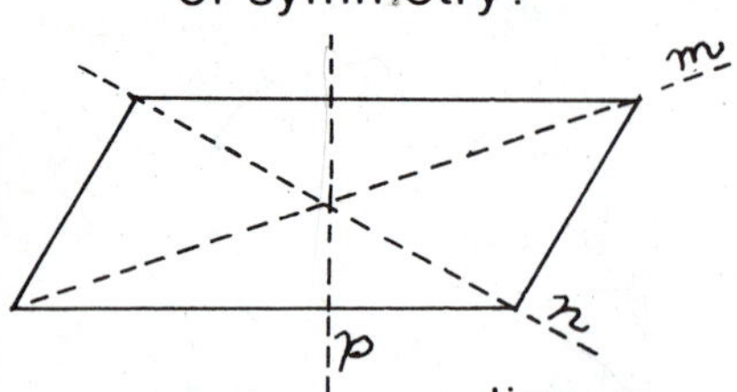

   - a. line *m*
   - b. line *n*
   - c. line *p*
   - d. none of these

12. Which figure is congruent to

   - a. 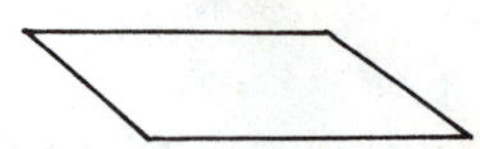
   - b. 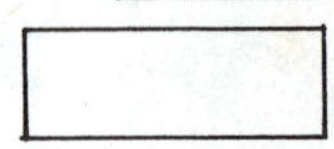
   - c. 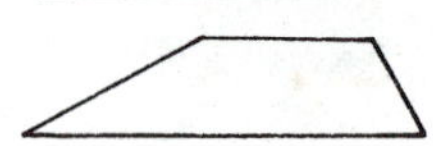
   - d. none of these

5 Fractions—Addition and Subtraction

READY OR NOT ?

Give the fraction that is shaded.

1. 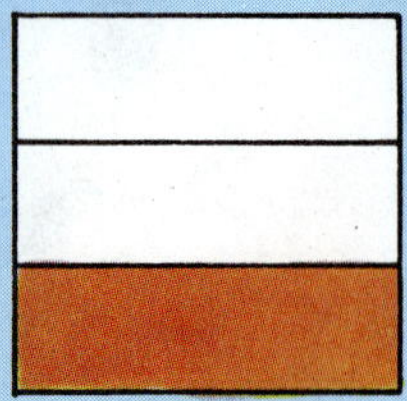

2. 

3. What is the numerator of $\frac{3}{8}$?
4. What is the numerator of $\frac{8}{3}$?
5. What is the denominator of $\frac{2}{5}$?

What is the least common multiple?

6. 2, 3 7. 3, 6 8. 4, 1

9. 6, 7 10. 4, 6 11. 6, 9

What is the greatest common factor?

12. 4, 6 13. 10, 15 14. 21, 28

15. 24, 18 16. 5, 6 17. 8, 13

Give the length of each segment.

18.

? inches

19.

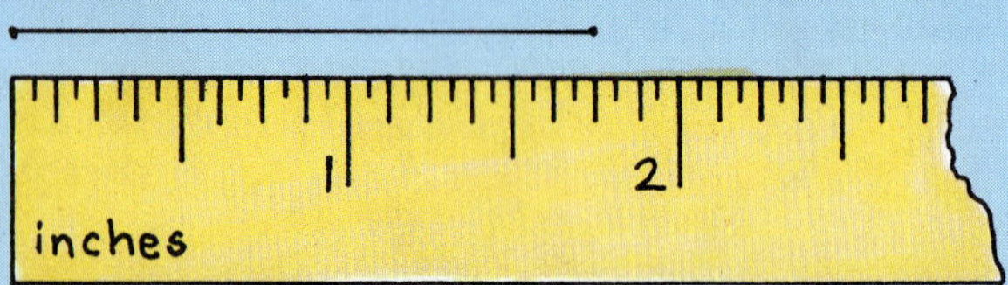

? inches

Equivalent fractions

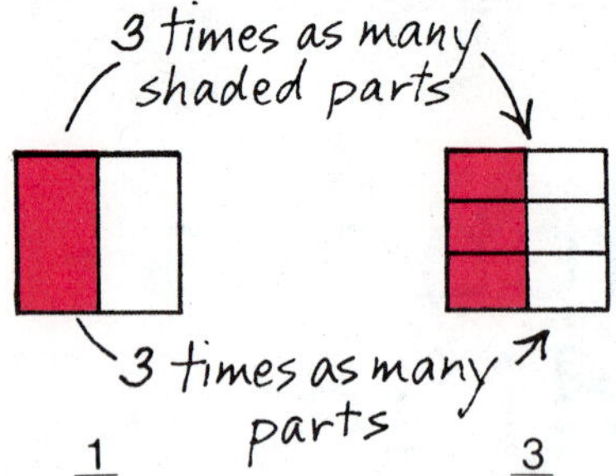

$\frac{1}{2}$ $\frac{3}{6}$

Notice that both squares have the same amount shaded. This means that the fractions $\frac{1}{2}$ and $\frac{3}{6}$ stand for the same number.

Also notice that the square on the right has 3 times as many parts **and** 3 times as many shaded parts.

The picture helps us understand why we can change from one fraction to an **equivalent fraction** by multiplying both numerator and denominator by the same number.

$$\frac{1}{2} \underset{\times 3}{\overset{\times 3}{=}} \frac{3}{6}$$

We have made the **terms** (numerator and denominator) larger. We have raised the fraction to **higher terms.**

We can do the reverse to **reduce** the fraction to **lower terms.** We can divide both terms by the same number and get an equivalent fraction.

$$\frac{3}{6} \underset{\div 3}{\overset{\div 3}{=}} \frac{1}{2}$$

EXERCISES

Complete to get an equivalent fraction.

1. $\frac{2}{3} = \frac{?}{12}$ (×4, ×4)

2. $\frac{1}{3} = \frac{?}{6}$ (×2, ×2)

3. $\frac{15}{20} = \frac{?}{4}$ (÷5, ÷5)

4. $\frac{1}{2} = \frac{?}{8}$ (×?, ×?)

5. $\frac{2}{8} = \frac{?}{4}$ (÷?, ÷?)

6. $\frac{9}{24} = \frac{?}{8}$ (÷?, ÷?)

7. $\frac{1}{6} = \frac{?}{18}$
8. $\frac{3}{5} = \frac{?}{15}$
9. $\frac{5}{8} = \frac{?}{16}$
10. $\frac{2}{5} = \frac{?}{35}$
11. $\frac{6}{8} = \frac{?}{4}$
12. $\frac{6}{9} = \frac{?}{3}$
13. $\frac{9}{12} = \frac{?}{4}$
14. $\frac{8}{12} = \frac{?}{3}$
15. $\frac{12}{15} = \frac{?}{5}$
16. $\frac{3}{8} = \frac{?}{32}$
17. $\frac{5}{7} = \frac{?}{21}$
18. $\frac{4}{5} = \frac{?}{25}$
19. $\frac{1}{4} = \frac{?}{16}$
20. $\frac{2}{3} = \frac{?}{21}$
21. $\frac{4}{9} = \frac{?}{45}$

Give the next three equivalent fractions.

22. $\frac{1}{3}, \frac{2}{6}, \frac{3}{9}, \frac{?}{?}, \frac{?}{?}, \frac{?}{?}$
23. $\frac{3}{4}, \frac{6}{8}, \frac{9}{12}, \frac{?}{?}, \frac{?}{?}, \frac{?}{?}$
24. $\frac{1}{2}, \frac{2}{4}, \frac{?}{?}, \frac{?}{?}, \frac{?}{?}$
25. $\frac{3}{5}, \frac{?}{?}, \frac{?}{?}, \frac{?}{?}$
26. $\frac{1}{4}, \frac{?}{?}, \frac{?}{?}, \frac{?}{?}$
27. $\frac{5}{6}, \frac{?}{?}, \frac{?}{?}, \frac{?}{?}$

Keeping Skills Sharp

Give the greatest common factor.

1. 12, 18
2. 6, 15
3. 10, 15
4. 9, 12
5. 6, 7
6. 24, 36
7. 30, 40
8. 15, 22
9. 32, 36
10. 32, 40

Give the least common multiple.

11. 3, 5
12. 2, 5
13. 4, 8
14. 2, 8
15. 3, 7
16. 5, 10
17. 3, 10
18. 6, 7
19. 7, 8
20. 8, 12

Reducing fractions to lowest terms

A fraction is in **lowest terms** if the greatest common factor of the numerator and denominator is 1.

To reduce a fraction to lowest terms:

1. Find the greatest common factor of the numerator and denominator. $\frac{12}{18}$

2. Divide both terms by their greatest common factor. $\frac{12}{18} \overset{\div 6}{\underset{\div 6}{=}} \frac{2}{3}$

$\frac{\overset{2}{\cancel{12}}}{\underset{3}{\cancel{18}}}$ $\frac{\overset{2}{\cancel{6}}}{\underset{5}{\cancel{15}}}$ $\frac{\overset{3}{\cancel{15}}}{\underset{1}{\cancel{5}}} = 3$

EXERCISES

Reduce to lowest terms.

1. $\frac{10}{15}$ 2. $\frac{9}{12}$ 3. $\frac{6}{7}$ 4. $\frac{24}{36}$ 5. $\frac{40}{30}$ 6. $\frac{15}{22}$ 7. $\frac{32}{36}$

8. $\frac{18}{6}$ 9. $\frac{15}{45}$ 10. $\frac{18}{14}$ 11. $\frac{24}{32}$ 12. $\frac{28}{16}$ 13. $\frac{16}{24}$ 14. $\frac{20}{24}$

Give answers in lowest terms.

15. What fraction of a foot is 4 inches?

16. What fraction of a yard is 24 inches?

17. What fraction of a pound is 12 ounces?

18. What fraction of an hour is 15 minutes?

Skill Game

Equivalent Fractions Bingo

1. Pick 16 fractions from the list below. Fill in a card like this one, putting your fractions in any order you wish.

$\frac{1}{2}$, $\frac{3}{2}$, $\frac{1}{3}$, $\frac{2}{3}$, $\frac{1}{4}$, $\frac{3}{4}$, $\frac{5}{4}$,

$\frac{1}{5}$, $\frac{2}{5}$, $\frac{3}{5}$, $\frac{4}{5}$, $\frac{1}{6}$, $\frac{5}{6}$, $\frac{1}{8}$, $\frac{3}{8}$,

$\frac{5}{8}$, $\frac{7}{8}$, $\frac{9}{8}$, $\frac{1}{10}$, $\frac{3}{10}$, $\frac{7}{10}$, $\frac{9}{10}$

$\frac{7}{8}$	$\frac{1}{4}$	$\frac{1}{10}$	$\frac{5}{4}$
$\frac{1}{6}$	$\frac{1}{8}$	$\frac{3}{5}$	$\frac{5}{6}$
$\frac{5}{8}$	$\frac{4}{5}$	$\frac{1}{2}$	$\frac{3}{4}$
$\frac{3}{2}$	$\frac{2}{5}$	$\frac{3}{10}$	$\frac{3}{8}$

2. Your teacher will hold up a card like this one:
3. Reduce that fraction to lowest terms.
4. If that fraction is on your card, put a marker on it.
5. Repeat until someone wins by getting four markers in a row.

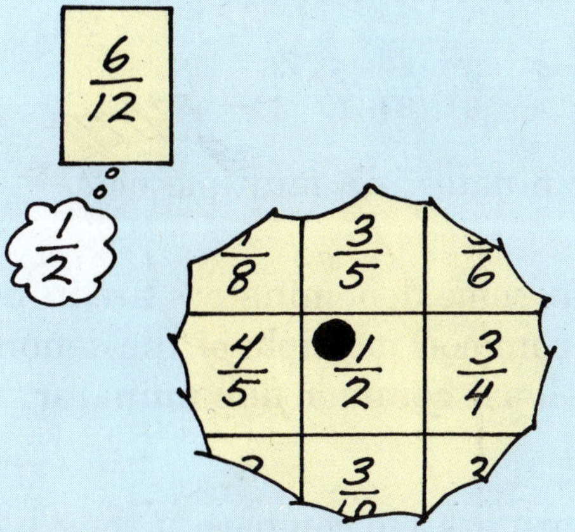

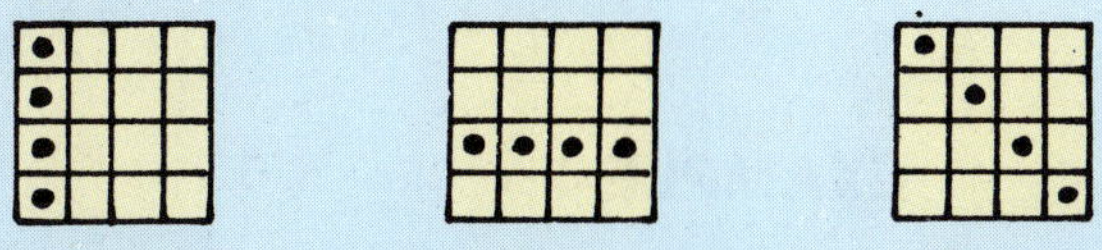

Art and Betty made these cards.

Art

$\frac{5}{6}$	$\frac{1}{10}$	$\frac{5}{4}$	$\frac{1}{5}$
$\frac{1}{2}$	$\frac{3}{4}$	$\frac{7}{10}$	$\frac{1}{8}$
$\frac{4}{5}$	$\frac{1}{6}$	$\frac{3}{2}$	$\frac{9}{10}$
$\frac{1}{4}$	$\frac{3}{8}$	$\frac{9}{8}$	$\frac{2}{5}$

Betty

$\frac{4}{5}$	$\frac{2}{3}$	$\frac{3}{4}$	$\frac{5}{6}$
$\frac{1}{6}$	$\frac{3}{10}$	$\frac{1}{2}$	$\frac{5}{4}$
$\frac{1}{3}$	$\frac{5}{8}$	$\frac{3}{8}$	$\frac{1}{5}$
$\frac{3}{5}$	$\frac{7}{10}$	$\frac{2}{5}$	$\frac{1}{8}$

The fraction cards were picked in the following order. Who won?

$\frac{8}{10}$, $\frac{5}{10}$, $\frac{2}{12}$, $\frac{3}{12}$, $\frac{12}{16}$,

$\frac{6}{20}$, $\frac{10}{16}$, $\frac{8}{20}$, $\frac{9}{15}$, $\frac{14}{20}$, $\frac{9}{6}$, $\frac{25}{20}$, $\frac{9}{24}$, $\frac{4}{32}$, $\frac{28}{32}$, $\frac{4}{20}$,

$\frac{5}{15}$, $\frac{8}{12}$, $\frac{25}{30}$, $\frac{27}{30}$, $\frac{5}{50}$, $\frac{36}{32}$.

Comparing fractions

Ralph washed $\frac{1}{2}$ of the windows on the 46th floor.

Some fractions equivalent to $\frac{1}{2}$:

$$\frac{1}{2}, \frac{2}{4}, \frac{3}{6}, \frac{4}{8}, \frac{5}{10}, \frac{6}{12}$$

Each denominator is a multiple of 2.

Alex washed $\frac{2}{3}$ of the windows on the 47th floor.

Some fractions equivalent to $\frac{2}{3}$:

$$\frac{2}{3}, \frac{4}{6}, \frac{6}{9}, \frac{8}{12}, \frac{10}{15}, \frac{12}{18}$$

Each denominator is a multiple of 3.

Notice that some denominators are in both lists. The least common multiple of the denominators is called the **least common denominator.**

Fractions are easy to compare if they have the same denominator.

EXAMPLE 1.

Compare $\frac{1}{2}$ and $\frac{2}{3}$.

Think of equivalent fractions having the least common denominator 6.

$$\left(\frac{3}{6}\right) \quad \frac{1}{2} < \frac{2}{3} \quad \left(\frac{4}{6}\right)$$

Ralph washed fewer windows than Alex did.

EXAMPLE 2.

Compare $\frac{3}{4}$ and $\frac{5}{8}$.

The least common denominator is 8.

$$\left(\frac{6}{8}\right) \quad \frac{3}{4} > \frac{5}{8} \quad \left(\frac{5}{8}\right)$$

EXERCISES

Give the least common denominator.

1. $\frac{1}{2}, \frac{1}{3}$ 2. $\frac{3}{2}, \frac{2}{3}$ 3. $\frac{1}{4}, \frac{2}{3}$ 4. $\frac{1}{2}, \frac{1}{4}$ 5. $\frac{1}{3}, \frac{5}{6}$ 6. $\frac{3}{4}, \frac{2}{5}$

7. $\frac{1}{3}, \frac{3}{5}$ 8. $\frac{1}{2}, \frac{3}{5}$ 9. $\frac{3}{4}, \frac{5}{8}$ 10. $\frac{1}{4}, \frac{1}{8}$ 11. $\frac{1}{2}, \frac{5}{8}$ 12. $\frac{1}{3}, \frac{2}{7}$

13. $\frac{1}{5}, \frac{3}{10}$ 14. $\frac{1}{3}, \frac{7}{10}$ 15. $\frac{5}{6}, \frac{3}{7}$ 16. $\frac{2}{7}, \frac{3}{8}$ 17. $\frac{1}{8}, \frac{5}{12}$ 18. $\frac{2}{3}, \frac{1}{6}$

Find the least common denominator. Change to equivalent fractions having that denominator.

19. $\frac{2}{3} = \frac{8}{12}$, $\frac{1}{4} = \frac{3}{12}$ 20. $\frac{1}{2}$, $\frac{1}{3}$ 21. $\frac{3}{2}$, $\frac{2}{3}$ 22. $\frac{3}{4}$, $\frac{2}{3}$ 23. $\frac{1}{2}$, $\frac{1}{4}$

24. $\frac{1}{3}$, $\frac{5}{6}$ 25. $\frac{3}{4}$, $\frac{2}{5}$ 26. $\frac{1}{3}$, $\frac{3}{5}$ 27. $\frac{1}{2}$, $\frac{3}{5}$ 28. $\frac{3}{4}$, $\frac{5}{8}$

29. $\frac{1}{4}$, $\frac{1}{8}$ 30. $\frac{1}{2}$, $\frac{5}{8}$ 31. $\frac{1}{3}$, $\frac{2}{7}$ 32. $\frac{1}{5}$, $\frac{3}{10}$ 33. $\frac{1}{3}$, $\frac{7}{10}$

<, =, or >?

34. $\frac{2}{8} \bigcirc \frac{5}{8}$ 35. $\frac{3}{4} \bigcirc \frac{1}{4}$ 36. $\frac{1}{2} \bigcirc \frac{1}{3}$ 37. $\frac{3}{8} \bigcirc \frac{1}{2}$

38. $\frac{3}{4} \bigcirc \frac{2}{3}$ 39. $\frac{5}{6} \bigcirc \frac{3}{4}$ 40. $\frac{5}{9} \bigcirc \frac{3}{8}$ 41. $\frac{4}{5} \bigcirc \frac{2}{3}$

42. $\frac{4}{7} \bigcirc \frac{3}{5}$ 43. $\frac{2}{3} \bigcirc \frac{6}{9}$ 44. $\frac{5}{9} \bigcirc \frac{1}{2}$ 45. $\frac{3}{8} \bigcirc \frac{5}{9}$

46. $\frac{3}{4} \bigcirc \frac{3}{5}$ 47. $\frac{7}{8} \bigcirc \frac{5}{6}$ 48. $\frac{4}{9} \bigcirc \frac{3}{8}$ 49. $\frac{3}{5} \bigcirc \frac{12}{20}$

Solve.

50. Jerry was $\frac{1}{2}$ of the way to the top of the building and Tim was $\frac{1}{3}$ of the way. Who was higher?

51. Jerry worked $\frac{3}{8}$ of a day and Tim worked $\frac{1}{2}$ of a day. Who worked less?

Adding fractions

To add two fractions that have the same denominator:

1. Add the numerators to get the numerator of the sum.
2. Use the common denominator for the denominator of the sum.

$$\frac{3}{5} + \frac{1}{5} = \frac{4}{5}$$

To add fractions that have different denominators:

1. Find the least common denominator.
2. Change to equivalent fractions.
3. Add.

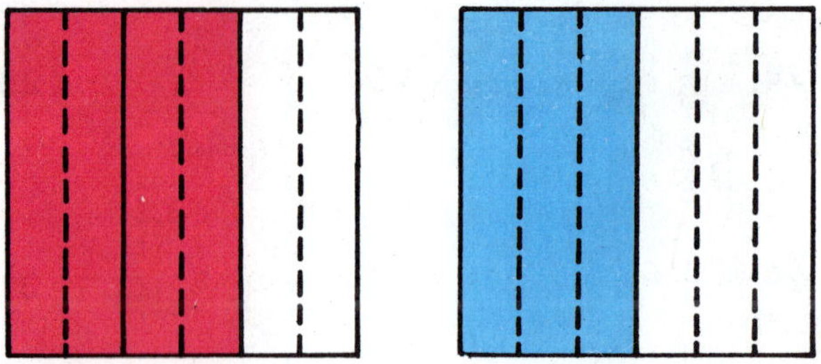

$$\frac{2}{3} + \frac{1}{2} = \frac{4}{6} + \frac{3}{6}$$
$$= \frac{7}{6}$$

EXERCISES

Give each sum in lowest terms.

1. $\frac{1}{3} + \frac{1}{3}$	**2.** $\frac{1}{4} + \frac{1}{4}$	**3.** $\frac{2}{4} + \frac{1}{4}$	**4.** $\frac{1}{2} + \frac{3}{4}$
5. $\frac{5}{8} + \frac{1}{4}$	**6.** $\frac{5}{6} + \frac{2}{3}$	**7.** $\frac{0}{2} + \frac{3}{4}$	**8.** $\frac{5}{6} + \frac{2}{9}$
9. $\frac{7}{8} + \frac{3}{4}$	**10.** $\frac{5}{9} + \frac{1}{6}$	**11.** $\frac{3}{7} + \frac{1}{2}$	**12.** $\frac{1}{3} + \frac{5}{8}$
13. $\frac{7}{8} + \frac{2}{3}$	**14.** $\frac{1}{4} + \frac{2}{3}$	**15.** $\frac{2}{3} + \frac{3}{4}$	**16.** $\frac{8}{9} + \frac{2}{1}$
17. $\frac{3}{5} + \frac{5}{3}$	**18.** $\frac{7}{3} + \frac{3}{7}$	**19.** $\frac{1}{2} + \frac{2}{5}$	**20.** $\frac{1}{5} + \frac{5}{11}$

Add. Give each sum in lowest terms.

21. $\frac{3}{4} = \frac{6}{8}$ $+\frac{5}{8} = \frac{5}{8}$	22. $\frac{5}{9}$ $+\frac{2}{3}$	23. $\frac{1}{2}$ $+\frac{1}{6}$	24. $\frac{1}{3}$ $+\frac{1}{4}$	25. $\frac{1}{3}$ $+\frac{1}{6}$	
26. $\frac{2}{5}$ $+\frac{3}{10}$	27. $\frac{5}{6}$ $+\frac{4}{9}$	28. $\frac{2}{3}$ $+\frac{3}{8}$	29. $\frac{2}{3}$ $+\frac{1}{9}$	30. $\frac{1}{8}$ $+\frac{2}{3}$	31. $\frac{1}{3}$ $+\frac{5}{6}$
32. $\frac{5}{6}$ $+\frac{4}{3}$	33. $\frac{5}{16}$ $+\frac{1}{8}$	34. $\frac{3}{8}$ $+\frac{9}{16}$	35. $\frac{1}{4}$ $+\frac{1}{10}$	36. $\frac{2}{3}$ $+\frac{2}{5}$	37. $\frac{5}{3}$ $+\frac{1}{2}$
38. $\frac{1}{2}$ $+\frac{1}{3}$	39. $\frac{3}{4}$ $+\frac{2}{2}$	40. $\frac{0}{5}$ $+\frac{3}{10}$	41. $\frac{1}{2}$ $+\frac{5}{16}$	42. $\frac{1}{9}$ $+\frac{7}{3}$	43. $\frac{3}{4}$ $+\frac{1}{6}$
44. $\frac{2}{3}$ $+\frac{5}{6}$	45. $\frac{3}{5}$ $+\frac{5}{4}$	46. $\frac{3}{4}$ $+\frac{3}{8}$	47. $\frac{3}{16}$ $+\frac{7}{8}$	48. $\frac{5}{8}$ $+\frac{1}{3}$	49. $\frac{1}{2}$ $+\frac{4}{15}$

Solve.

50. Charles studied $\frac{1}{2}$ hour before dinner and $\frac{3}{4}$ hour after dinner. How many hours did he study?

51. Joan stood on one leg for $\frac{1}{6}$ hour on her first try and for $\frac{1}{4}$ hour on her second try. How many hours is that?

Add across. Add down.

52.

		⊕ →
$\frac{1}{2}$	$\frac{1}{3}$	?
$\frac{2}{3}$	$\frac{3}{4}$	?
?	?	?

53.

		⊕ →
$\frac{5}{8}$	$\frac{1}{2}$	?
$\frac{1}{4}$	$\frac{3}{8}$	?
?	?	?

54.

		⊕ →
$\frac{5}{9}$	$\frac{1}{3}$	?
$\frac{1}{2}$	$\frac{5}{6}$	?
?	?	?

Subtracting fractions

Amy had $\frac{3}{4}$ of a cup of syrup. She used $\frac{2}{3}$ of a cup to make caramel apples. How much did she have left?

To solve the problem, we subtract. Notice that we first write equivalent fractions having the least common denominator.

$$\begin{aligned} \frac{3}{4} - \frac{2}{3} &= \frac{9}{12} - \frac{8}{12} \\ &= \frac{1}{12} \end{aligned}$$

She had $\frac{1}{12}$ of a cup left.

EXERCISES

Give each difference in lowest terms.

1. $\frac{2}{3} - \frac{1}{3}$	**2.** $\frac{3}{4} - \frac{1}{4}$	**3.** $\frac{5}{8} - \frac{3}{8}$	**4.** $\frac{3}{4} - \frac{1}{2}$
5. $\frac{5}{6} - \frac{1}{3}$	**6.** $\frac{3}{2} - \frac{3}{4}$	**7.** $\frac{7}{8} - \frac{1}{3}$	**8.** $\frac{2}{3} - \frac{5}{9}$
9. $\frac{9}{10} - \frac{3}{5}$	**10.** $\frac{7}{12} - \frac{1}{3}$	**11.** $\frac{3}{4} - \frac{1}{3}$	**12.** $\frac{5}{6} - \frac{2}{3}$
13. $\frac{3}{4} - \frac{3}{8}$	**14.** $\frac{2}{3} - \frac{5}{12}$	**15.** $\frac{7}{8} - \frac{3}{4}$	**16.** $\frac{3}{2} - \frac{7}{8}$
17. $\frac{5}{6} - \frac{0}{3}$	**18.** $\frac{3}{4} - \frac{2}{3}$	**19.** $\frac{1}{2} - \frac{2}{5}$	**20.** $\frac{1}{2} - \frac{3}{7}$

Subtract. Give each difference in lowest terms.

21. $\frac{5}{6} - \frac{1}{4}$	**22.** $\frac{3}{8} - \frac{1}{3}$	**23.** $\frac{5}{9} - \frac{1}{6}$	**24.** $\frac{7}{4} - \frac{3}{4}$	**25.** $\frac{9}{5} - \frac{3}{10}$	**26.** $\frac{7}{8} - \frac{5}{6}$
27. $\frac{5}{8} - \frac{1}{6}$	**28.** $\frac{1}{2} - \frac{1}{3}$	**29.** $\frac{7}{10} - \frac{1}{5}$	**30.** $\frac{5}{9} - \frac{1}{3}$	**31.** $\frac{3}{4} - \frac{5}{16}$	**32.** $\frac{3}{5} - \frac{1}{2}$
33. $\frac{3}{4} - \frac{3}{8}$	**34.** $\frac{4}{4} - \frac{3}{3}$	**35.** $\frac{1}{3} - \frac{1}{5}$	**36.** $\frac{3}{2} - \frac{5}{6}$	**37.** $\frac{1}{6} - \frac{1}{8}$	**38.** $\frac{5}{12} - \frac{1}{3}$
39. $\frac{8}{9} - \frac{1}{6}$	**40.** $\frac{5}{6} - \frac{5}{8}$	**41.** $\frac{4}{3} - \frac{5}{6}$	**42.** $\frac{5}{8} - \frac{1}{3}$	**43.** $\frac{7}{4} - \frac{0}{8}$	**44.** $\frac{3}{4} - \frac{1}{5}$
45. $\frac{11}{12} - \frac{2}{3}$	**46.** $\frac{3}{4} - \frac{3}{10}$	**47.** $\frac{3}{4} - \frac{1}{7}$	**48.** $\frac{5}{6} - \frac{1}{3}$	**49.** $\frac{9}{10} - \frac{1}{4}$	**50.** $\frac{7}{6} - \frac{3}{4}$

Solve.

51. Bill has $\frac{3}{4}$ of a cup of brown sugar. How much more is needed for the recipe?

52. Before making the caramel apples, Bill had 2 cans of condensed milk. How much did he have after making the caramel apples?

Practice exercises

Add or subtract.

1. $\frac{3}{4} - \frac{1}{4}$	2. $\frac{1}{4} + \frac{1}{2}$	3. $\frac{1}{2} - \frac{1}{4}$	4. $\frac{1}{2} + \frac{1}{3}$	5. $\frac{5}{9} - \frac{1}{3}$	6. $\frac{5}{4} - \frac{1}{2}$
7. $\frac{5}{6} + \frac{2}{9}$	8. $\frac{7}{8} - \frac{3}{4}$	9. $\frac{5}{6} + \frac{2}{3}$	10. $\frac{1}{2} - \frac{3}{8}$	11. $\frac{6}{5} + \frac{2}{10}$	12. $\frac{5}{6} - \frac{2}{3}$
13. $\frac{1}{2} + \frac{5}{6}$	14. $\frac{1}{2} - \frac{1}{3}$	15. $\frac{3}{4} - \frac{3}{7}$	16. $\frac{0}{5} + \frac{2}{3}$	17. $\frac{3}{10} - \frac{1}{5}$	18. $\frac{3}{2} + \frac{2}{3}$
19. $\frac{5}{8} + \frac{1}{2}$	20. $\frac{2}{3} - \frac{1}{4}$	21. $\frac{2}{3} + \frac{2}{2}$	22. $\frac{2}{3} + \frac{3}{4}$	23. $\frac{4}{5} + \frac{2}{3}$	24. $\frac{3}{4} - \frac{1}{5}$
25. $\frac{6}{10} - \frac{3}{5}$	26. $\frac{3}{7} + \frac{1}{5}$	27. $\frac{5}{12} - \frac{1}{3}$	28. $\frac{5}{12} + \frac{2}{3}$	29. $\frac{9}{8} - \frac{2}{3}$	30. $\frac{3}{2} + \frac{3}{8}$
31. $\frac{1}{4} + \frac{5}{8}$	32. $\frac{3}{8} + \frac{1}{6}$	33. $\frac{3}{4} - \frac{3}{10}$	34. $\frac{5}{12} + \frac{1}{4}$	35. $\frac{5}{6} + \frac{7}{10}$	36. $\frac{9}{16} - \frac{1}{4}$
37. $\frac{9}{10} + \frac{3}{5}$	38. $\frac{13}{16} - \frac{3}{8}$	39. $\frac{5}{6} + \frac{3}{3}$	40. $\frac{13}{12} - \frac{5}{6}$	41. $\frac{3}{5} + \frac{1}{3}$	42. $\frac{7}{8} + \frac{1}{3}$
43. $\frac{2}{3} - \frac{3}{5}$	44. $\frac{1}{2} + \frac{3}{10}$	45. $\frac{1}{2} + \frac{5}{7}$	46. $\frac{1}{4} + \frac{7}{10}$	47. $\frac{3}{10} - \frac{1}{4}$	48. $\frac{2}{9} - \frac{1}{6}$

Compute. Remember to work inside the grouping symbols first.

49. $\left(\frac{3}{5} + \frac{1}{2}\right) - \frac{3}{10}$

50. $\frac{3}{5} + \left(\frac{1}{2} - \frac{3}{10}\right)$

51. $\left(\frac{7}{8} - \frac{1}{2}\right) - \frac{1}{4}$

52. $\frac{7}{8} - \left(\frac{1}{2} - \frac{1}{4}\right)$

53. $\left(\frac{5}{9} + \frac{1}{3}\right) + \frac{1}{2}$

54. $\frac{5}{9} + \left(\frac{1}{3} + \frac{1}{2}\right)$

55. $\frac{5}{6} - \left(\frac{1}{3} + \frac{1}{2}\right)$

56. $\left(\frac{5}{6} - \frac{1}{3}\right) + \frac{1}{2}$

57. $\left(\frac{7}{8} - \frac{1}{4}\right) + \frac{1}{2}$

58. $\frac{7}{8} - \left(\frac{1}{4} + \frac{1}{2}\right)$

59. $\left(\frac{5}{12} + \frac{2}{3}\right) + \frac{1}{4}$

60. $\frac{5}{12} + \left(\frac{2}{3} + \frac{1}{4}\right)$

61. $\left(\frac{5}{16} + \frac{3}{8}\right) - \frac{1}{4}$

62. $\frac{5}{16} + \left(\frac{3}{8} - \frac{1}{4}\right)$

63. $\left(\frac{3}{8} + \frac{5}{12}\right) + \frac{2}{3}$

64. $\frac{3}{8} + \left(\frac{5}{12} + \frac{2}{3}\right)$

65. $\left(\frac{5}{2} - \frac{5}{6}\right) - \frac{1}{2}$

66. $\frac{5}{2} - \left(\frac{5}{6} - \frac{1}{2}\right)$

Add across. Subtract down.

67. (+ across, − down)

$\frac{3}{2}$	$\frac{3}{4}$	?
$\frac{1}{4}$	$\frac{5}{8}$	?
?	?	?

68. (+ across, − down)

$\frac{7}{4}$	$\frac{5}{6}$	?
$\frac{2}{3}$	$\frac{5}{12}$	?
?	?	?

69. (+ across, − down)

$\frac{9}{5}$	$\frac{1}{4}$	?
$\frac{3}{10}$	$\frac{1}{10}$	?
?	?	?

Work through the path to find the missing number.

70.

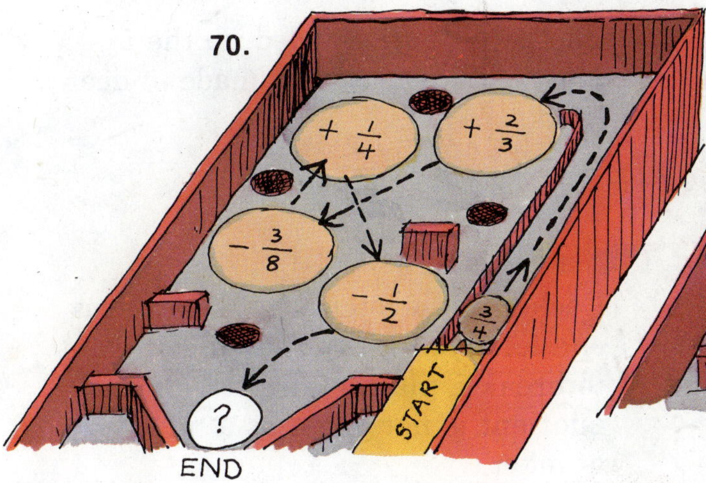

71.

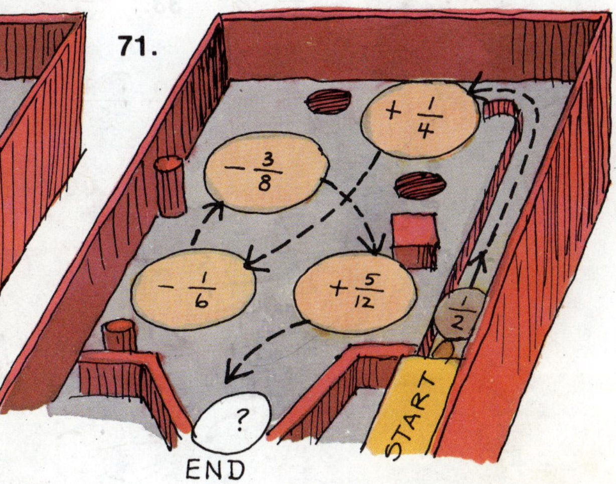

Division and mixed numbers

Four hikers divided 3 oranges.

$$3 \div 4 = \frac{3}{4}$$

Each got $\frac{3}{4}$ of an orange.

Two hikers divided 5 granola bars.

The extra bar was cut in half.

Mixed number → $2\frac{1}{2}$ ← This is a short way of writing $2 + \frac{1}{2}$

$$\begin{array}{r} 2\tfrac{1}{2} \\ 2\overline{)5} \\ -4 \\ \hline 1 \end{array}$$

Each got $2\frac{1}{2}$ bars. Notice that a mixed number has a whole number part and a fraction part.

The two hikers divided up the 5 plaster casts they had made of deer tracks.

$$\begin{array}{r} 2 \text{ R}1 \\ 2\overline{)5} \\ -4 \\ \hline 1 \end{array}$$

Each got 2 plaster casts. There was one left over. Since they didn't want to break a plaster cast in half, the quotient is not given as a mixed number.

EXERCISES

Give each quotient. Reduce to lowest terms.

1. $1 \div 3$
2. $2 \div 3$
3. $1 \div 4$
4. $3 \div 4$
5. $4 \div 5$
6. $3 \div 5$
7. $3 \div 7$
8. $4 \div 6$
9. $6 \div 9$
10. $1 \div 6$
11. $5 \div 6$
12. $7 \div 10$

Divide. Write your answers as whole numbers or mixed numbers. Reduce the fraction part to lowest terms.

13. $3\overline{)9}$
14. $3\overline{)11}$
15. $5\overline{)8}$
16. $2\overline{)9}$
17. $3\overline{)10}$
18. $4\overline{)28}$
19. $4\overline{)29}$
20. $4\overline{)30}$
21. $4\overline{)31}$
22. $4\overline{)32}$
23. $2\overline{)12}$
24. $2\overline{)13}$
25. $2\overline{)14}$
26. $2\overline{)15}$
27. $2\overline{)16}$
28. $3\overline{)43}$
29. $5\overline{)97}$
30. $8\overline{)93}$
31. $7\overline{)85}$
32. $6\overline{)75}$

Solve. Give your answer as a mixed number if the objects can be divided.

33. 8 pieces of bacon
3 people
How many pieces to each?

34. 5 baseballs
3 people
How many baseballs to each?

35. 3 pancakes
2 people
How many pancakes to each person?

36. 7 apples
2 people
How many apples to each person?

37. 15 sandwiches
6 people
How many sandwiches to each person?

38. 13 bus tokens
2 people
How many tokens to each person?

Keeping Skills Sharp

1. 42×9
2. 345×7
3. 1628×4
4. 52341×8
5. 730650×5
6. 25×12
7. 138×21
8. 5206×35
9. 783×152
10. 596×847

Mixed numbers and fractions

Many numbers can be written both as fractions and as mixed numbers. You must be able to change from one form to another.

Mixed Numbers to Fractions

Add the whole number part and the fraction part.

$$5\frac{2}{3} = \frac{5}{1} + \frac{2}{3} = \frac{15}{3} + \frac{2}{3} = \frac{17}{3}$$

Here is a shortcut.

$$5\frac{2}{3} = \frac{17}{3} \qquad 3\frac{1}{2} = \frac{7}{2}$$

Fractions to Mixed Numbers

Divide the numerator by the denominator.

$$\frac{17}{3} = 17 \div 3 = 5\frac{2}{3} \qquad \frac{29}{8} = 3\frac{5}{8}$$

EXERCISES

Change to fractions.

1. $1\frac{1}{2}$
2. $2\frac{1}{2}$
3. $3\frac{1}{2}$
4. $4\frac{1}{2}$
5. $5\frac{1}{2}$
6. $6\frac{1}{2}$
7. $1\frac{2}{3}$
8. $4\frac{1}{3}$
9. $6\frac{2}{3}$
10. $8\frac{1}{3}$
11. $3\frac{1}{4}$
12. $3\frac{3}{4}$

13. $5\frac{1}{6}$ 14. $7\frac{3}{4}$ 15. $5\frac{2}{5}$ 16. $4\frac{3}{4}$ 17. $6\frac{5}{6}$ 18. $3\frac{5}{8}$

19. $4\frac{7}{10}$ 20. $3\frac{3}{10}$ 21. $5\frac{9}{10}$ 22. $6\frac{3}{8}$ 23. $4\frac{1}{7}$ 24. $3\frac{1}{9}$

25. $5\frac{2}{5}$ 26. $7\frac{3}{8}$ 27. $9\frac{2}{3}$ 28. $15\frac{1}{2}$ 29. $19\frac{2}{3}$ 30. $25\frac{3}{4}$

Change to mixed numbers or whole numbers. Fraction parts should be in lowest terms.

31. $\frac{5}{2}$ 32. $\frac{6}{2}$ 33. $\frac{7}{2}$ 34. $\frac{10}{3}$ 35. $\frac{12}{5}$ 36. $\frac{14}{6}$

37. $\frac{15}{8}$ 38. $\frac{15}{7}$ 39. $\frac{15}{6}$ 40. $\frac{15}{5}$ 41. $\frac{15}{4}$ 42. $\frac{15}{3}$

43. $\frac{27}{5}$ 44. $\frac{39}{6}$ 45. $\frac{48}{7}$ 46. $\frac{56}{9}$ 47. $\frac{54}{7}$ 48. $\frac{38}{5}$

49. $\frac{37}{3}$ 50. $\frac{68}{5}$ 51. $\frac{73}{5}$ 52. $\frac{47}{4}$ 53. $\frac{27}{2}$ 54. $\frac{38}{3}$

55. $\frac{50}{4}$ 56. $\frac{60}{7}$ 57. $\frac{95}{8}$ 58. $\frac{43}{10}$ 59. $\frac{98}{10}$ 60. $\frac{203}{10}$

61. Kareem Abdul-Jabbar is about $\frac{43}{6}$ feet tall.

62. Bill Walton is about $\frac{83}{12}$ feet tall.

63. Dwight Stones has high jumped about $\frac{15}{2}$ feet.

64. Brenda Morehead has run the 100-meter dash in about $\frac{23}{2}$ seconds.

Problem solving

Remember:
To find a unit fractional part of a quantity (when the numerator is 1), divide by the denominator.

To find a fractional part of a quantity, divide by the denominator and multiply by the numerator.

EXAMPLE. How many eggs are in 2½ dozen?

$$2 \text{ dozen} = 2 \times 12 = 24$$
$$\frac{1}{2} \text{ dozen} = \frac{1}{2} \text{ of } 12 = 6$$
$$2\frac{1}{2} \text{ dozen} = 24 + 6 = 30$$

There are 30 eggs in $2\frac{1}{2}$ dozen.

EXERCISES

Solve.

1. $1\frac{1}{4}$ dozen doughnuts
   How many doughnuts is that?

2. Doughnuts cost $2.00 per dozen.
   How much will $1\frac{1}{2}$ dozen cost?
   How much will $2\frac{1}{4}$ dozen cost?

3. Hamburger costs $2.20 per pound. What do $2\frac{1}{2}$ pounds of hamburger cost?

4. A car can be driven 20 miles on one gallon of gasoline. How many miles can be driven on $3\frac{3}{4}$ gallons?

5. How many inches are there in $2\frac{3}{4}$ feet?

6. How many feet are there in $4\frac{2}{3}$ yards?

7.

How much for $10\frac{9}{10}$ gallons?

8.

How many egg rolls will $2\frac{1}{2}$ recipes make?

9. An indoor track has 12 laps to the mile. How many laps make $3\frac{3}{4}$ miles?

10. Cookies cost 84¢ per dozen. How much do 30 cookies cost?

11.

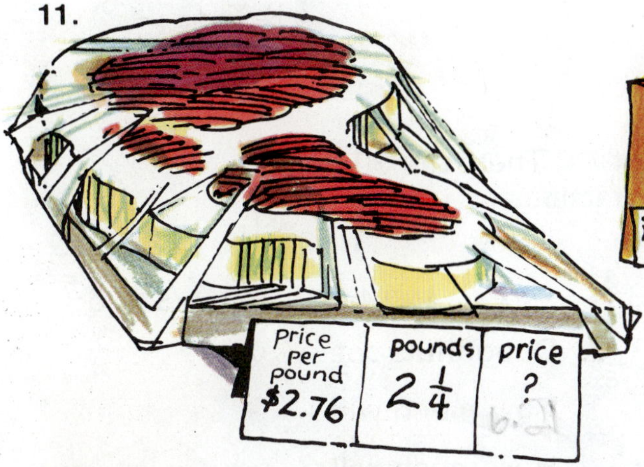

12.

How much for $3\frac{1}{6}$ six-packs?

13. Jenny ran 1 mile in 6 minutes. At that rate, how long would it take her to run $2\frac{1}{4}$ miles?

14. Bill drove 60 miles per hour for $2\frac{3}{4}$ hours. How far did he drive?

15. How many inches are in $5\frac{5}{6}$ feet?

16. How many ounces are in $6\frac{1}{4}$ pounds?

Adding and subtracting mixed numbers

Just add fractions and add whole numbers.

$$\begin{array}{r} 3\frac{1}{5} \\ +4\frac{3}{5} \\ \hline 7\frac{4}{5} \end{array}$$

Just subtract fractions and subtract whole numbers.

$$\begin{array}{r} 5\frac{4}{7} \\ -2\frac{3}{7} \\ \hline 3\frac{1}{7} \end{array}$$

Sometimes you will have to change to equivalent fractions with the same denominator.

$$\begin{array}{rcr} 6\frac{1}{2} & = & 6\frac{3}{6} \\ +4\frac{1}{3} & = & +4\frac{2}{6} \\ \hline & & 10\frac{5}{6} \end{array}$$

$$\begin{array}{rcr} 8\frac{3}{4} & = & 8\frac{3}{4} \\ -2\frac{1}{2} & = & -2\frac{2}{4} \\ \hline & & 6\frac{1}{4} \end{array}$$

EXERCISES

Add or subtract. Watch the signs. Reduce fractions to lowest terms.

1. $3\frac{1}{3} + 1\frac{1}{3}$
2. $4\frac{2}{5} + 7$
3. $3\frac{3}{5} + \frac{1}{5}$
4. $4\frac{3}{7} - 3\frac{1}{7}$
5. $5\frac{5}{6} - 4$
6. $3\frac{4}{5} - \frac{3}{5}$
7. $4\frac{3}{8} + 2\frac{1}{8}$
8. $4\frac{3}{8} - 2\frac{1}{8}$
9. $7\frac{5}{9} - 4\frac{2}{9}$
10. $3\frac{1}{7} + 4\frac{4}{7}$
11. $3\frac{1}{2} + 2\frac{1}{4}$
12. $3\frac{1}{2} - 2\frac{1}{4}$
13. $5\frac{3}{4} - 2\frac{1}{2}$
14. $4\frac{3}{8} + 1\frac{1}{2}$
15. $6\frac{7}{8} - 2\frac{3}{4}$

16. $4\frac{1}{3} + 2\frac{1}{2}$

17. $3\frac{2}{3} - 1\frac{1}{2}$

18. $2\frac{3}{4} + 1\frac{1}{6}$

19. $4\frac{3}{4} - 2\frac{2}{5}$

20. $12\frac{7}{8} - 6\frac{1}{3}$

21. $6\frac{3}{5} - 2\frac{1}{4}$

22. $5\frac{1}{4} + 6\frac{1}{6}$

23. $8\frac{3}{8} + 5\frac{1}{6}$

24. $9\frac{7}{8} - 4\frac{5}{6}$

25. $15\frac{3}{4} - 12\frac{6}{8}$

Solve.

26. Cindy ran $2\frac{1}{2}$ miles in the morning and $3\frac{1}{4}$ miles in the afternoon. How far did she run in all?

27. Bill ran $2\frac{1}{2}$ miles in the morning and $3\frac{3}{4}$ miles in the afternoon. How much farther did he run in the afternoon?

28. Jim drove for $2\frac{1}{4}$ hours and Joan drove for $3\frac{2}{3}$ hours. How much longer did Joan drive?

29. Carol ate $2\frac{1}{5}$ oranges and Carrie ate $1\frac{3}{4}$ oranges. How many oranges did they eat?

Add across. Add down.

30. (+ across, + down)

3	$\frac{1}{3}$	?
2	$\frac{1}{3}$	?
?	?	?

31. (+ across, + down)

4	$\frac{1}{2}$	?
5	$\frac{1}{4}$	?
?	?	?

32. (+ across, + down)

2	$\frac{1}{3}$	?
3	$\frac{1}{2}$	?
?	?	?

Add across. Subtract down.

33. (+ across, − down)

4	$\frac{2}{3}$	?
1	$\frac{1}{3}$	?
?	?	?

34. (+ across, − down)

5	$\frac{3}{4}$	?
2	$\frac{1}{2}$	?
?	?	?

35. (+ across, − down)

6	$\frac{2}{3}$	?
4	$\frac{1}{2}$	?
?	?	?

Mixed numbers—addition with regrouping

Sometimes when you add mixed numbers the fraction in the sum will be equal to 1 or more. When that happens you will have to regroup to make the fraction less than 1.

EXAMPLE 1.

$$\begin{array}{r} 2\frac{2}{3} \\ +1\frac{2}{3} \\ \hline 3\frac{4}{3} = 4\frac{1}{3} \end{array}$$

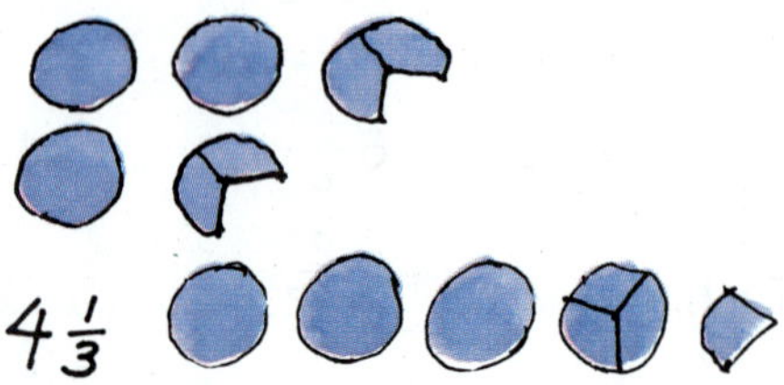

EXAMPLE 2.

$$\begin{array}{rcl} 3\frac{3}{4} & = & 3\frac{3}{4} \\ +2\frac{1}{2} & = & 2\frac{2}{4} \\ \hline & & 5\frac{5}{4} = 6\frac{1}{4} \end{array}$$

EXAMPLE 3.

$$\begin{array}{r} {\scriptstyle 1} \\ 3\frac{3}{5} \\ +8\frac{4}{5} \\ \hline 12\frac{2}{5} \end{array} \qquad \frac{7}{5} = 1\frac{2}{5}$$

EXERCISES

Add. The fraction should be in lowest terms and less than 1.

1. $4\frac{3}{5} + \frac{2}{5}$
2. $6\frac{1}{3} + \frac{2}{3}$
3. $10\frac{7}{8} + \frac{1}{8}$
4. $2\frac{5}{6} + \frac{1}{6}$
5. $3\frac{2}{5} + 1\frac{3}{5}$
6. $6\frac{3}{4} + 2\frac{3}{4}$
7. $5\frac{3}{7} + 4\frac{6}{7}$
8. $2\frac{3}{5} + 1\frac{4}{5}$
9. $3\frac{2}{3} + 5\frac{2}{3}$
10. $4\frac{7}{8} + 3\frac{5}{8}$

11. $4\frac{3}{4} + 1\frac{1}{2}$

12. $3\frac{2}{3} + 4\frac{1}{2}$

13. $7\frac{3}{8} + 4\frac{3}{4}$

14. $5\frac{5}{6} + 2\frac{1}{3}$

15. $3\frac{7}{8} + 1\frac{1}{2}$

16. $4\frac{3}{5} + 1\frac{2}{3}$

17. $5\frac{4}{5} + 6\frac{1}{4}$

18. $12\frac{3}{8} + 6\frac{2}{3}$

19. $5\frac{5}{6} + 3\frac{1}{2}$

20. $2\frac{7}{8} + 8\frac{3}{4}$

21. $4\frac{5}{8} + 3\frac{1}{8}$

22. $2\frac{3}{5} + 4\frac{1}{3}$

23. $6\frac{5}{6} + 3\frac{2}{3}$

24. $14\frac{3}{10} + 17\frac{3}{4}$

25. $13\frac{11}{12} + 4\frac{5}{6}$

Solve.

26.

How many pounds of apples in all?

27.

How many ounces of tuna in all?

Follow the path.

28.

29.

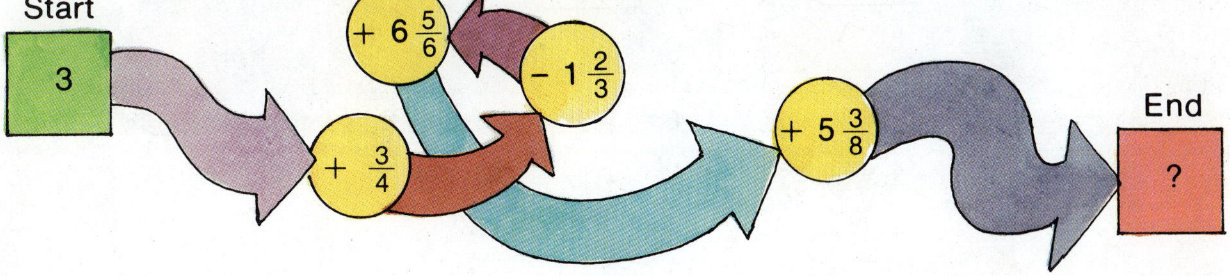

Mixed numbers–subtraction with regrouping

Just as you sometimes have to regroup in order to subtract whole numbers, you sometimes have to regroup to subtract mixed numbers.

EXAMPLE 1.

Step 1.

Not enough sixths.

$$\begin{array}{r} 5\frac{1}{6} \\ -\ 2\frac{5}{6} \\ \hline \end{array}$$

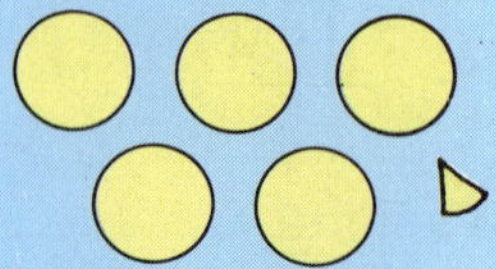

Step 2.

Regroup 1 whole for 6 sixths.

$$\begin{array}{rcl} 5\frac{1}{6} & = & 4\frac{7}{6} \\ -\ 2\frac{5}{6} & = & -2\frac{5}{6} \\ \hline \end{array}$$

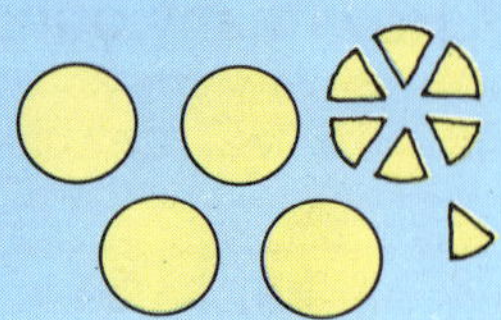

Step 3.

Subtract and reduce.

$$\begin{array}{rcl} 5\frac{1}{6} & = & 4\frac{7}{6} \\ -\ 2\frac{5}{6} & = & -2\frac{5}{6} \\ \hline & & 2\frac{2}{6} = 2\frac{1}{3} \end{array}$$

EXAMPLE 2.

	Change to common denominator.	Regroup.
$8\frac{3}{8}$ =	$8\frac{3}{8}$ =	$7\frac{11}{8}$
$-\ 1\frac{3}{4}$ =	$-1\frac{6}{8}$ =	$-1\frac{6}{8}$
		$6\frac{5}{8}$

EXAMPLE 3.

change 1 to $\frac{4}{4}$

$$\begin{array}{r} \overset{5}{\cancel{6}}\,\frac{\overset{5}{\cancel{1}}}{4} \\ -\ 2\frac{3}{4} \\ \hline 3\frac{2}{4} = 3\frac{1}{2} \end{array}$$

EXERCISES

Subtract.

1. $5\frac{1}{5} - 2\frac{3}{5}$
2. $4\frac{3}{7} - 1\frac{5}{7}$
3. $2\frac{1}{8} - \frac{7}{8}$
4. $7\frac{1}{4} - 3\frac{3}{4}$
5. $8\frac{2}{5} - 6\frac{4}{5}$
6. $3\frac{1}{9} - 1\frac{5}{9}$
7. $6\frac{7}{9} - 2\frac{8}{9}$
8. $3 - 1\frac{1}{2}$
9. $5 - 3\frac{3}{4}$
10. $7 - 1\frac{5}{8}$
11. $9\frac{1}{4} - 2\frac{1}{2}$
12. $5\frac{3}{8} - 1\frac{1}{2}$
13. $6\frac{1}{6} - 3\frac{1}{3}$
14. $5\frac{3}{8} - 2\frac{1}{2}$
15. $4\frac{1}{8} - 2\frac{3}{4}$
16. $4\frac{1}{3} - 1\frac{1}{2}$
17. $6\frac{1}{4} - 2\frac{2}{3}$
18. $5\frac{3}{8} - 4\frac{5}{6}$
19. $2\frac{1}{8} - 1\frac{3}{4}$
20. $5\frac{7}{8} - 2\frac{1}{4}$
21. $6\frac{5}{7} - 1\frac{3}{7}$
22. $9\frac{1}{4} - 2\frac{1}{3}$
23. $6\frac{1}{6} - 4\frac{3}{8}$
24. $9\frac{3}{5} - \frac{7}{8}$
25. $4\frac{3}{10} - 1\frac{5}{6}$
26. $12\frac{1}{5} - 8\frac{2}{9}$
27. $25\frac{3}{4} - 18\frac{7}{8}$
28. $11\frac{1}{6} - 9\frac{2}{5}$
29. $31\frac{4}{9} - 18\frac{11}{12}$
30. $29\frac{1}{10} - 17\frac{4}{5}$
31. $47\frac{2}{3} - 19\frac{5}{9}$
32. $36\frac{3}{10} - 18\frac{4}{5}$
33. $23\frac{1}{2} - 17\frac{3}{5}$
34. $52\frac{2}{15} - 14\frac{1}{3}$
35. $71\frac{1}{3} - 19\frac{4}{6}$

Solve.

36. 

How much heavier is the larger box?

37. 

How much does she want in all?

CHAPTER CHECKUP

Complete. [pages 122–123]

1. $\frac{3}{4} = \frac{?}{20}$
2. $\frac{5}{6} = \frac{?}{12}$
3. $\frac{1}{3} = \frac{?}{12}$
4. $\frac{5}{7} = \frac{?}{35}$

Reduce to lowest terms. [pages 124–127]

5. $\frac{12}{18}$
6. $\frac{15}{20}$
7. $\frac{32}{40}$
8. $\frac{24}{36}$
9. $\frac{28}{42}$
10. $\frac{16}{20}$

What is the correct sign, < or >? [pages 128–129]

11. $\frac{3}{8}$ ● $\frac{7}{8}$
12. $\frac{3}{5}$ ● $\frac{3}{7}$
13. $\frac{2}{3}$ ● $\frac{3}{5}$
14. $\frac{3}{8}$ ● $\frac{1}{4}$

Add or subtract. Watch the signs. [pages 130–131]

15. $\frac{3}{5} + \frac{1}{5}$
16. $\frac{5}{8} - \frac{2}{8}$
17. $\frac{1}{3} + \frac{1}{2}$
18. $\frac{3}{4} - \frac{2}{3}$
19. $\frac{3}{8} + \frac{1}{4}$
20. $\frac{5}{6} - \frac{2}{3}$

Complete. [pages 132–137]

	21.	22.	23.	24.	25.	26.
Fraction	$\frac{5}{2}$		$\frac{11}{4}$	$\frac{17}{5}$		
Mixed number		$2\frac{2}{3}$			$3\frac{1}{5}$	$1\frac{3}{4}$

Add. [pages 140–143]

27. $2\frac{3}{8} + 1\frac{1}{8}$
28. $4\frac{2}{3} + 7\frac{1}{3}$
29. $3\frac{4}{5} + 1\frac{1}{10}$
30. $5\frac{3}{4} + 2\frac{1}{2}$
31. $4\frac{4}{5} + 2\frac{1}{3}$

Subtract. [pages 140–141, 144–145]

32. $4\frac{5}{7} - 1\frac{1}{7}$
33. $2\frac{3}{8} - \frac{3}{8}$
34. $5\frac{2}{3} - 1\frac{1}{6}$
35. $4\frac{1}{2} - 1\frac{3}{4}$
36. $7\frac{1}{3} - 2\frac{3}{4}$

Project

1. Take a survey. Ask classmates what drink they prefer with lunch: white milk, chocolate milk, orange drink, cola, or ginger ale.
2. Record the results of your survey on a bar graph.
3. What fraction of the students prefer white milk? Chocolate milk? Orange drink? Cola? Ginger ale?
4. If you had to order enough drinks for 1000 servings, how many of each kind would you order?

LUNCH DRINK SURVEY

NUMBER	WHITE MILK	CHOC. MILK	ORANGE DRINK	COLA	GINGER ALE

CHAPTER REVIEW

Complete.

1. $\frac{2}{3} = \frac{?}{12}$ (×4 on top and bottom)

2. $\frac{1}{6} = \frac{?}{18}$ (×? on top and bottom)

3. $\frac{3}{4} = \frac{?}{20}$

4. $\frac{1}{2} = \frac{?}{10}$

5. $\frac{3}{6} = \frac{?}{2}$ (÷3 on top and bottom)

6. $\frac{6}{8} = \frac{?}{4}$ (÷? on top and bottom)

7. $\frac{12}{15} = \frac{?}{5}$

8. $\frac{12}{18} = \frac{?}{3}$

Reduce to lowest terms.

9. $\frac{9}{12}$ (Divide terms by 3)
10. $\frac{12}{16}$
11. $\frac{20}{25}$
12. $\frac{25}{35}$
13. $\frac{28}{32}$
14. $\frac{18}{24}$

Change to equivalent fractions with the least common denominator.

15. $\frac{1}{2}, \frac{2}{3}$
16. $\frac{1}{6}, \frac{1}{3}$
17. $\frac{1}{2}, \frac{3}{8}$
18. $\frac{5}{6}, \frac{1}{5}$
19. $\frac{3}{4}, \frac{2}{3}$

Add or subtract.

20. Change to common denominator. Add.
$$\frac{1}{6} = \frac{2}{12}$$
$$+\frac{3}{4} = +\frac{9}{12}$$

21. Change to common denominator. Subtract.
$$\frac{5}{8} = \frac{5}{8}$$
$$-\frac{1}{4} = -\frac{2}{8}$$

22. $\frac{4}{5} - \frac{1}{10}$

23. $\frac{3}{8} + \frac{1}{2}$

24. $\frac{4}{7} + \frac{1}{7}$

25. $\frac{2}{3} - \frac{1}{2}$

26. $\frac{3}{4} - \frac{1}{3}$

27. $\frac{1}{4} + \frac{2}{5}$

28. $\frac{2}{3} - \frac{1}{4}$

29. $\frac{1}{3} + \frac{2}{5}$

30. $2\frac{1}{5} + 1\frac{3}{5}$

31. $4\frac{3}{5} - 2\frac{2}{5}$

32. $4\frac{1}{4} + 6\frac{1}{2}$

33. $5\frac{7}{8} + 1\frac{1}{8}$

34. $6\frac{1}{5} - 4\frac{3}{5}$

35. $7\frac{1}{2} + 4\frac{3}{4}$

CHAPTER CHALLENGE

Here is a die: 

If it is thrown, it can land with any one of 6 numbers up. The chance of any one number coming up is the same as the chance of any other number. This means that there are 6 **equally likely outcomes.**

We say:

The probability of each outcome is $\frac{1}{6}$.

EXAMPLE .

The probability of 4 coming up $= \frac{1}{6}$

$$P(4) = \frac{1}{6}$$

$$P(\text{either 2 } \mathbf{or} \text{ 3}) = \frac{1}{6} + \frac{1}{6} = \frac{1}{3}$$

Can you find these probabilities?

1. $P(6) =$?
2. $P(1) =$?
3. $P(5) =$?
4. ★ $P(7) =$?
5. $P(3 \text{ or } 4) =$?
6. $P(5 \text{ or } 6) =$?
7. $P(2 \text{ and } 3 \text{ at the same time}) =$?
8. $P(2, 4, \text{ or } 6) =$?
9. $P(\text{odd number}) =$?
10. $P(\text{prime number}) =$?
11. $P(\text{number greater than } 4) =$?
12. $P(\text{number less than } 4) =$?
13. $P(1, 2, 3, 4, 5, \text{ or } 6) =$?

Think about tossing a coin.

14. How many outcomes are there?
15. What is $P(\text{heads})$?
16. What is $P(\text{tails})$?

MAJOR CHECKUP
Standardized Format

Choose the correct letter.

1. The standard numeral for two billion, seventy-two thousand is

a. 2,072,000
b. 2,000,072,000
c. 2,000,000,072
d. none of these

2. Round 247,499 to the nearest thousand.

a. 247,000
b. 248,000
c. 247,500
d. none of these

3. Add.

$$\begin{array}{r} 375 \\ 968 \\ +421 \\ \hline \end{array}$$

a. 1654
b. 1764
c. 1754
d. none of these

4. Subtract.

$$\begin{array}{r} 2003 \\ -149 \\ \hline \end{array}$$

a. 2146
b. 1854
c. 1954
d. none of these

5. Multiply.

$$\begin{array}{r} 1735 \\ \times 3 \\ \hline \end{array}$$

a. 5195
b. 4205
c. 5205
d. none of these

6. Multiply.

$$\begin{array}{r} 421 \\ \times 206 \\ \hline \end{array}$$

a. 86,726
b. 10,946
c. 844,526
d. none of these

7. Divide.

$4\overline{)6432}$

a. 168
b. 1358
c. 16,008
d. none of these

8. Divide.

$38\overline{)2914}$

a. 76 R26
b. 66 R14
c. 76 R14
d. none of these

9. Add.

$$\begin{array}{r} \frac{1}{5} \\ +\frac{2}{3} \\ \hline \end{array}$$

a. $\frac{3}{8}$
b. $\frac{11}{15}$
c. $\frac{13}{15}$
d. none of these

10. Subtract.

$$\begin{array}{r} \frac{5}{8} \\ -\frac{1}{2} \\ \hline \end{array}$$

a. $\frac{1}{8}$
b. $\frac{2}{3}$
c. $\frac{9}{8}$
d. none of these

11. Add.

$$\begin{array}{r} 4\frac{2}{3} \\ +1\frac{2}{3} \\ \hline \end{array}$$

a. $5\frac{2}{3}$
b. $6\frac{1}{3}$
c. $5\frac{1}{3}$
d. none of these

12. Subtract.

$$\begin{array}{r} 3\frac{1}{8} \\ -1\frac{5}{8} \\ \hline \end{array}$$

a. $2\frac{1}{2}$
b. $1\frac{1}{2}$
c. $1\frac{3}{4}$
d. none of these

6 Fractions—Multiplication and Division

READY OR NOT ?

Give the fraction that is shaded.

1. 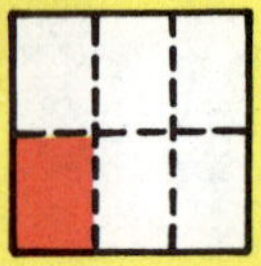

2. 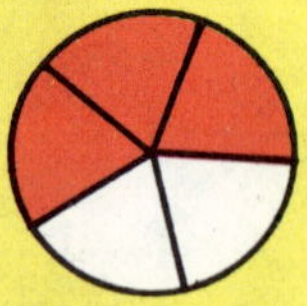

Reduce to lowest terms.

3. $\frac{6}{8}$ 4. $\frac{9}{12}$

5. $\frac{12}{18}$ 6. $\frac{24}{18}$

7. $\frac{30}{25}$ 8. $\frac{25}{30}$

Complete.

9. $\frac{1}{2}$ of \$12 = \$?

10. $\frac{2}{3}$ of 24¢ = ? ¢

11. $\frac{3}{5}$ of 15 = ?

Change to fractions.

12. $3\frac{1}{2}$ 13. $4\frac{3}{5}$

14. $5\frac{2}{3}$

Change to mixed numbers.

15. $\frac{8}{3}$ 16. $\frac{9}{2}$ 17. $\frac{20}{7}$

Multiplication of fractions

If you were just learning to multiply whole numbers and wanted to know the product of 2 and 3, you might do this:

1. Pick a unit square. 

2. Make a region that is as long as 3 squares and as wide as 2 squares.

3. Count the total number of unit squares.

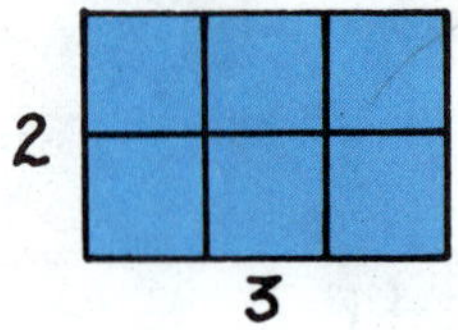

$2 \times 3 = 6$

You can do the same thing to find the product of $\frac{1}{2}$ and $\frac{1}{3}$.

1. Pick a unit square. 

2. Make a region that is $\frac{1}{3}$ as long as the unit square and $\frac{1}{2}$ as wide.

3. Find what fraction of the unit square that region is.

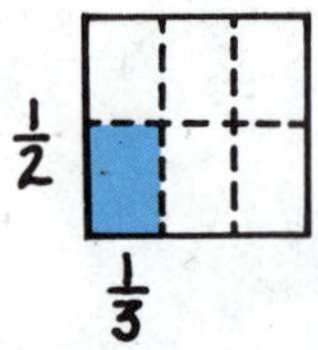

$\frac{1}{2} \times \frac{1}{3} = \frac{1}{6}$

Here are other examples.

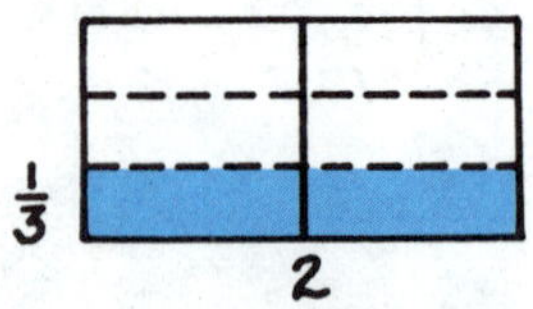

$\frac{1}{3} \times 2 = \frac{2}{3}$

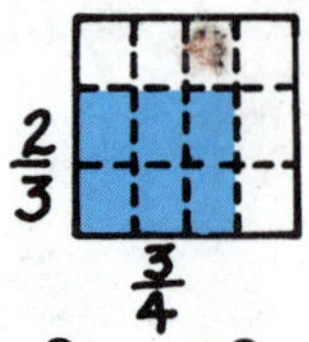

$\frac{2}{3} \times \frac{3}{4} = \frac{6}{12}$

EXERCISES

Study the pictures of unit squares. Give the products.

1.

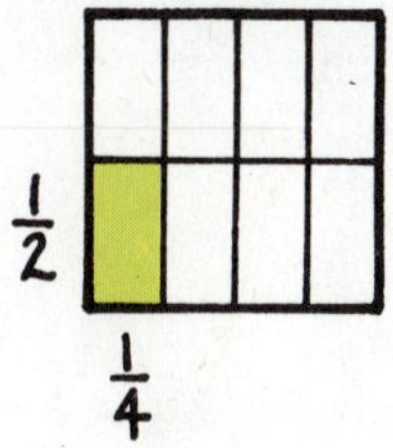

$\frac{1}{2} \times \frac{1}{4} = \underline{?}$

2.

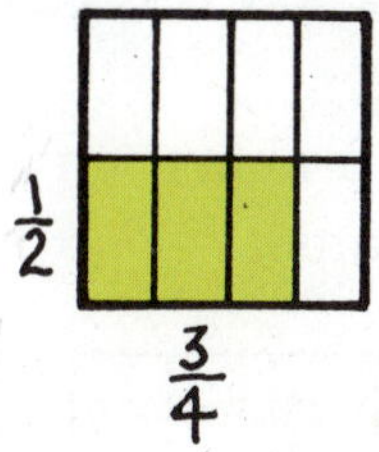

$\frac{1}{2} \times \frac{3}{4} = \underline{?}$

3.

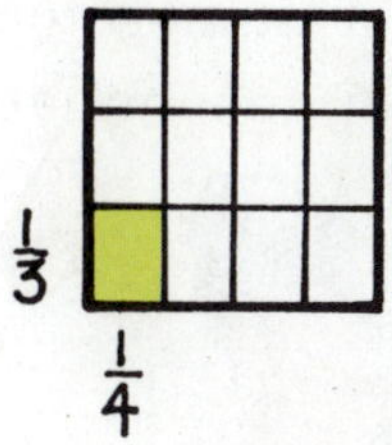

$\frac{1}{3} \times \frac{1}{4} = \underline{?}$

4.

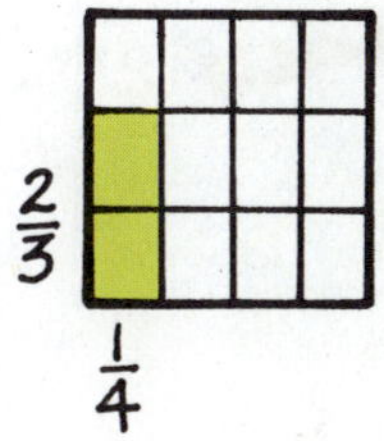

$\frac{2}{3} \times \frac{1}{4} = \underline{?}$

5.

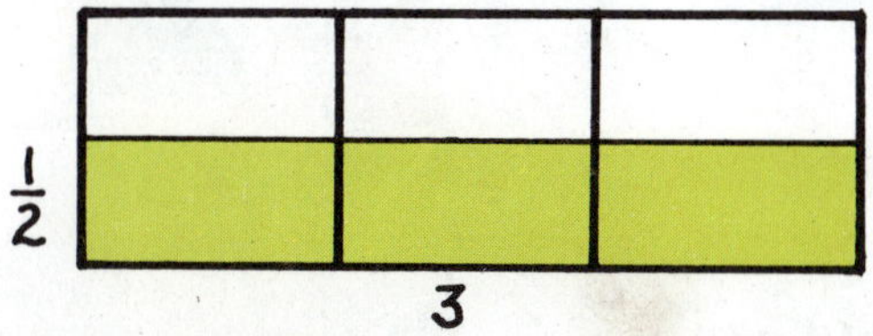

$\frac{1}{2} \times 3 = \underline{?}$

6.

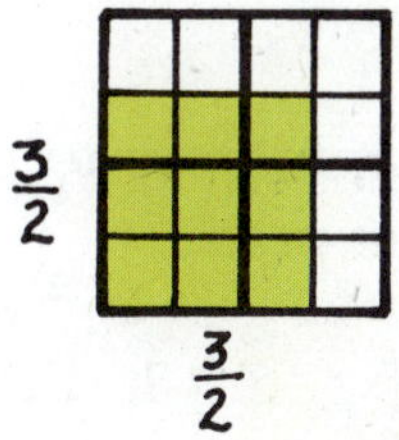

$\frac{3}{2} \times \frac{3}{2} = \underline{?}$

Write the multiplication equations.

7.

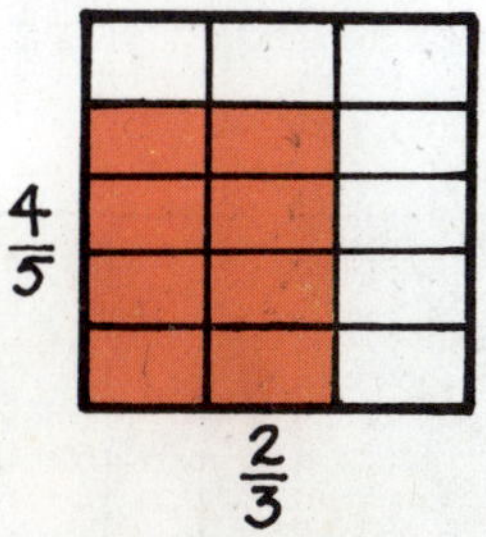

8.

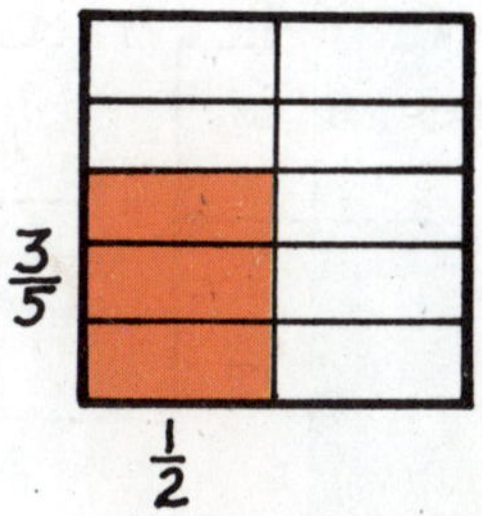

9.

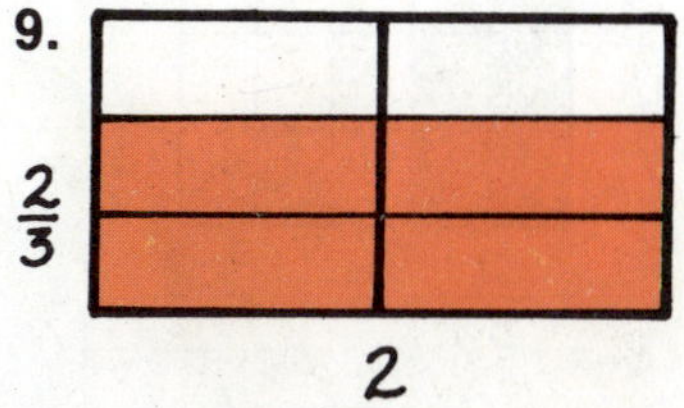

10.

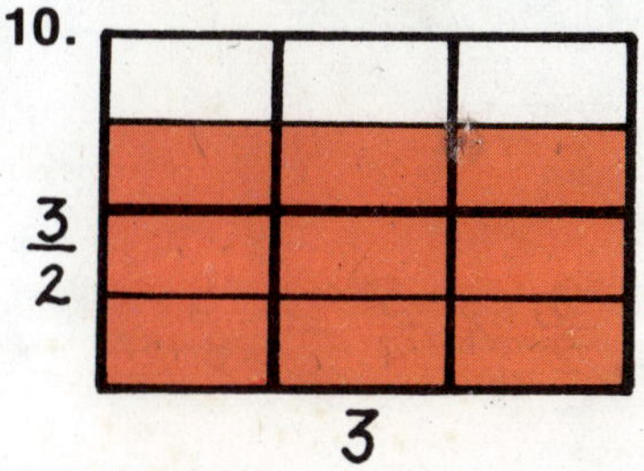

11.

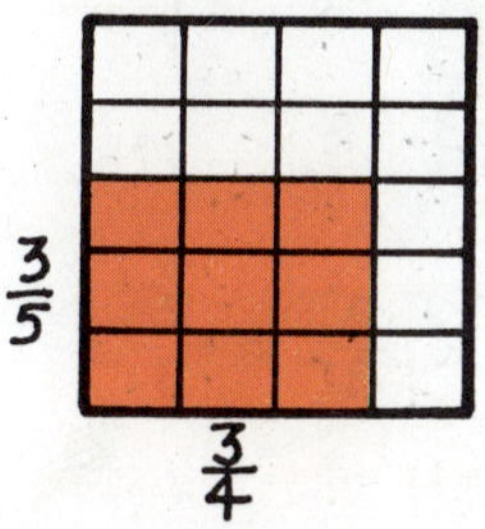

12.

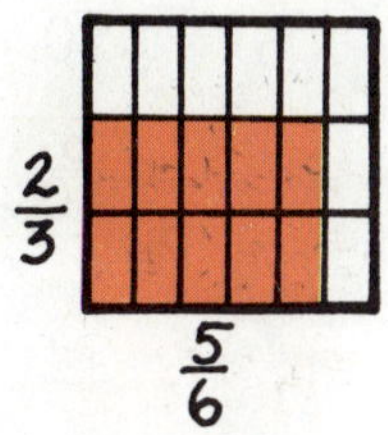

More about multiplication of fractions

This example shows a short way to find the product of two fractions.

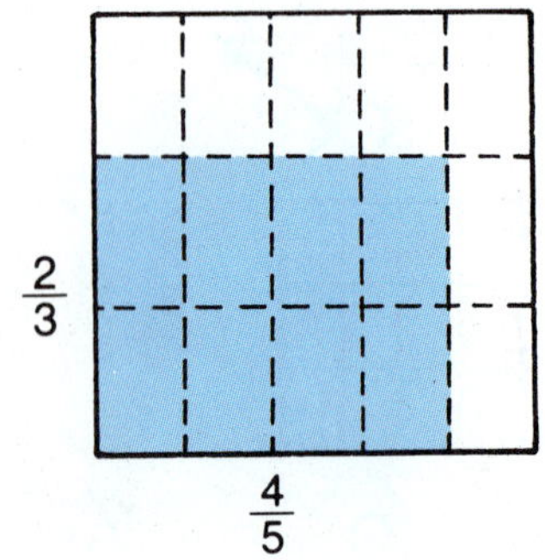

$$\frac{2}{3} \times \frac{4}{5} = \frac{8}{15}$$

product of numerators

product of denominators

1. Multiply the numerators to get the numerator of the product.
2. Multiply the denominators to get the denominator of the product.

Study these examples.

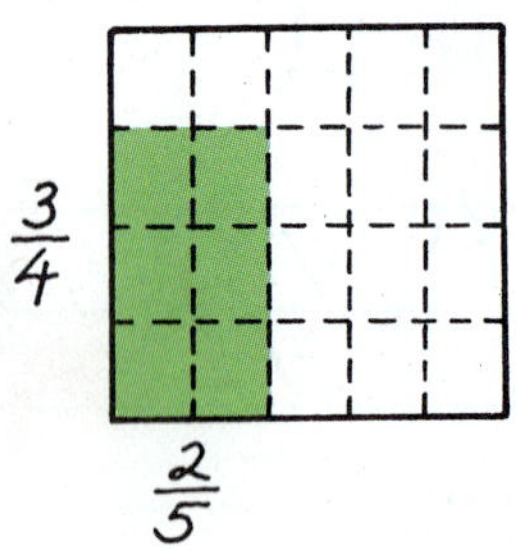

$$\frac{3}{4} \times \frac{2}{5} = \frac{6}{20}$$
$$= \frac{3}{10}$$

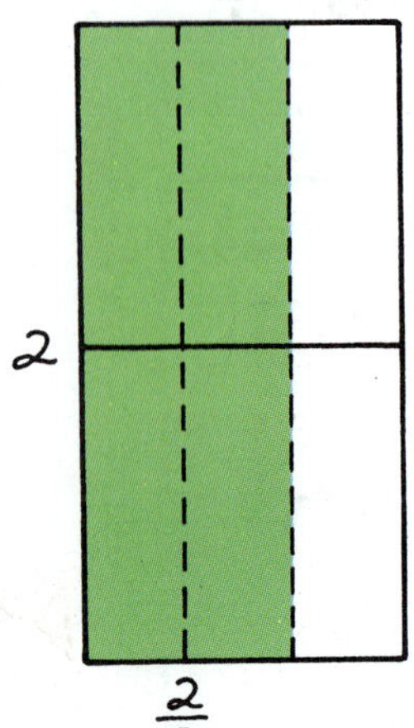

$$\frac{2}{1} \times \frac{2}{3} = \frac{4}{3}$$

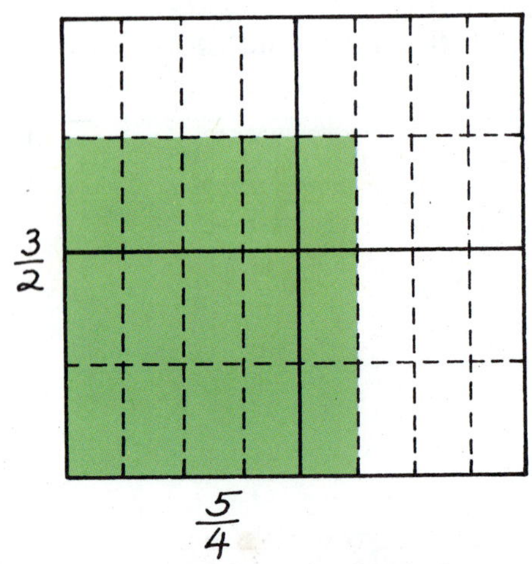

$$\frac{3}{2} \times \frac{5}{4} = \frac{15}{8}$$

EXERCISES

Multiply. Reduce products to lowest terms.

1. $\frac{1}{2} \times \frac{1}{3}$
2. $\frac{1}{2} \times \frac{3}{4}$
3. $\frac{1}{2} \times \frac{5}{6}$
4. $\frac{1}{3} \times \frac{2}{5}$
5. $\frac{2}{3} \times \frac{3}{4}$
6. $\frac{3}{4} \times \frac{1}{3}$
7. $\frac{3}{5} \times \frac{1}{4}$
8. $\frac{3}{5} \times \frac{3}{4}$
9. $\frac{3}{8} \times \frac{1}{2}$
10. $\frac{5}{8} \times \frac{3}{4}$
11. $2 \times \frac{1}{3}$
12. $2 \times \frac{2}{3}$
13. $5 \times \frac{3}{4}$
14. $5 \times \frac{3}{5}$
15. $7 \times \frac{1}{2}$
16. $\frac{3}{4} \times 7$
17. $\frac{4}{3} \times 5$
18. $\frac{2}{5} \times \frac{4}{7}$
19. $\frac{3}{8} \times \frac{5}{3}$
20. $\frac{2}{7} \times \frac{3}{4}$
21. $\frac{3}{8} \times \frac{2}{2}$
22. $\frac{5}{6} \times \frac{3}{3}$
23. $\frac{1}{3} \times \frac{4}{4}$
24. $\frac{3}{4} \times \frac{3}{3}$
25. $\frac{3}{8} \times \frac{8}{3}$
26. $\frac{2}{5} \times \frac{5}{2}$
27. $\frac{3}{4} \times \frac{4}{3}$
28. $5 \times \frac{1}{5}$
29. $3 \times \frac{1}{3}$
30. $\frac{1}{2} \times 2$
31. $8 \times \frac{1}{8}$
32. $\frac{5}{6} \times \frac{6}{5}$
33. $\frac{9}{4} \times \frac{4}{9}$

If the product of two factors is 1, the factors are called **reciprocals.**

$\frac{3}{5} \times \frac{5}{3} = 1$ — $\frac{3}{5}$ and $\frac{5}{3}$: Reciprocals

$4 \times \frac{1}{4} = 1$ — 4 and $\frac{1}{4}$: Reciprocals

34. What is the reciprocal of 5?
35. What is the reciprocal of $\frac{1}{3}$?
36. What is the reciprocal of $\frac{2}{3}$?
37. What is the reciprocal of 1?

Keeping Skills Sharp

Estimate each sum or difference.

1. $87 + 98$

$$\begin{array}{r} 87 \\ +98 \\ \hline \end{array}$$

2.

$$\begin{array}{r} 32 \\ 49 \\ +58 \\ \hline \end{array}$$

3.

$$\begin{array}{r} 605 \\ -197 \\ \hline \end{array}$$

4.

$$\begin{array}{r} 502 \\ -288 \\ \hline \end{array}$$

5.

$$\begin{array}{r} 5003 \\ -1977 \\ \hline \end{array}$$

6.

$$\begin{array}{r} 685 \\ 796 \\ +888 \\ \hline \end{array}$$

7.

$$\begin{array}{r} 998 \\ 976 \\ +988 \\ \hline \end{array}$$

8.

$$\begin{array}{r} 2217 \\ -1988 \\ \hline \end{array}$$

9.

$$\begin{array}{r} 6135 \\ -4987 \\ \hline \end{array}$$

10.

$$\begin{array}{r} 98974 \\ +83516 \\ \hline \end{array}$$

Fraction of a number

In Chapter 5 you learned to find a fraction of a number.

$\frac{2}{3}$ of \$21 = \$14 ... $3\overline{)21}$ with quotient 7, $7 \times 2 = 14$

That is exactly the same as multiplying.

$\frac{2}{3} \times \$21 = \$14$ ... $\frac{2}{3} \times 21 = \frac{42}{3} = 14$

Finding a fractional part of a number is the same as multiplying the fraction and the number.

$\frac{1}{2}$ of $12 = \frac{1}{2} \times 12$

One way:

$$\frac{2}{3} \times \frac{3}{4} = \frac{6}{12} = \frac{1}{2}$$

Another way:

Step 1. Cancel to divide by the common factor 3.

$$\frac{2}{\cancel{3}_1} \times \frac{\cancel{3}^1}{4}$$

Step 2. Cancel to divide by the common factor 2.

$$\frac{\cancel{2}^1}{\cancel{3}_1} \times \frac{\cancel{3}^1}{\cancel{4}_2}$$

Step 3. Multiply.

$$\frac{\cancel{2}^1}{\cancel{3}_1} \times \frac{\cancel{3}^1}{\cancel{4}_2} = \frac{1}{2}$$

EXAMPLES.

$$\frac{5}{\cancel{6}_3} \times \frac{\cancel{2}^1}{3} = \frac{5}{9}$$

$$\cancel{6}^3 \times \frac{3}{\cancel{4}_2} = \frac{9}{2}$$

$$\frac{\cancel{3}^1}{\cancel{8}_2} \times \frac{\cancel{4}^1}{\cancel{9}_3} = \frac{1}{6}$$

EXERCISES

Solve. Use canceling if you wish.

1. $\frac{3}{4}$ of 28	**2.** $\frac{2}{3}$ of 18	**3.** $\frac{3}{5}$ of 60	**4.** $\frac{2}{3}$ of $\frac{1}{2}$
5. $\frac{5}{6}$ of 24	**6.** $\frac{3}{8}$ of 48	**7.** $\frac{5}{4}$ of 20	**8.** $\frac{3}{2}$ of 30
9. $\frac{5}{7}$ of 35	**10.** $\frac{3}{4}$ of $\frac{1}{2}$	**11.** $\frac{1}{2}$ of $\frac{1}{2}$	**12.** $\frac{2}{3}$ of $\frac{4}{5}$
13. $\frac{3}{5}$ of $\frac{3}{4}$	**14.** $\frac{2}{3}$ of \$15	**15.** $\frac{3}{8}$ of 32¢	**16.** $\frac{3}{10}$ of \$40

Solve.

17. A baker made 100 dozen cookies. She had sold $\frac{3}{5}$ of them before noon. How many dozen were sold?

18. The baker's recipe for angel food cake made enough batter for 30 cakes. Today she used only $\frac{2}{3}$ of the recipe. How many cakes did that make?

19. One recipe called for $\frac{3}{4}$ cup of salt. The baker doubled the recipe. How much salt was used?

20. Left-over baked items are sold the next day at $\frac{2}{5}$ off the regular price. If the regular price of a dozen doughnuts is \$2.00, what is the sale price?

Follow the path.

21.

Multiplying mixed numbers

Here are two ways to multiply mixed numbers.

METHOD 1. Change to fractions and then multiply.

$$2 \times 3\frac{1}{4} = \overset{1}{\cancel{2}} \times \frac{13}{\underset{2}{\cancel{4}}} = \frac{13}{2} = 6\frac{1}{2}$$

METHOD 2. Multiply just as you do whole numbers.

Step 1.

$$\begin{array}{r} 3\frac{1}{4} \\ \times 2 \\ \hline \frac{2}{4} \end{array}$$

Step 2.

$$\begin{array}{r} 3\frac{1}{4} \\ \times 2 \\ \hline 6\frac{2}{4} \end{array} = 6\frac{1}{2}$$

The first method is better when both factors are mixed numbers.

$$3\frac{1}{2} \times 1\frac{2}{3} = \frac{7}{2} \times \frac{5}{3} = \frac{35}{6} = 5\frac{5}{6}$$

The second method is better when one factor is a whole number.

$$\begin{array}{r} 4\frac{2}{3} \\ \times 5 \\ \hline 20\frac{10}{3} \end{array} = 23\frac{1}{3} \qquad \left(\frac{10}{3} = 3\frac{1}{3} \right)$$

EXERCISES

Multiply.

1. $3\frac{1}{2} \times 2$	**2.** $4\frac{1}{4} \times 3$	**3.** $2\frac{3}{4} \times 2$	**4.** $5\frac{1}{3} \times 4$
5. $2\frac{1}{5} \times 3$	**6.** $2\frac{2}{5} \times 3$	**7.** $2\frac{4}{5} \times 3$	**8.** $2\frac{3}{8} \times 4$
9. $3\frac{1}{4} \times 6$	**10.** $6\frac{1}{4} \times 3$	**11.** $3\frac{3}{4} \times 7$	**12.** $7\frac{3}{4} \times 3$
13. $1\frac{1}{2} \times 1\frac{1}{3}$	**14.** $1\frac{2}{3} \times 2\frac{1}{2}$	**15.** $3\frac{1}{2} \times 1\frac{3}{4}$	**16.** $2\frac{1}{4} \times 3\frac{1}{2}$

Solve.

17. How many dozen cookies does the recipe make?

18. By what should the recipe be multiplied to make 75 cookies?

19. How much flour is used if the recipe is doubled?

20. How much baking powder is used if the recipe is multiplied by $2\frac{1}{2}$?

Carrot Cookies
Makes 30 cookies

1 egg

$\frac{1}{3}$ cup cooking oil

$\frac{1}{3}$ cup sugar

$\frac{3}{4}$ cup flour

$\frac{2}{3}$ teaspoon baking powder

$\frac{1}{4}$ teaspoon salt

1 teaspoon grated orange peel

$\frac{1}{3}$ cup cooked carrots

4 tablespoons raisins

$\frac{1}{2}$ teaspoon ginger

21. How much ginger is used if the recipe is cut in half?

22. How many cookies will $3\frac{1}{2}$ recipes make?

23. How much flour is used in $\frac{1}{2}$ a recipe?

24. How much sugar is used in $\frac{1}{2}$ a recipe?

25. How many cups of carrots are used if the recipe is tripled?

26. How much more flour than sugar is used?

★27. A tablespoon is $\frac{1}{16}$ of a cup. What fraction of a cup of raisins is used?

★28. How many tablespoons of cooked carrots are used? (See Exercise 27.)

Problem solving

Sometimes there is not enough information given to solve a problem. In the exercises below, tell what else you would need to know to solve the problem.

EXERCISES

1. Gail wants to buy a stereo. She has $85.65. How much more money does she need?

2. Alex rode in a bike tour. He completed the tour in 5 hours. How many kilometers did he average each hour?

3. Adult tickets for the school play sell for $.60. The sixth grade sold 48 adult tickets and 35 children's tickets. How much money did they make?

4. Mrs. Smith drove from Toledo to Chicago at an average rate of 80 kilometers per hour. How long did the trip take?

5. Sarah had $32.48. She earned $1.50 an hour for working one Saturday. How much money did she have then?

6. Mr. Allison bought 8 long-playing records for $6.95 each. He also bought some singles for $2.15 each. How much did he spend for the records?

7. Kevin bought 500 grams of peanuts for $1.25 and some pecans for $1.75. How many grams of nuts did he buy?

8. Gail and her father went backpacking for 7 days. Gail's pack weighed $\frac{1}{2}$ as much as her father's. How much did Gail's pack weigh?

9.

About how much does each apple weigh?

10.

How much does 1 gram of cereal cost?

11. Miss Johnson drove 480 kilometers on a tank of gas. How many kilometers did she average per liter of gas?

12. Mr. Schultz increased a recipe that would make 15 cupcakes. How many cupcakes did he make?

13.

How many hamburgers can be made from this package?

14.

How much does 1 liter of gas cost?

Dividing by a fraction

If you were just learning about division and wanted to know what $6 \div 2$ was, you could do this.

Take 6 objects and count the number of groups of 2.

$$6 \div 2 = 3$$

You can do the same thing when dividing by a fraction. Suppose you wish to find how many $\frac{1}{2}$s there are in 3.

Take 3 objects and count the number of $\frac{1}{2}$s.

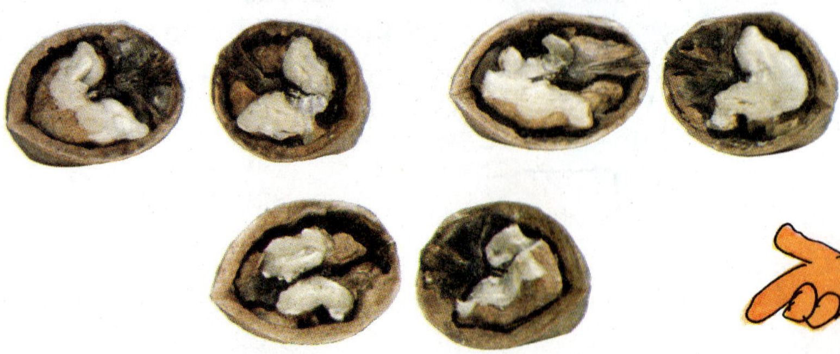

$3 \div \frac{1}{2} = 6$ $\qquad$ $3 \times 2 = 6$

EXERCISES

Complete. Compare your answers.

1. a. $2 \div \frac{1}{2} = \underline{?}$ 

   b. $2 \times 2 = \underline{?}$

2. a. $2 \div \frac{1}{4} = \underline{?}$ 

   b. $2 \times 4 = \underline{?}$

3. a. $2 \div \frac{1}{3} = \underline{?}$

b. $2 \times 3 = \underline{?}$

4. a. $3 \div \frac{1}{4} = \underline{?}$

b. $3 \times 4 = \underline{?}$

5. a. $\frac{3}{2} \div \frac{1}{2} = \underline{?}$

b. $\frac{3}{2} \times 2 = \underline{?}$

6. a. $\frac{3}{4} \div \frac{1}{4} = \underline{?}$

b. $\frac{3}{4} \times 4 = \underline{?}$

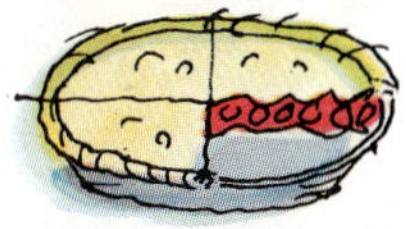

7. a. $2 \div \frac{2}{3} = \underline{?}$

b. $2 \times \frac{3}{2} = \underline{?}$

8. a. $3 \div \frac{3}{4} = \underline{?}$

b. $3 \times \frac{4}{3} = \underline{?}$

9. a. $\frac{4}{5} \div \frac{2}{5} = \underline{?}$

b. $\frac{4}{5} \times \frac{5}{2} = \underline{?}$

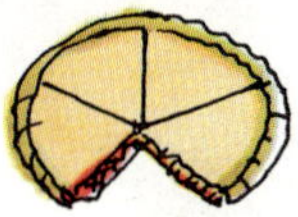

10. a. $\frac{6}{7} \div \frac{3}{7} = \underline{?}$

b. $\frac{6}{7} \times \frac{7}{3} = \underline{?}$

11. Dividing by $\frac{1}{2}$ is just like multiplying by $\underline{?}$.

12. Dividing by $\frac{1}{3}$ is just like multiplying by $\underline{?}$.

13. Dividing by 4 is just like multiplying by $\underline{?}$.

14. Dividing by $\frac{3}{4}$ is just like multiplying by $\underline{?}$.

Keeping Skills Sharp

Remember that two numbers are reciprocals if their product is 1. Give each reciprocal.

1. $\frac{1}{2}$ ($\frac{1}{2} \times 2 = 1$)
2. $\frac{2}{3}$ ($\frac{2}{3} \times \frac{3}{2} = 1$)
3. 6 ($6 \times \frac{1}{6} = 1$)
4. $\frac{1}{3}$
5. $\frac{3}{4}$
6. $\frac{5}{2}$
7. $\frac{4}{7}$
8. $\frac{5}{3}$
9. 8
10. 12
11. 1
12. $\frac{2}{5}$
13. 9
14. $\frac{7}{3}$
15. $\frac{5}{6}$

More about dividing by fractions

You can use pictures to divide a fraction by a fraction.

$\frac{5}{2} \div \frac{3}{4}$ ... *How many $\frac{3}{4}$s in $\frac{5}{2}$?*

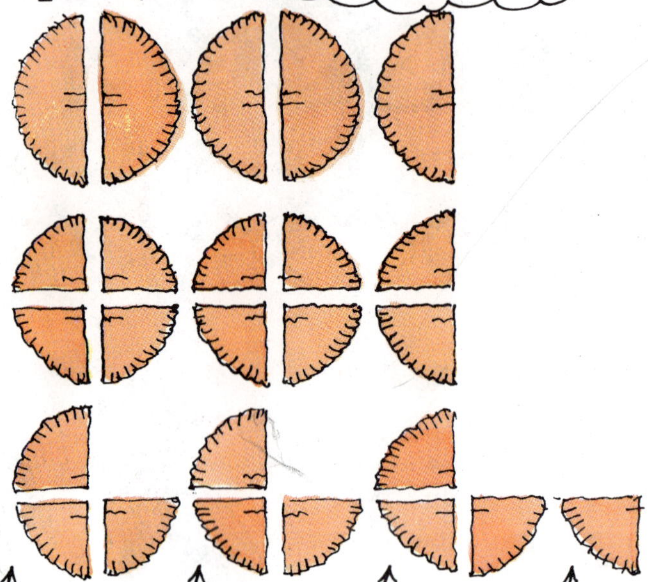

Draw $\frac{5}{2}$ circles:

Divide into $\frac{1}{4}$s:

Separate into $\frac{3}{4}$s:

Count groups of $\frac{3}{4}$s: 1 group of $\frac{3}{4}$ | 1 group of $\frac{3}{4}$ | 1 group of $\frac{3}{4}$ | $\frac{1}{3}$ of a group of $\frac{3}{4}$

$$\frac{5}{2} \div \frac{3}{4} = 3\frac{1}{3}$$

You would have gotten the same answer if you had multiplied $\frac{5}{2}$ by $\frac{4}{3}$.

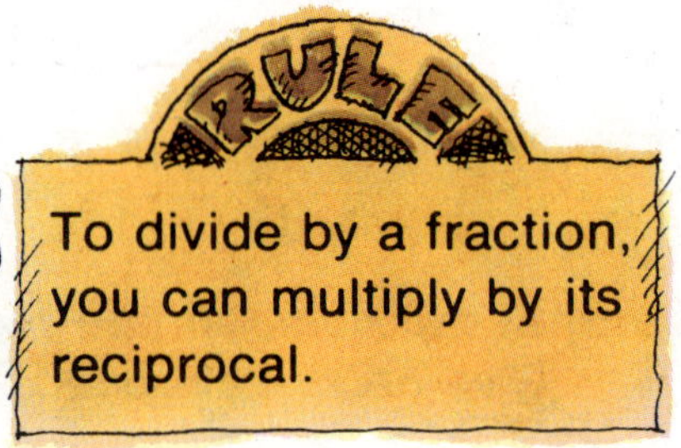

EXAMPLE. Divide: $\frac{3}{5} \div \frac{2}{3}$

$$\frac{3}{5} \div \frac{2}{3} = \frac{3}{5} \times \frac{3}{2} = \frac{9}{10}$$

EXERCISES

Complete.

1. $3 \div \frac{1}{2} = 3 \times \underline{?}$
2. $\frac{1}{2} \div \frac{2}{3} = \frac{1}{2} \times \underline{?}$
3. $\frac{3}{4} \div 2 = \frac{3}{4} \times \underline{?}$
4. $\frac{4}{5} \div \frac{3}{8} = \underline{?} \times \underline{?}$
5. $\frac{3}{7} \div \frac{2}{5} = \underline{?} \times \underline{?}$
6. $6 \div \frac{3}{4} = \underline{?} \times \underline{?}$
7. $\frac{4}{5} \div 3 = \underline{?} \times \underline{?}$
8. $\frac{6}{7} \div \frac{5}{3} = \underline{?} \times \underline{?}$
9. $\frac{4}{3} \div \frac{5}{2} = \underline{?} \times \underline{?}$

Divide. Give answers in lowest terms.

10. $\frac{3}{4} \div \frac{5}{2}$ ∘∘ ($\frac{3}{4} \times \frac{2}{5}$)
11. $\frac{4}{5} \div \frac{4}{3}$
12. $\frac{3}{8} \div 2$
13. $\frac{7}{8} \div \frac{3}{4}$
14. $\frac{4}{5} \div \frac{3}{4}$
15. $\frac{3}{5} \div \frac{2}{5}$
16. $9 \div \frac{3}{2}$
17. $\frac{4}{5} \div \frac{2}{3}$
18. $\frac{5}{8} \div \frac{5}{7}$
19. $\frac{6}{5} \div 3$
20. $6 \div \frac{3}{4}$
21. $\frac{9}{4} \div \frac{3}{2}$
22. $\frac{7}{10} \div \frac{3}{5}$
23. $\frac{5}{9} \div \frac{2}{3}$
24. $\frac{7}{8} \div \frac{5}{6}$
25. $\frac{3}{8} \div \frac{1}{2}$
26. $\frac{5}{6} \div \frac{5}{6}$
27. $\frac{3}{4} \div \frac{3}{4}$
28. $\frac{6}{7} \div 1$
29. $1 \div \frac{1}{5}$
30. $1 \div \frac{2}{5}$

Follow the path.

31.

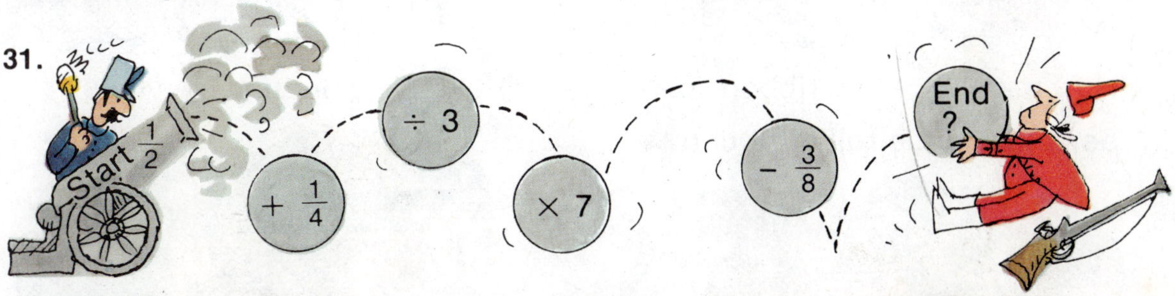

32.

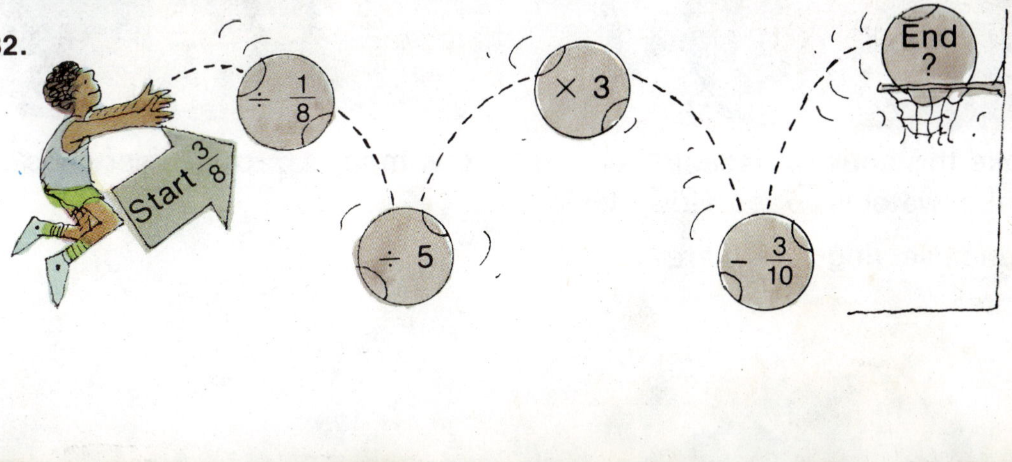

Dividing mixed numbers

The easiest way to divide mixed numbers is to change them to fractions first.

$$2\frac{1}{2} \div 1\frac{3}{4} = \frac{5}{2} \div \frac{7}{4} = \frac{5}{\overset{}{\underset{1}{\cancel{2}}}} \times \frac{\overset{2}{\cancel{4}}}{7} = \frac{10}{7} = 1\frac{3}{7}$$

Change to fractions → Change to multiplication → Multiply → Change to mixed number.

EXERCISES

Divide.

1. $3\frac{1}{2} \div 1\frac{3}{4}$
2. $7\frac{1}{2} \div 2\frac{1}{2}$
3. $1\frac{1}{4} \div 2\frac{1}{2}$
4. $11\frac{1}{2} \div 2\frac{7}{8}$
5. $5\frac{1}{4} \div 3$
6. $7 \div 3\frac{1}{2}$
7. $7 \div 2\frac{1}{3}$
8. $6\frac{1}{3} \div 2$
9. $4\frac{1}{4} \div 3\frac{1}{8}$
10. $2\frac{3}{4} \div 5\frac{2}{3}$
11. $4\frac{7}{8} \div 6\frac{1}{4}$
12. $12\frac{3}{8} \div 2\frac{3}{4}$

Solve.

13. 

How many $5\frac{1}{3}$-inch ribbons can be cut from the roll of ribbon?

14. 

How many 6-ounce servings?

15. 

To make this soup, an equal amount of water is added. How many $6\frac{1}{2}$-ounce servings are there?

16. 

How many $1\frac{1}{4}$-ounce servings?

FRACTION EXERCISES

Add across. Add down.

1.

3	$\frac{1}{4}$	?
4	$\frac{1}{2}$	?
?	?	?

2.

2	$\frac{1}{3}$	?
5	$\frac{1}{4}$	?
?	?	?

3.

6	$\frac{3}{8}$	?
4	$\frac{1}{3}$	?
?	?	?

Add across. Subtract down.

4.

5	$\frac{2}{3}$	?
2	$\frac{1}{2}$	?
?	?	?

5.

6	$\frac{3}{4}$	?
4	$\frac{1}{3}$	?
?	?	?

6.

9	$\frac{5}{8}$	?
2	$\frac{2}{5}$	?
?	?	?

Follow the path.

7. Start

8. Start

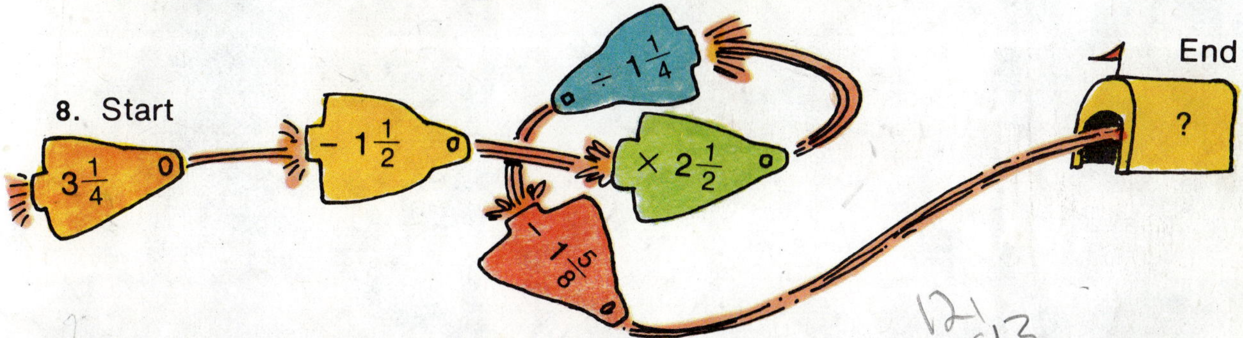

9.

10. Start

Problem solving

Solve.

1. The Harper family grows fruit and vegetables and sells them at a roadside stand. Jeff Harper works at the stand $3\frac{1}{2}$ hours a day, 7 days a week. How many hours does he work each week?

2. Tomatoes sell for 78¢ per pound. How much do $2\frac{1}{2}$ pounds of tomatoes cost?

3. Peaches cost $5.60 per half-bushel. How much does a quarter-bushel cost?

4. A head of cauliflower weighs about $\frac{3}{4}$ pound. How many heads would it take to make 3 pounds?

5. One morning the Harpers picked $5\frac{1}{2}$ bushels of sweet corn. They sold $\frac{3}{4}$ of the corn that day. How much corn was left at the end of the day?

6. The Harpers sell large apples at 2 for 35¢. That is $3\frac{1}{2}$¢ less per apple than at a nearby grocery store. How much do 2 apples cost at the grocery store?

7. Mrs. Harper made dill pickles from some small cucumbers. She canned 22 pounds of pickles. Each jar contained $10\frac{2}{3}$ ounces. How many jars did she can?

8. She also canned 45 jars of sweet pickles. Each jar contained $8\frac{3}{4}$ ounces of cucumbers. How many pounds of cucumbers did she use?

CHAPTER CHECKUP

Multiply. Give each answer as a fraction in lowest terms. [pages 152–155]

1. $\frac{3}{4} \times \frac{1}{2}$ **2.** $\frac{4}{5} \times \frac{3}{2}$ **3.** $\frac{5}{6} \times \frac{9}{10}$ **4.** $\frac{3}{8} \times \frac{7}{4}$

5. $\frac{7}{3} \times \frac{6}{5}$ **6.** $\frac{10}{3} \times \frac{6}{5}$ **7.** $\frac{3}{8} \times \frac{5}{8}$ **8.** $\frac{5}{4} \times \frac{12}{5}$

Multiply. Give each answer as a mixed number with the fraction in lowest terms. [pages 158–159]

9. $3\frac{1}{2} \times \frac{2}{3}$ **10.** $5\frac{1}{4} \times 3$ **11.** $3\frac{2}{3} \times 7$ **12.** $9 \times 2\frac{1}{2}$

13. $3\frac{1}{2} \times 2\frac{1}{2}$ **14.** $2\frac{3}{4} \times 1\frac{5}{6}$ **15.** $4\frac{2}{3} \times 3\frac{7}{8}$ **16.** $4\frac{3}{8} \times 5\frac{2}{3}$

Divide. Give each answer as a fraction in lowest terms. [pages162–165]

17. $\frac{4}{5} \div 2$ **18.** $2 \div \frac{4}{5}$ **19.** $\frac{3}{8} \div 6$ **20.** $\frac{3}{4} \div \frac{1}{4}$

21. $\frac{3}{5} \div \frac{2}{3}$ **22.** $\frac{3}{5} \div \frac{3}{2}$ **23.** $\frac{5}{8} \div \frac{3}{4}$ **24.** $\frac{5}{3} \div \frac{5}{3}$

Divide. Give each answer as a mixed number with the fraction in lowest terms. [pages 166–167]

25. $2\frac{1}{2} \div 5$ **26.** $3\frac{1}{4} \div \frac{1}{4}$ **27.** $3\frac{1}{4} \div 2\frac{1}{8}$ **28.** $5\frac{2}{3} \div 3\frac{1}{4}$

29. $6\frac{3}{8} \div 3$ **30.** $12 \div 2\frac{1}{2}$ **31.** $16 \div 3\frac{1}{3}$ **32.** $3\frac{1}{5} \div 2\frac{3}{10}$

Solve. [pages 168–169]

33.

How many $4\frac{1}{2}$-inch pieces?

34.

How many ounces in 3 cans?

Project

Making a circle graph

Here is a circle graph that shows how many points were scored by each player in a basketball game. Answer these questions about the graph.

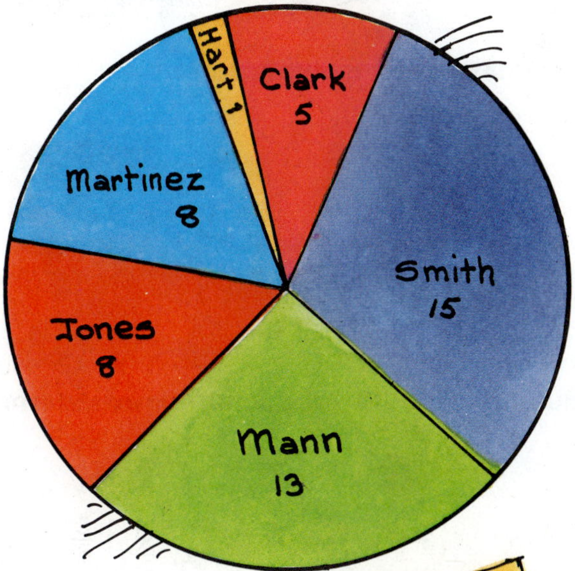

1. How many points were scored in all by the team?
2. What fraction of the team's points did Smith score?
3. Multiply the fraction in exercise 2 by 360°.
4. Measure the central angle in the part of the graph that shows Smith's points. Are the answers to exercises 3 and 4 the same?

Now let's make a circle graph.

1. Get a list of the individual scoring in one game played by your favorite basketball team.
2. Find what fraction of the team's points each player scored.
3. Multiply each fraction by 360°. Round to the nearest degree. This tells how many degrees are in the central angle for each part of the graph.
4. Draw a circle. Measure and draw each central angle.
5. Label and color your graph.

CHAPTER REVIEW

Complete each multiplication.

1. 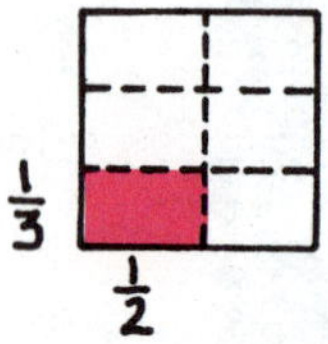

$\frac{1}{3} \times \frac{1}{2} = \underline{?}$

2. 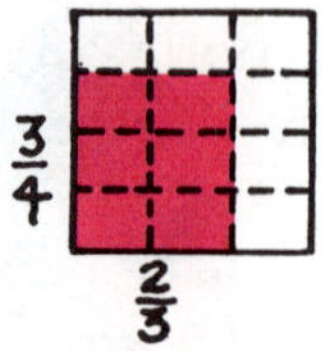

$\frac{3}{4} \times \frac{2}{3} = \underline{?}$

3. 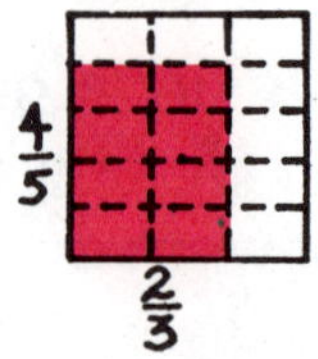

$\frac{4}{5} \times \frac{2}{3} = \underline{?}$

Multiply. Reduce products to lowest terms.

$\frac{3}{4} \times \frac{5}{7} = \frac{15}{28}$

4. $\frac{2}{3} \times \frac{5}{8}$

5. $\frac{5}{8} \times 2$

6. $\frac{4}{5} \times 6$

7. $\frac{5}{6} \times \frac{3}{10}$

8. $\frac{3}{4} \times \frac{5}{6}$

9. $\frac{4}{7} \times \frac{3}{7}$

10. $\frac{5}{8} \times \frac{4}{5}$

Multiply.

$2\frac{1}{3} \times 5$ — $10\frac{5}{3} = 11\frac{2}{3}$

11. $4\frac{1}{2} \times 3$

12. $3\frac{2}{3} \times 6$

13. $5\frac{3}{8} \times 4$

14. $2\frac{3}{7} \times 5$

Complete.

15. a. $2 \div \frac{2}{3} = \underline{?}$
    b. $2 \times \frac{3}{2} = \underline{?}$

16. a. $\frac{1}{2} \div \frac{1}{4} = \underline{?}$
    b. $\frac{1}{2} \times \frac{4}{1} = \underline{?}$

17. a. $\frac{4}{5} \div \frac{2}{5} = \underline{?}$
    b. $\frac{4}{5} \times \frac{5}{2} = \underline{?}$

Divide.

$\frac{3}{8} \div \frac{1}{2} = \frac{3}{8} \times \frac{2}{1} = \frac{6}{8} = \frac{3}{4}$

18. $6 \div \frac{1}{2}$

19. $\frac{3}{4} \div \frac{2}{3}$

20. $\frac{7}{8} \div 2$

21. $\frac{2}{3} \div \frac{5}{6}$

22. $\frac{4}{5} \div \frac{2}{3}$

23. $\frac{7}{3} \div \frac{5}{2}$

Divide.

$2\frac{1}{2} \div 1\frac{2}{3} = \frac{5}{2} \div \frac{5}{3} = \frac{5}{2} \times \frac{3}{5} = \frac{3}{2} = 1\frac{1}{2}$

24. $3\frac{1}{2} \div 2$

25. $7 \div 2\frac{1}{3}$

26. $4\frac{1}{2} \div 3\frac{1}{3}$

27. $2\frac{5}{6} \div 4\frac{1}{5}$

28. $3\frac{2}{3} \div 1\frac{1}{3}$

CHAPTER CHALLENGE

A die can land in any one of 6 ways. All ways are equally likely.

Probability of 2 up $= \frac{1}{6}$

$P(2) = \frac{1}{6}$

A coin can land in one of 2 ways. Both ways are equally likely.

$P(\text{H}) = \frac{1}{2}$

This spinner can land in any one of 4 ways. All ways are equally likely.

$P(\text{yellow}) = \frac{1}{4}$

Suppose we are going to roll a die *twice* and wish to know the probability of getting 6 on the first throw and 2 on the second. We can multiply.

$P(\text{6 on first, 2 on second}) = P(\text{6 on first}) \times P(\text{2 on second}) = \frac{1}{6} \times \frac{1}{6} = \frac{1}{36}$

Give the probability.

1.

P(3 on first, 4 on second)

2.

P(head on first, tail on second)

3.

P(tail on first, head on second)

4.

P(H, H)

5.

P(yellow, green)

6.

P(tail, red)

7.

P(5, H)

8.

P(H, H, H)

9.

P(1, green)

MAJOR CHECKUP
Standardized Format

Choose the correct letter.

1. 99,999 rounded to the nearest thousand is
   - **a.** 99,000
   - **b.** 10,000
   - **c.** 100,000
   - **d.** none of these

2. Line ℓ is a line of symmetry. Which segment is congruent to $\overline{AB}$?

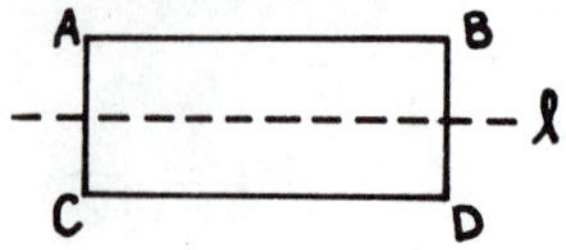

   - **a.** $\overline{BC}$
   - **b.** $\overline{CD}$
   - **c.** $\overline{DA}$
   - **d.** none of these

3. Which figure is congruent to this figure? 
   - **a.** 
   - **b.** 
   - **c.** 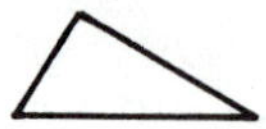
   - **d.** none of these

4. Subtract.
$$\begin{array}{r} 702 \\ -\ 143 \\ \hline \end{array}$$
   - **a.** 641
   - **b.** 559
   - **c.** 659
   - **d.** none of these

5. Multiply.
$$\begin{array}{r} 34 \\ \times 25 \\ \hline \end{array}$$
   - **a.** 850
   - **b.** 650
   - **c.** 228
   - **d.** none of these

6. Divide.
$6\overline{)5766}$
   - **a.** 966
   - **b.** 960
   - **c.** 961
   - **d.** none of these

7. Add.
$\frac{1}{2} + \frac{1}{3}$
   - **a.** $\frac{2}{5}$
   - **b.** $\frac{5}{6}$
   - **c.** $\frac{1}{5}$
   - **d.** none of these

8. Subtract.
$\frac{3}{4} - \frac{1}{3}$
   - **a.** $\frac{2}{1}$
   - **b.** $\frac{2}{7}$
   - **c.** $\frac{5}{12}$
   - **d.** none of these

9. Add.
$$\begin{array}{r} 3\frac{1}{2} \\ +2\frac{3}{4} \\ \hline \end{array}$$
   - **a.** $6\frac{1}{4}$
   - **b.** $6\frac{1}{2}$
   - **c.** $5\frac{4}{6}$
   - **d.** none of these

10. Subtract.
$$\begin{array}{r} 5\frac{1}{3} \\ -\ 2\frac{2}{3} \\ \hline \end{array}$$
    - **a.** $3\frac{1}{3}$
    - **b.** $2\frac{9}{3}$
    - **c.** $2\frac{2}{3}$
    - **d.** none of these

11. Multiply.
$\frac{3}{4} \times \frac{5}{3}$
    - **a.** $\frac{5}{4}$
    - **b.** $\frac{9}{20}$
    - **c.** $\frac{20}{9}$
    - **d.** none of these

12. Divide.
$\frac{3}{4} \div \frac{5}{3}$
    - **a.** $\frac{5}{4}$
    - **b.** $\frac{9}{20}$
    - **c.** $\frac{20}{9}$
    - **d.** none of these

7
Adding and Subtracting Decimals

READY OR NOT ?

What fraction of the square is shaded?

1. 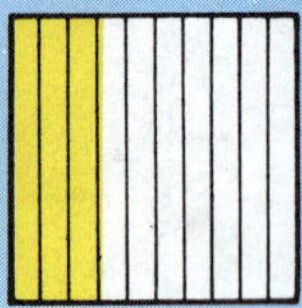

2. 

3. 

4. 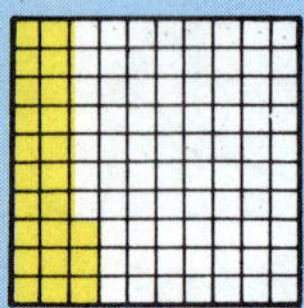

5. 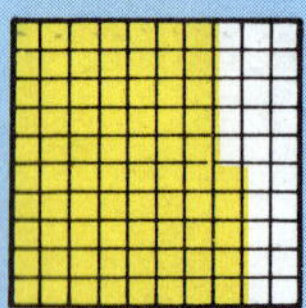

6. 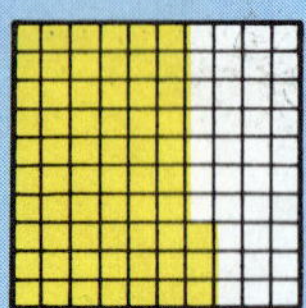

Tenths and hundredths

We can write about the number of unit squares here using fractions.

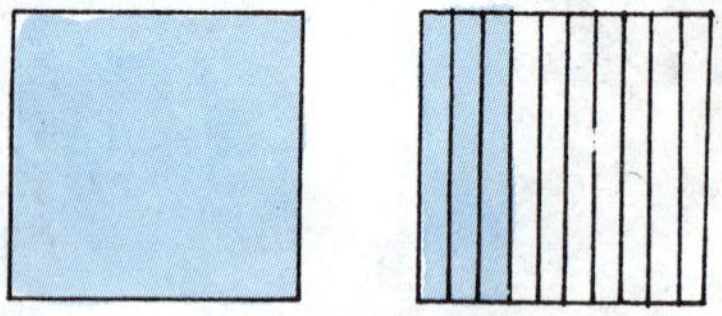

$1\frac{3}{10}$

We can also use place value and write **decimals.**

Ones	Tenths
1	3

1.3

Notice that a **decimal point** is written between the ones place and the tenths place.

Read "1.3" as "one and three tenths."

Here are some other examples of decimals.

EXAMPLE 1.

Ones	Tenths	Hundredths
1	0	7

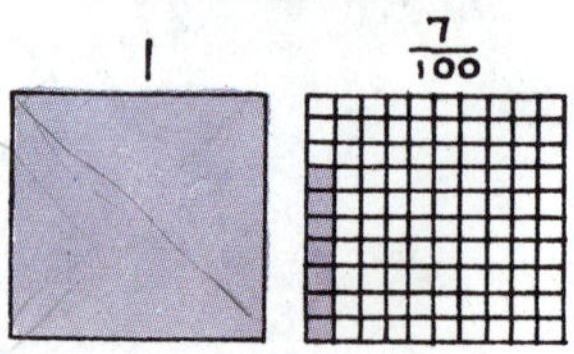

1.07

one and seven hundredths

EXAMPLE 2.

Ones	Tenths	Hundredths
1	3	7

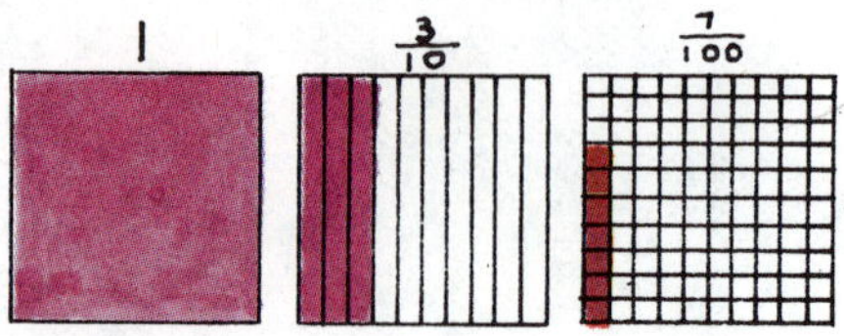

1.37

one and thirty-seven hundredths

To read the decimal, think of 3 tenths as 30 hundredths.

EXERCISES

Give a decimal for the number of shaded squares.

1.
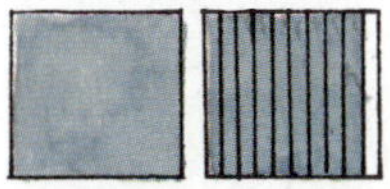

2.
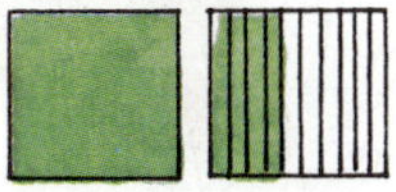

3.
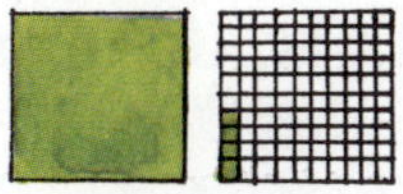

4.
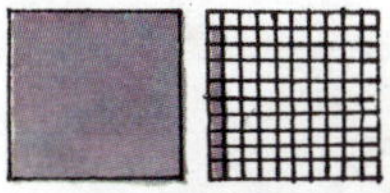

5.

6.
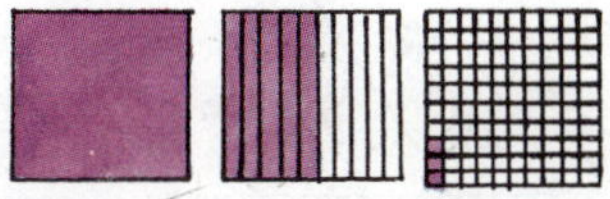

Tell what the red digit stands for.

7. 63.8
   8 tenths
8. 30.52
9. 82.16
10. 74.38
11. 66.66
12. 66.66
13. 66.66

Read each decimal aloud.

14. 42.6
15. 38.09
16. 6.5
17. 84.03
18. 35.24
19. 426.74

Read each fact aloud.

20. The distance around the earth at the equator is 40,074.98 kilometers.
21. The distance around the earth at a meridian is 40,008.01 kilometers.
22. The distance from the equator to the center of the earth is 6378.15 kilometers.

Write as decimals.

23. four and six tenths
24. nineteen and seven hundredths
25. twenty-two and five tenths
26. thirty-seven and twenty-one hundredths
27. two hundred and two hundredths
28. twelve and thirty-nine hundredths

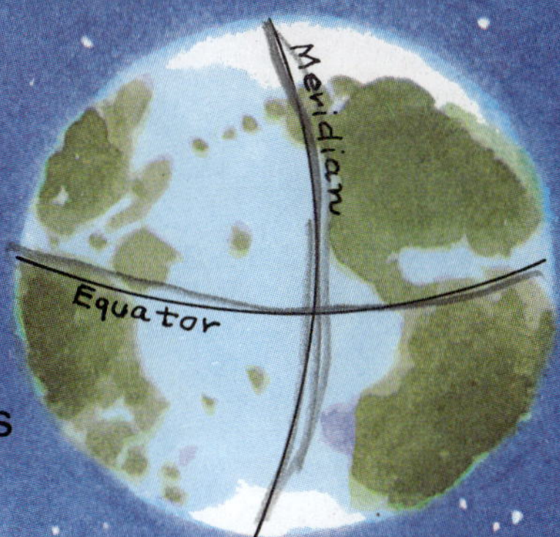

Thousandths

If we divide one hundredth into 10 equal parts, each part is one **thousandth.**

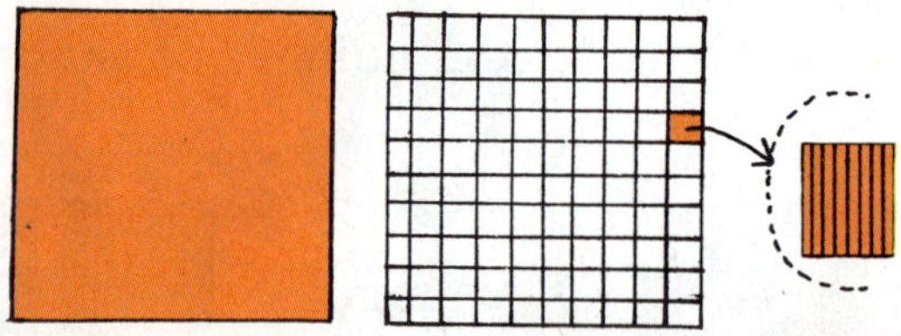

Ones	Tenths	Hundredths	Thousandths
1	0	0	8

1. 0 0 8

Read this as
"one and eight thousandths."

Here is another example:

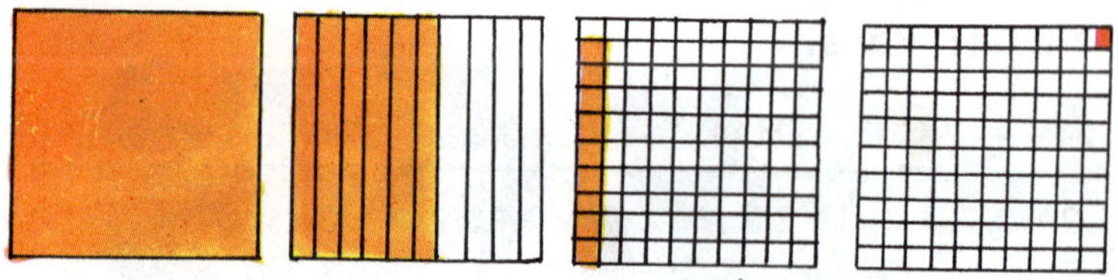

Ones	Tenths	Hundredths	Thousandths
1	6	9	7

1. 6 9 7

This is read as "one and six hundred ninety-seven thousandths."

EXAMPLE.
Read "23.047" as "twenty-three and forty-seven thousandths."

EXERCISES

Build a numeral.

1. 3 in the tenths place
   7 in the ones place
   4 in the thousandths place
   1 in the tens place
   9 in the hundredths place

2. 5 in the hundredths place
   6 in the ones place
   3 in the tenths place
   8 in the thousandths place
   6 in the hundreds place
   9 in the thousands place
   4 in the tens place

Read each decimal.

3. 1.8	**4.** 1.08	**5.** 1.008	**6.** 1.88	**7.** 1.888
8. 3.082	**9.** 5.261	**10.** 4.019	**11.** 200.002	**12.** 0.202

Give the segment in which the number "falls."

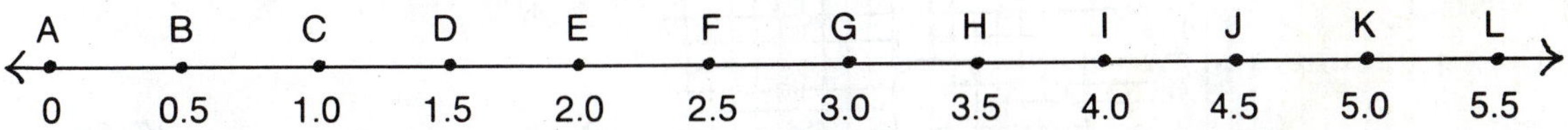

13. 2.67	**14.** 3.8	**15.** 2.4	**16.** 2.6	**17.** 4.3
FG	**18.** 2.95	**19.** 4.79	**20.** 3.21	**21.** 4.38
	22. 0.134	**23.** 1.34	**24.** 0.396	**25.** 3.96

Write as a decimal.

26. eight and six thousandths
27. ninety-three and forty-five thousandths
28. fifty-eight and two hundred thirty-six thousandths
29. four hundred seventy and three hundred eleven thousandths
30. nine thousand eighteen and sixty-three thousandths

More about decimals

A place-value table can be extended on and on in both directions.

Hundred thousands	Ten thousands	Thousands	Hundreds	Tens	Ones	Tenths	Hundredths	Thousandths	Ten thousandths	Hundred thousandths
4	9	2	7	8	1	0	5	3	6	1

If you extend the table, what is the next place to the left? What is the next place to the right?

3.67 three point six seven

45.079 forty-five point zero seven nine

EXERCISES

Read these decimals in two ways.

1. 6.08
2. 5.37
3. 8.724
4. 21.007
5. 35.168
6. 12.053
7. 72.865
8. 300.005
9. 0.305
10. 0.87
11. 120.007
12. 0.127
13. 0.0007
14. 0.00006
15. 0.3214

What digit is in the

16. tens place?
17. hundredths place?
18. ones place?
19. hundred thousands place?
20. thousandths place?
21. tenths place?
22. thousands place?

Complete, using decimals.

Remember: 1 dime = .1 dollar

23. 2 dimes = ? dollar
24. 5 dimes = ? dollar
25. 7 dimes = ? dollar
26. 10 dimes = ? dollar
27. 13 dimes = ? dollars
28. 47 dimes = ? dollars
29. 145 dimes = ? dollars
30. 1465 dimes = ? dollars

Remember: 1 centimeter (cm) = .01 meter (m)

31. 3 cm = ? m
32. 9 cm = ? m
33. 10 cm = ? m
34. 17 cm = ? m
35. 38 cm = ? m
36. 85 cm = ? m
37. 100 cm = ? m
38. 104 cm = ? m
39. 235 cm = ? m
40. 987 cm = ? m
41. 1000 cm = ? m

Remember: 1 millimeter (mm) = .001 meter

42. 2 mm = ? m
43. 9 mm = ? m
44. 10 mm = ? m
45. 18 mm = ? m
46. 58 mm = ? m
47. 100 mm = ? m
48. 352 mm = ? m
49. 875 mm = ? m
50. 999 mm = ? m
51. 1000 mm = ? m
52. 1001 mm = ? m
53. 2462 mm = ? m

The total number of tens in 346 is 34.6.

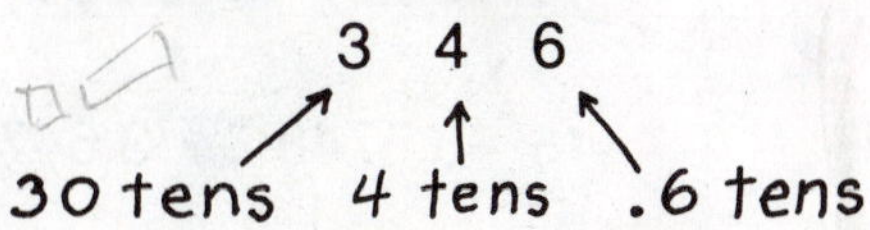

★ **Answer the questions.**

54. What is the total number of tens in 56?
55. What is the total number of hundreds in 287?
56. What is the total number of hundreds in 5421?

Comparing decimals

Thinking about shaded squares can help you compare decimals.

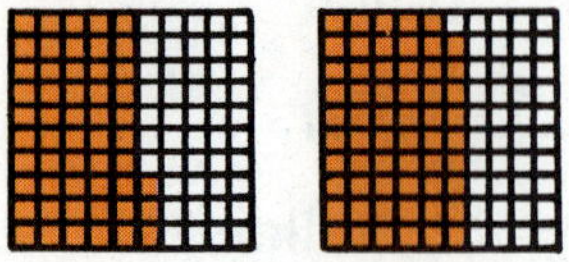

$0.53 < 0.59$
is less than

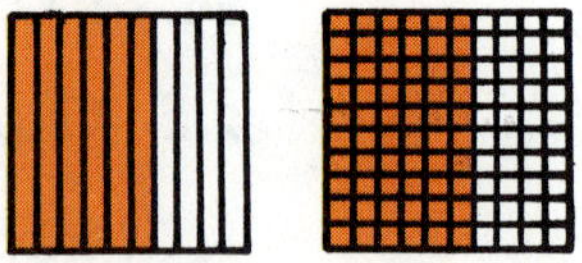

$0.6 = 0.60$
is equal to

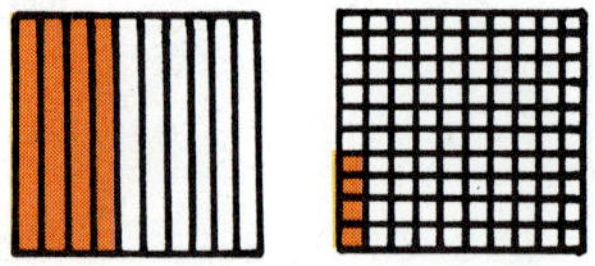

$0.4 > 0.04$
is greater than

These examples show how to compare decimals by comparing the digits that are in the same place.

Same

6.2043   6.2511

$0 < 5$

So, $6.2043 < 6.2511$

8.1   0.999

$8 > 0$

So, $8.1 > 0.999$

EXERCISES

$<$, $=$, or $>$?

1. 0.5 ● 0.50	**2.** 0.30 ● 0.3	**3.** 0.20 ● 0.200
4. 0.02 ● 0.2	**5.** 0.6 ● 0.06	**6.** 0.007 ● 0.07
7. 1 ● 0.9	**8.** 0.99 ● 1	**9.** 0.999 ● 1.0
10. 0.53 ● 0.54	**11.** 0.78 ● 0.76	**12.** 0.257 ● 0.261
13. 0.3984 ● 0.3964	**14.** 5.263 ● 5.291	**15.** 3.6748 ● 3.6478
16. 0.193 ● 0.26	**17.** 0.583 ● 0.49	**18.** 0.7823 ● 0.94
19. 0.526 ● 5.26	**20.** 78.3 ● 7.83	**21.** 9.5 ● 9.50
22. 0.4295 ● 0.4106	**23.** 3.782 ● 3.871	**24.** 4.2 ● 0.9658

Skill Game

Give a decimal that is between the two decimals.

1. 6.3, 6.6
2. 52.5, 52.6
3. 48.9, 49.0
4. 4.51, 4.53
5. 6.84, 6.9
6. 12.3, 12.42
7. 15.82, 15.825
8. 9.745, 9.75
9. 1.0, 1.04
10. 4.1361, 4.1365
11. 3.8219, 3.822
12. 63.425, 63.45

Play the game.

1. Choose a game leader and divide the class into two teams, team A and team B.
2. The game leader writes two whole numbers on the chalkboard.
3. Team A earns 1 point if a player from team A can go to the chalkboard and write a number that is between the two numbers.
4. The game leader erases one of the "end numbers." A player from team B can earn his team 1 point by writing a number that is between the two numbers that are left on the chalkboard.
5. Play continues until each player has had a turn. The team with the greater total wins.

Rounding decimals

Joe Leonard was the first winner of the California 500 (in 1971). He averaged 152.354 miles per hour.

Here is how to round his average speed.

Rounded to the nearest whole number:

152.354 rounds to 152.

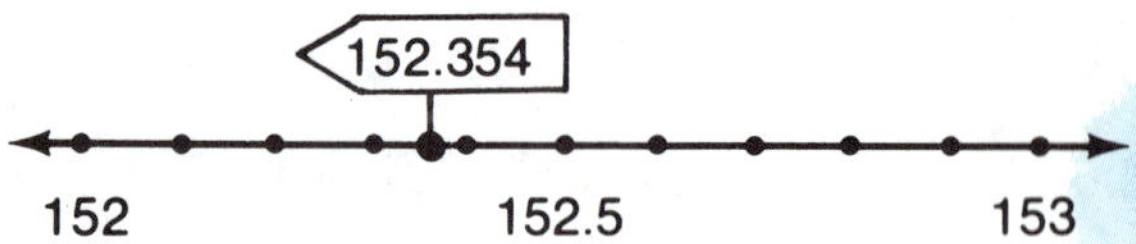

Rounded to the nearest tenth:

152.354 rounds to 152.4.

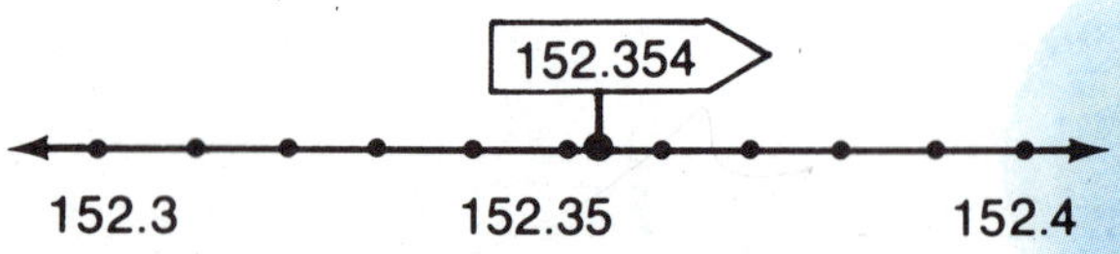

Rounded to the nearest hundredth:

152.354 rounds to 152.35.

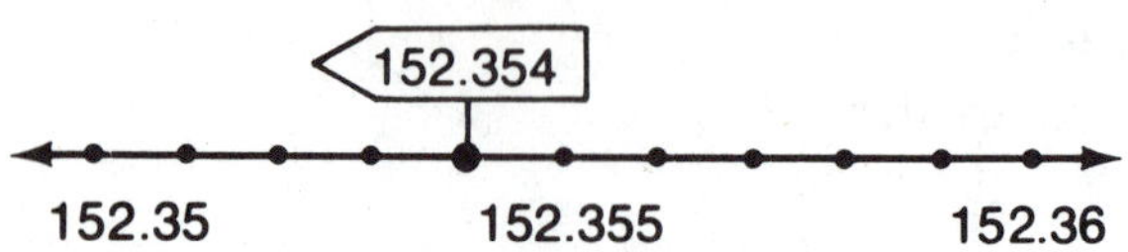

EXERCISES

Round to the nearest whole number.

1. 15.8 **2.** 37.3 **3.** 94.5 **4.** 53.43 **5.** 62.29

6. 74.928 **7.** 59.523 **8.** 421.05 **9.** 315.083 **10.** 529.564

Round to the nearest tenth.

11. 2.635 **12.** 8.093 **13.** 7.284 **14.** 0.552 **15.** 3.470

16. 89.269 **17.** 35.044 **18.** 92.835 **19.** 765.46 **20.** 923.81

Round to the nearest hundredth.

21. 35.216 **22.** 48.108 **23.** 92.543 **24.** 71.825 **25.** 63.581

26. 5.1039 **27.** 7.4265 **28.** 9.3815 **29.** 6.3452 **30.** 5.9305

Round the number to the nearest

31. tenth. **32.** hundredth.

33. ten thousandth. **34.** thousandth.

Round to the nearest thousandth.

35. A sheet of paper from a paperback book is about 0.0032 millimeter thick.

36. The smallest spider is about 0.0395 centimeter long.

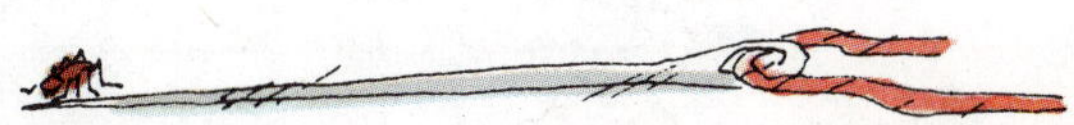

Keeping Skills Sharp

Give each product in lowest terms.

1. $\frac{3}{4} \times \frac{1}{2}$ **2.** $\frac{1}{4} \times \frac{1}{3}$ **3.** $\frac{3}{2} \times \frac{4}{3}$ **4.** $\frac{5}{8} \times \frac{4}{5}$

5. $5 \times \frac{2}{5}$ **6.** $\frac{5}{6} \times 3$ **7.** $\frac{4}{3} \times \frac{3}{4}$ **8.** $4 \times \frac{5}{2}$

9. $\frac{7}{8} \times \frac{2}{3}$ **10.** $\frac{5}{6} \times \frac{3}{8}$ **11.** $\frac{4}{5} \times \frac{10}{3}$ **12.** $\frac{3}{4} \times \frac{9}{12}$

Adding decimals

You add decimals the same way you add whole numbers.

EXAMPLE 1. 5.634 + 2.909

Step 1. Add thousandths and regroup.

```
  5.634
+ 2.909
-------
      3
```

Step 2. Add hundredths.

```
  5.634
+ 2.909
-------
     43
```

Step 3. Add tenths and regroup.

```
  5.634
+ 2.909
-------
   .543
```

Step 4. Add ones.

```
  5.634
+ 2.909
-------
  8.543
```

REMEMBER!

Before adding decimals, make sure that the decimal points are lined up.

EXAMPLE 2. 8.86 + 0.9 + 15.352

Step 1. Add thousandths.

```
   8.86
   0.9
+ 15.352
--------
       2
```

Step 2. Add hundredths and regroup.

```
   8.86
   0.9
+ 15.352
--------
      12
```

Step 3. Add tenths and regroup.

```
   8.86
   0.9
+ 15.352
--------
    .112
```

Step 4. Add ones and regroup.

```
   8.86
   0.9
+ 15.352
--------
   5.112
```

Step 5. Add tens.

```
   8.86
   0.9
+ 15.352
--------
  25.112
```

EXERCISES

First estimate the sum. Then add.

1. 3.54 + 0.63
2. 6.82 + 0.09
3. 6.3 + 8.2
4. 5.34 + 2.96
5. 7.42 + 8.95
6. 0.976 + 0.003
7. 0.752 + 0.396
8. 0.5384 + 0.2574
9. 0.3896 + 0.74
10. 0.56 + 2.489
11. 3.6421 + 1.2538
12. 7.8396 + 0.7482
13. 96.783 + 2.59
14. 4.96 + 3.8594
15. 7.236 + 15.974
16. $1.86 + .73 + 5.25
17. $5.38 + 2.92 + 4.26
18. $3.95 + 2.84 + 5.63
19. $18.74 + 23.52 + 9.74
20. $35.28 + 36.25 + 17.09
21. 3.8 + 2.09
22. 5.6 + 9.48
23. 15.3 + 2.95
24. 6.8 + 3.04 + 5
25. 6 + 2.89 + 3.5
26. 7.62 + 4.8 + 2.359

Solve.

27. What is the normal precipitation during January and February in Atlanta? In New York?
28. In Los Angeles, what is the normal rainfall during the first six months of the year?
29. What is the normal yearly precipitation in Chicago?

Monthly Normal Precipitation in Inches

City	Jan.	Feb.	Mar.	Apr.	May	June	July	Aug.	Sept.	Oct.	Nov.	Dec.
Atlanta	4.4	4.5	5.4	4.5	3.2	3.8	4.7	3.6	3.3	2.4	3.0	4.4
Chicago	1.9	1.6	2.7	3.0	3.7	4.1	3.4	3.2	2.7	2.8	2.2	1.9
Los Angeles	3.1	3.3	2.3	1.2	.2	.1	.0	.0	.2	.4	1.1	2.9
Miami	2.0	1.9	2.3	3.9	6.4	7.4	6.8	7.0	9.5	8.2	2.8	1.7
New York	3.3	2.8	4.0	3.4	3.7	3.3	3.7	4.4	3.9	3.1	3.4	3.3

Subtracting decimals

You subtract decimals the same way you subtract whole numbers.

When subtracting decimals, make sure the decimal points are lined up.

EXAMPLE 1. 6.402 – 1.674

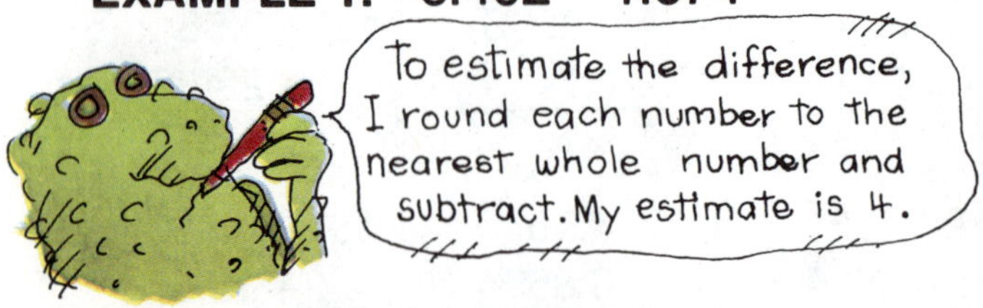

Step 1. Regroup and subtract thousandths.

$$\begin{array}{r} \overset{3\,9\,1}{6.402} \\ -\ 1.674 \\ \hline 8 \end{array}$$

Step 2. Subtract hundredths.

$$\begin{array}{r} \overset{3\,9\,1}{6.402} \\ -\ 1.674 \\ \hline 28 \end{array}$$

Step 3. Regroup and subtract tenths.

$$\begin{array}{r} \overset{5\,13\,9\,1}{6.402} \\ -\ 1.674 \\ \hline .728 \end{array}$$

Step 4. Subtract ones.

$$\begin{array}{r} \overset{5\,13\,9\,1}{6.402} \\ -\ 1.674 \\ \hline 4.728 \end{array}$$

EXAMPLE 2. 5.2 – 3.48

Step 1. Since .2 = .20, write a zero in the hundredths place.

$$\begin{array}{r} 5.20 \\ -\ 3.48 \\ \hline \end{array}$$

Step 2. Regroup and subtract hundredths.

$$\begin{array}{r} \overset{1\,1}{5.20} \\ -\ 3.48 \\ \hline 2 \end{array}$$

Step 3. Regroup and subtract tenths.

$$\begin{array}{r} \overset{4\,11\,1}{5.20} \\ -\ 3.48 \\ \hline .72 \end{array}$$

Step 4. Subtract ones.

$$\begin{array}{r} \overset{4\,11\,1}{5.20} \\ -\ 3.48 \\ \hline 1.72 \end{array}$$

Find each difference.

1. 8.6 − 3.4
2. 0.95 − 0.21
3. 6.48 − 0.03
4. 25.6 − 19.9
5. 24.6 − 18.9
6. 3.741 − 1.685
7. 3.592 − 1.648
8. 9.784 − 2.659
9. 3.017 − 1.854
10. 69.356 − 7.281
11. 8.001 − 3.456
12. 12.935 − 4.62
13. 36.521 − 15.934
14. 6.323 − 1.784
15. 5.2964 − 3.18
16. 5.3 − 2.64
17. 7.9 − 3.748
18. 9.02 − 5.644
19. 12.04 − 6.3815
20. 7.43 − 2.6975
21. 14.6 − 13.8
22. 23.4 − 2.89
23. 7 − 6.52
24. 5.1 − 2.658
25. 14.6 − 3.826
26. 5.43 − 1.009

Solve.

27. What is the population density (people per square kilometer) of Delaware?
28. Which state is the most densely populated?
29. Can you tell from this table which state has the greatest population?
30. How many more people per square kilometer does Connecticut have than Indiana?
31. Which state has a population density of 87.3 people more than Massachusetts?
32. Which state has the fourth greatest density?
33. Rhode Island is how much more densely populated than New York?

The 16 most densely populated states

State	Population Density (people per square km)
California	49.27
Massachusetts	280.69
Connecticut	240.81
Michigan	60.31
Delaware	106.76
New Jersey	367.99
Florida	48.46
New York	147.22
Hawaii	46.25
Ohio	100.39
Illinois	76.99
Pennsylvania	101.27
Indiana	55.56
Rhode Island	349.61
Maryland	153.13
Virginia	45.14

Practice exercises

Add.

1. 43.89 + 20.75
2. 9.682 + 7.395
3. 563.4 + 87.9
4. 0.8634 + 0.2963
5. 0.59648 + 0.2364
6. 0.73 + 0.59 + 0.62
7. 0.83 + 0.92 + 0.74
8. 0.64 + 0.82 + 0.91
9. 7.41 + 8.35 + 2.63
10. 8.249 + 17.6 + 9.05
11. 2.01 + 7.8 + 3.95 + 2.06
12. 3.5 + 5.28 + 2.09 + 15.3
13. 0.006 + 3.891 + 0.74 + 0.358
14. 38.21 + 61 + 5.78 + 96.3
15. 8.421 + 26.7 + 18 + 9.56

Subtract.

16. 274.3 − 139.4
17. 5.962 − 2.588
18. 74.38 − 36.95
19. 0.9106 − 0.0482
20. 5.342 − 0.799
21. 73.8 − 38.42
22. 5.26 − 1.567
23. 93.8 − 5.93
24. 91.5 − 78.21
25. 47.6 − 26.54
26. 520.3 − 43.67
27. 36.01 − 3.958
28. 29 − 2.834
29. 15 − 1.781
30. 38 − 2.563

Solve.

31. 

What is the total weight?

32. 

How much more is the larger box?

33. On her first try, Susan threw the softball 17.78 meters. On her second try she threw it 19.05 meters. How much farther did she throw it on her second try?

34. John ran the 100-meter dash in 14 seconds. David ran it in 12.7 seconds. How much faster was David's time?

35. A gasoline tank holds 60 liters. If 46.7 liters of gasoline filled the tank, how much gasoline was already in the tank?

36. Normal body temperature is 37 degrees Celsius. A patient's temperature is 38.6 degrees Celsius. How much is this above normal?

37. Robert wants to buy a tennis racket that costs $15.39. He has saved $8.25 and has a Saturday job that will pay him $6.50. How much more money will he need after doing the Saturday job?

38. The Levy family planned to drive 560 kilometers in one day. Before breakfast they drove 153.8 kilometers. After breakfast they drove 255.9 kilometers before stopping for lunch. How much farther did they have to drive after lunch?

Who am I?

39. *If you add me to 16.3, you get 30.74.*

40. *If you subtract 5.94 from me, you get 9.58.*

Keeping Skills Sharp

Give each quotient in lowest terms.

1. $\frac{2}{3} \div \frac{3}{4}$	**2.** $\frac{3}{4} \div \frac{2}{3}$	**3.** $3 \div \frac{5}{6}$	**4.** $\frac{3}{8} \div 2$
5. $\frac{3}{5} \div \frac{1}{4}$	**6.** $\frac{4}{3} \div \frac{5}{6}$	**7.** $\frac{3}{2} \div \frac{6}{5}$	**8.** $\frac{7}{8} \div \frac{3}{4}$
9. $\frac{5}{9} \div \frac{2}{3}$	**10.** $\frac{7}{10} \div \frac{4}{5}$	**11.** $\frac{7}{12} \div \frac{5}{3}$	**12.** $\frac{5}{9} \div \frac{5}{4}$

Problem solving

1. What is the total price of a baseball glove and cap?
2. What is the total price of a tennis racket, a can of balls, a pair of socks, and a pair of shoes?
3. How much more does a football cost than a basketball?
4. Martha has $19.46. She wants to buy a warm-up jacket. How much money will she have left?

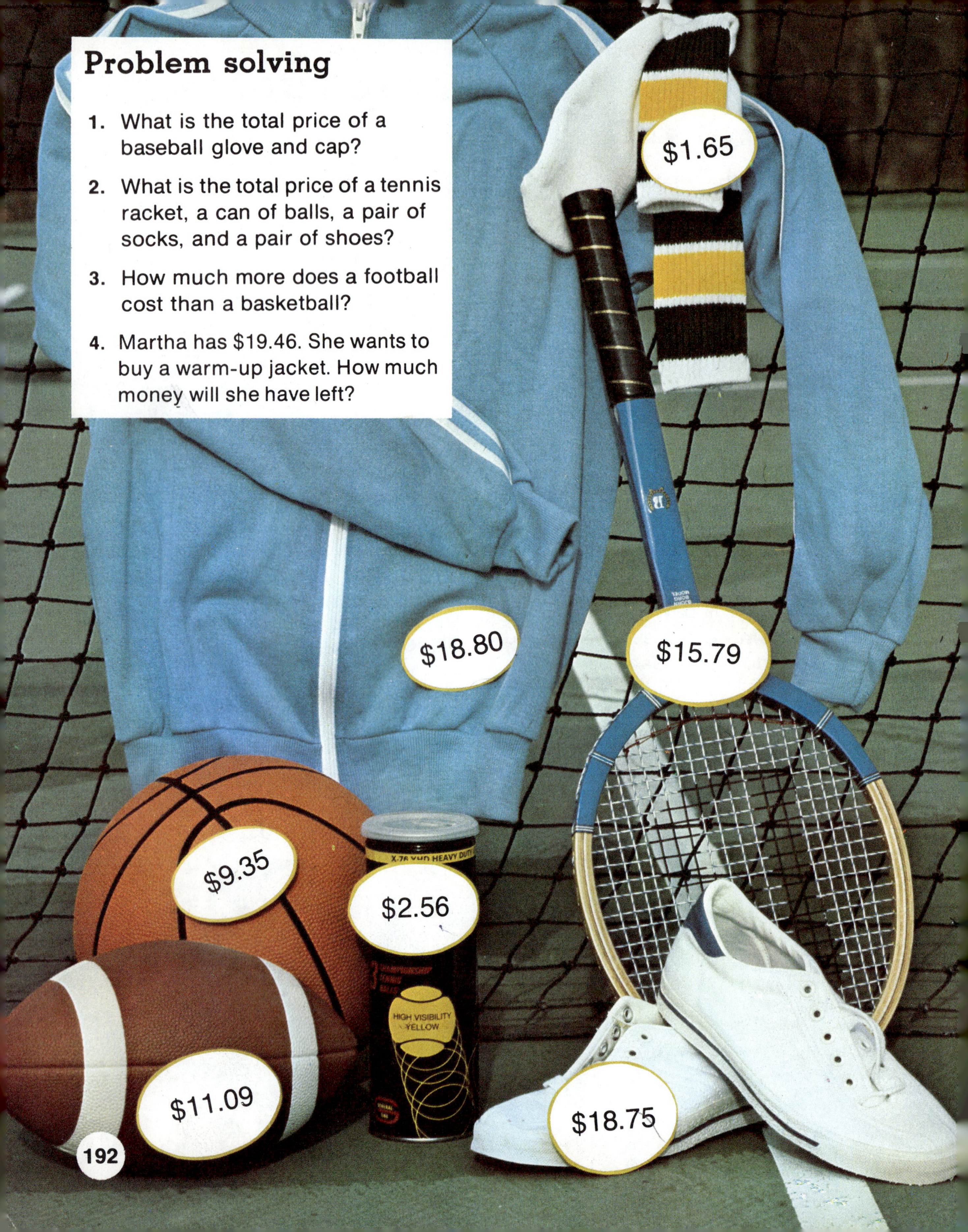

5. Sarah has $10. She wants to buy a softball and bat. How much money will she have left?

6. How much will 2 cans of tennis balls and a pair of socks cost?

7. Mark wants to buy a pair of tennis shoes and a jersey. He needs $5.85 more. How much money does he have?

8. Craig has $16.50 in the bank. He wants to buy a tennis racket, 2 pair of socks, a pair of shoes, and 2 cans of balls. How much more money does he need?

CHAPTER CHECKUP

Tell what the red digit stands for. [pages 176–181]

1. 4.**2**65
   2 tenths
2. 38.9**1**
3. 0.0**6**5
4. 3.**9**82
5. 0.47**3**6
6. 0.9**2**83
7. 0.28**7**4
8. 6.103**9**
9. 5.78**1**5

<, =, or >? [pages 182–183]

10. 0.8 ● 0.80
11. 0.06 ● 0.6
12. 0.9 ● 0.89
13. 1 ● 0.99
14. 2.02 ● 2
15. 3.5 ● 3.500
16. 0.429 ● 0.43
17. 0.3816 ● 0.3820
18. 7.3845 ● 7.3712

[pages 184–185]

Round the number to the nearest

19. tenth.
20. whole number.
21. thousandth.
22. hundredth.

Add. [pages 186–187, 190]

23. $\begin{array}{r} 59.8 \\ +23.7 \\ \hline \end{array}$
24. $\begin{array}{r} 6.43 \\ +0.28 \\ \hline \end{array}$
25. $\begin{array}{r} 5.028 \\ +3.746 \\ \hline \end{array}$
26. $\begin{array}{r} 3.9 \\ 4.75 \\ +1.993 \\ \hline \end{array}$
27. $\begin{array}{r} 14.6 \\ 8.342 \\ +16.48 \\ \hline \end{array}$

Subtract. [pages 188–190]

28. $\begin{array}{r} 7.21 \\ -3.75 \\ \hline \end{array}$
29. $\begin{array}{r} 5.804 \\ -1.298 \\ \hline \end{array}$
30. $\begin{array}{r} 917 \\ -25.4 \\ \hline \end{array}$
31. $\begin{array}{r} 605.3 \\ -58.97 \\ \hline \end{array}$
32. $\begin{array}{r} 2.0045 \\ -1.8329 \\ \hline \end{array}$

Solve [pages 191–193]

33. Before the Allens started on a trip, the odometer of their car read 38274.9 kilometers. After the trip it read 39109.4 kilometers. How many kilometers did they drive?

34. The Allens allowed $50 for car expenses. They paid $34.89 for gasoline, $2.18 for oil, and $2.75 for having a tire fixed. How much money did they have left?

Project

What is your throwing score?

In the punt-pass-kick competition, passing skill is judged on both distance and accuracy. The score is the distance the football is thrown down a line minus the distance from the line the ball lands. In this project you will test your football-throwing distance and accuracy.

1. When throwing, you must not step outside the parallel tapes that are 3 meters apart.
2. A "spotter" will mark where the ball lands. Your classmates will compute your score by using the method shown in the following example.

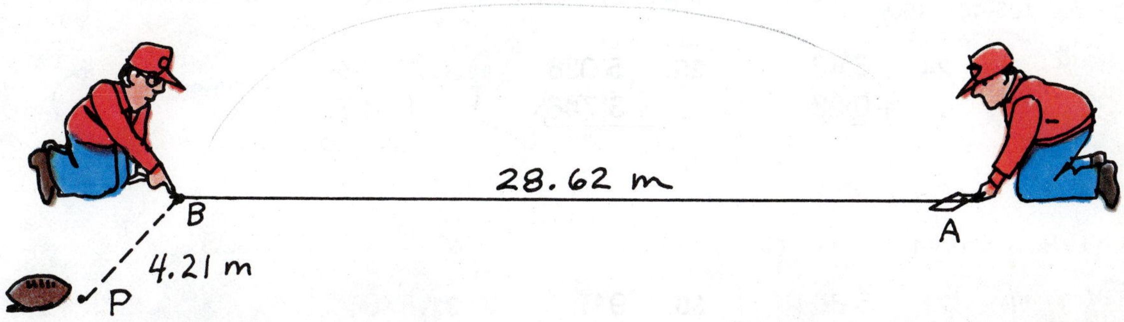

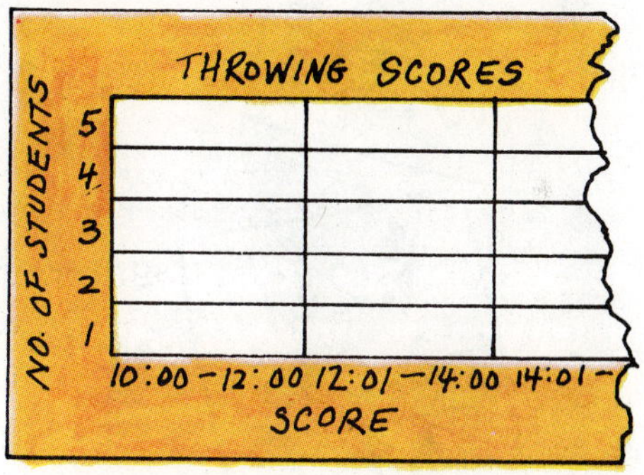

a. Measure $\overline{AB}$ to the nearest .01 meter (cm).
b. Measure $\overline{PB}$ to the nearest .01 meter (cm).
c. Subtract the measure of $\overline{PB}$ from the measure of $\overline{AB}$.

The score would be 28.62 − 4.21, or 24.41.

3. Make a bar graph of the class scores.

CHAPTER REVIEW

Which digit represents

1. tenths?
2. hundredths?
3. ones?
4. thousandths?

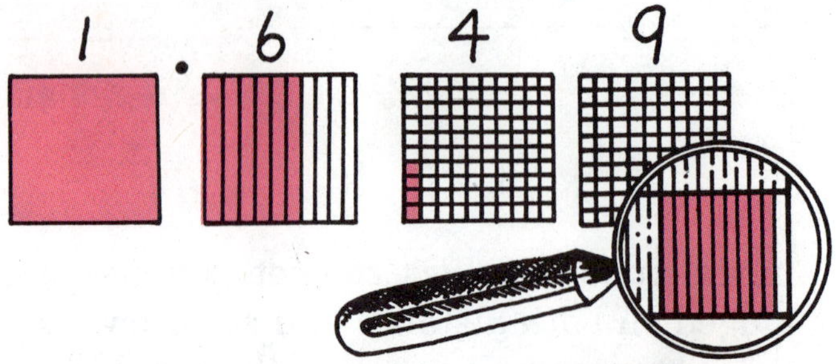

<, =, or >?

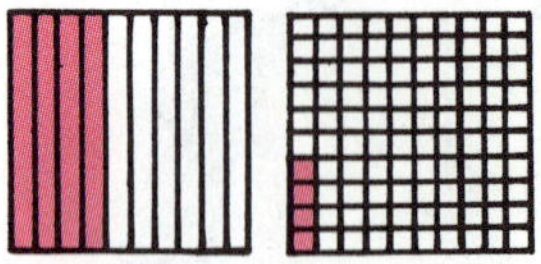

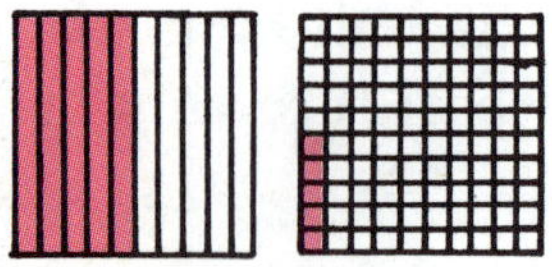

5. 0.4 ● 0.04
6. 0.5 ● 0.05
7. 0.38 ● 0.41
8. 0.521 ● 0.253
9. 4.6 ● 4.60
10. 0.3 ● 0.291

Round to the nearest tenth.

11. 3.26

12. 18.32

13. 6.758

Round to the nearest hundredth.

14. 3.6178
15. 15.4052
16. 24.3628

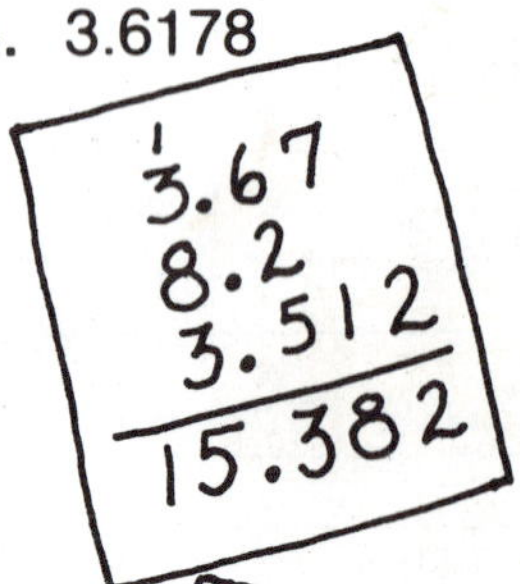

Add.

17. $\begin{array}{r} 3.782 \\ +2.599 \\ \hline \end{array}$

18. $\begin{array}{r} 8.567 \\ +4.45 \\ \hline \end{array}$

19. $\begin{array}{r} 7.42 \\ 8.3 \\ +9.052 \\ \hline \end{array}$

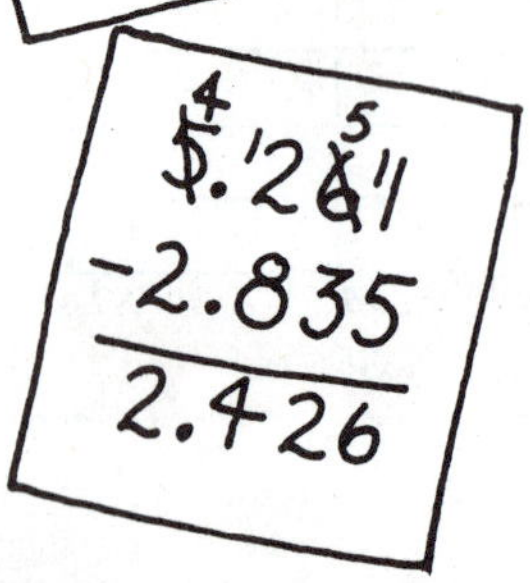

Subtract.

20. $\begin{array}{r} 6.415 \\ -2.378 \\ \hline \end{array}$

21. $\begin{array}{r} 53.04 \\ -29.78 \\ \hline \end{array}$

22. $\begin{array}{r} 6.20 \\ -2.84 \\ \hline \end{array}$

CHAPTER CHALLENGE

If the sum of the numbers in each row, column, and diagonal is the same, then the square is a magic square.

1. Is this a magic square?

2.24	2.11	2.1	2.21
2.13	2.18	2.19	2.16
2.17	2.14	2.15	2.2
2.12	2.23	2.22	2.09

2. Copy and complete this magic square.

3.7	3.57		
	3.64		3.62
3.63		3.61	
	3.69	3.68	3.55

3. Make a new square by adding the corresponding numbers in the magic squares above.

5.94			

Is it a magic square?

MAJOR CHECKUP

Standardized Format

Choose the correct letter.

1. 65,369,821 rounded to the nearest ten thousand is

a. 70,000,000
b. 65,000,000
c. 65,360,000
d. none of these

2. Add.

$$\begin{array}{r} 3916 \\ 287 \\ 1543 \\ +\ \ 95 \\ \hline \end{array}$$

a. 5841
b. 5621
c. 5821
d. none of these

3. Subtract.

$$\begin{array}{r} 10000 \\ -5824 \\ \hline \end{array}$$

a. 4186
b. 4076
c. 4176
d. none of these

4. Multiply.

$$\begin{array}{r} 5264 \\ \times 327 \\ \hline \end{array}$$

a. 1,718,908
b. 1,721,328
c. 1,600,228
d. none of these

5. Divide.

$37\overline{)3785}$

a. 12
b. 12 R10
c. 102 R10
d. none of these

6. If ΔABC is congruent to ΔSTR, then $\angle B$ is congruent to

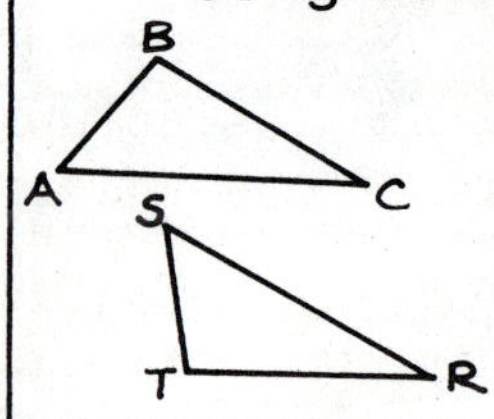

a. $\angle R$
b. $\angle S$
c. $\angle T$
d. none of these

7. $3\frac{3}{4} =$

a. $\frac{21}{4}$
b. $\frac{15}{4}$
c. $\frac{10}{4}$
d. none of these

8. $\frac{7}{3} =$

a. $2\frac{1}{3}$
b. $3\frac{1}{3}$
c. $1\frac{1}{3}$
d. none of these

9. Add.

$$\begin{array}{r} 3\frac{5}{6} \\ +4\frac{2}{3} \\ \hline \end{array}$$

a. $7\frac{1}{2}$
b. $7\frac{1}{6}$
c. $8\frac{1}{2}$
d. none of these

10. Subtract.

$$\begin{array}{r} 4\frac{1}{3} \\ -\ 2\frac{3}{4} \\ \hline \end{array}$$

a. $1\frac{7}{12}$
b. $2\frac{5}{12}$
c. $1\frac{5}{12}$
d. none of these

11. $\frac{3}{5} \times \frac{2}{3} =$

a. $\frac{1}{3}$
b. $\frac{2}{5}$
c. $\frac{9}{10}$
d. none of these

12. $\frac{5}{2} \div \frac{3}{4} =$

a. $\frac{10}{3}$
b. $\frac{3}{10}$
c. $\frac{8}{15}$
d. none of these

8 Multiplying and Dividing Decimals

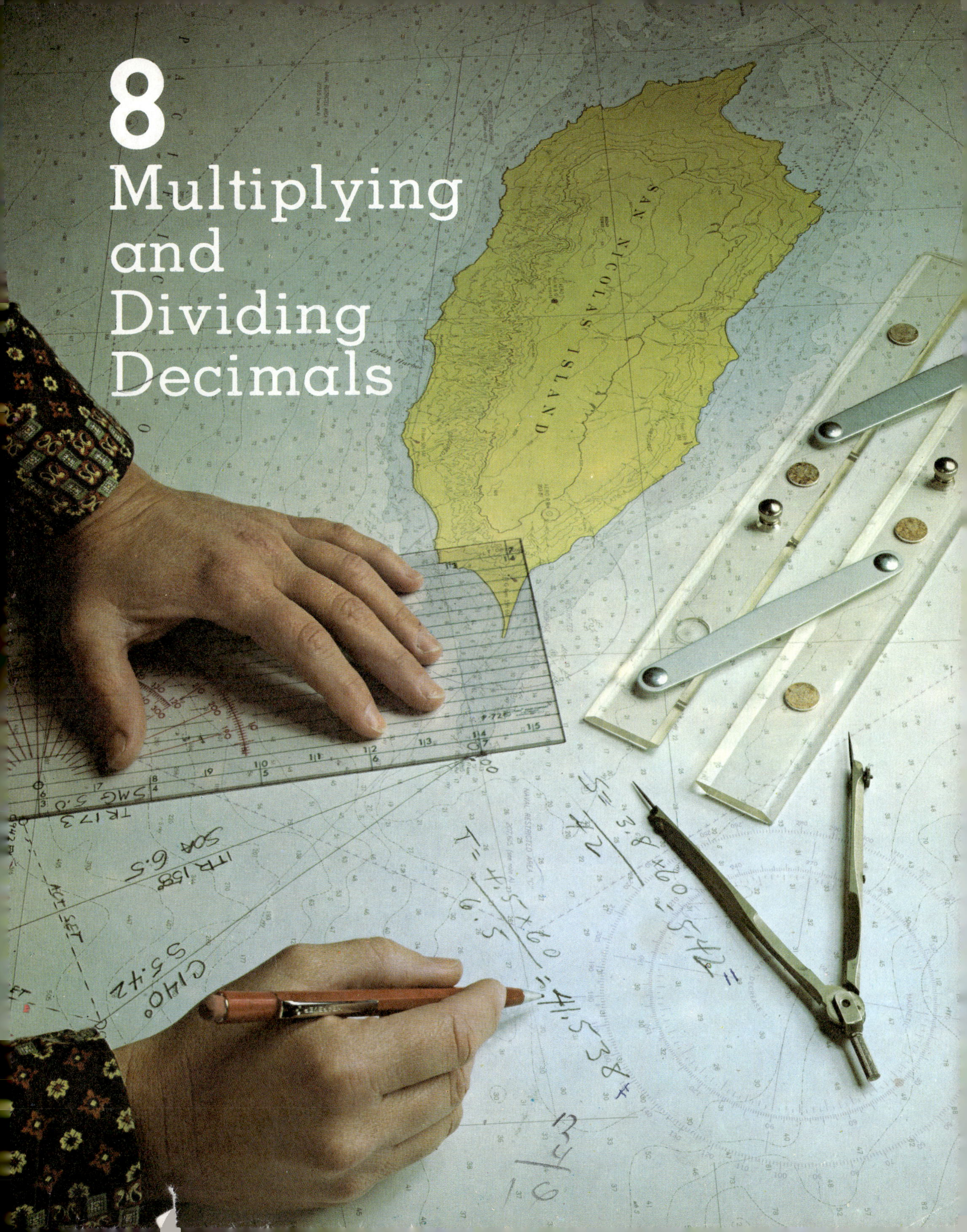

READY OR NOT ?

Give a decimal for the number of unit squares that are shaded yellow.

1. 

2. 

3. 

4. 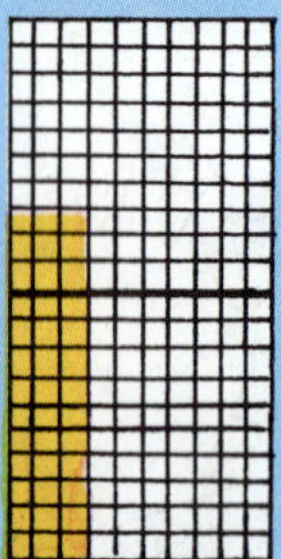

5. 

Multiplying decimals

We can use pictures to help us multiply decimals. Suppose that we wish to multiply 1.6 by .4.

Step 1. Pick a unit square and make a region that is as long as 1.6 unit squares and as wide as .4 unit squares.

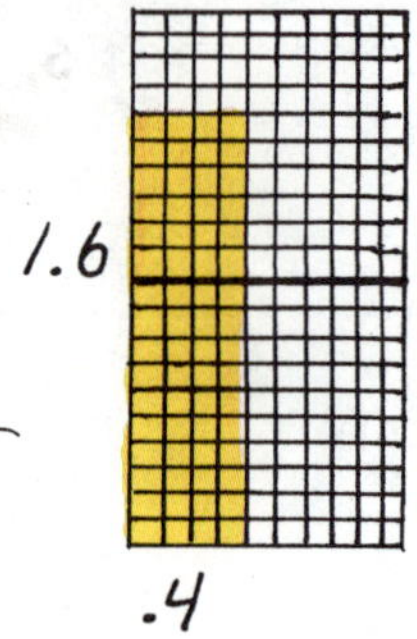

Step 2. Find what part of a unit square is shaded. In this case, 64 parts are shaded and each part is one hundredth of a unit square. So, .64 of a unit square is shaded.

$1.6 \times .4 = .64$

Here is how to do the problem without counting the shaded parts.

Step 1. Multiply the numbers as if they were whole numbers.

$$\begin{array}{r} 1.6 \\ \times\ .4 \\ \hline 64 \end{array}$$

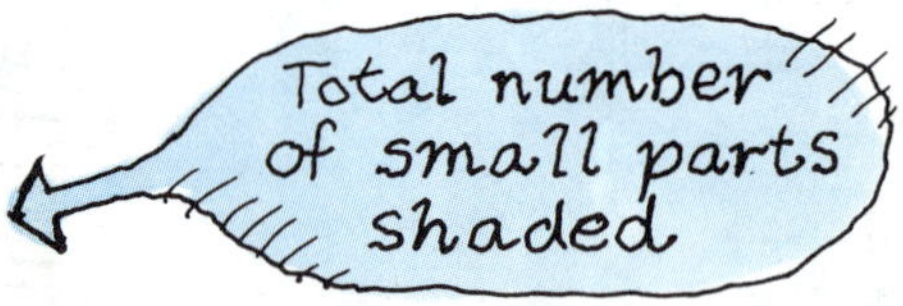

Step 2. Place the decimal point to show that the small parts are hundredths.

$$\begin{array}{r} 1.6 \\ \times\ .4 \\ \hline .64 \end{array}$$

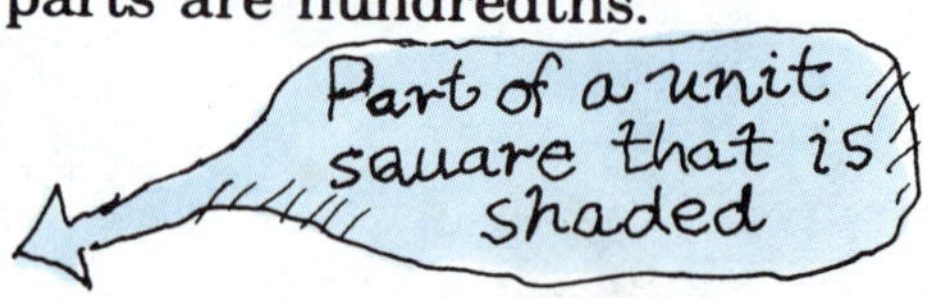

EXERCISES

Multiply. If you need to, look at the picture.

1. 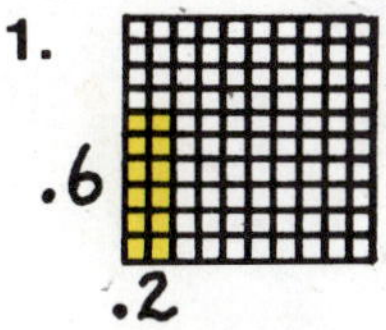

$$\begin{array}{r} .6 \\ \times .2 \\ \hline \end{array}$$

2. 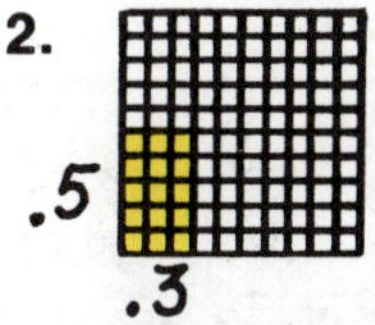

$$\begin{array}{r} .5 \\ \times .3 \\ \hline \end{array}$$

3. 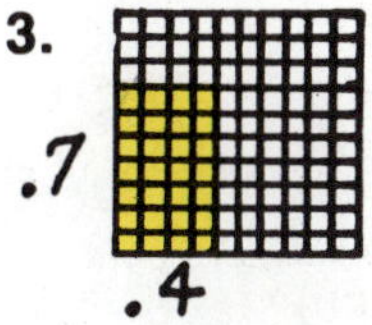

$$\begin{array}{r} .7 \\ \times .4 \\ \hline \end{array}$$

4. 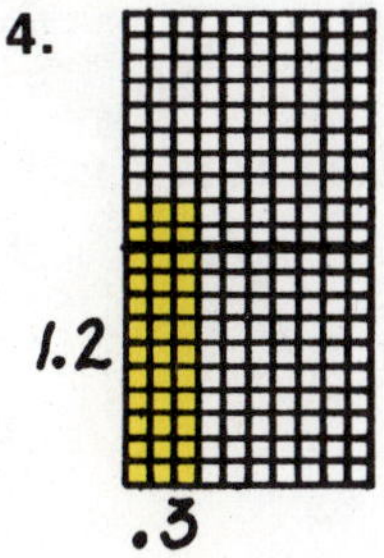

$$\begin{array}{r} 1.2 \\ \times\ .3 \\ \hline \end{array}$$

5. 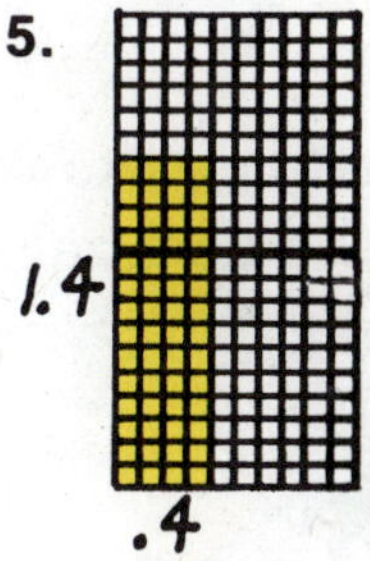

$$\begin{array}{r} 1.4 \\ \times\ .4 \\ \hline \end{array}$$

6. 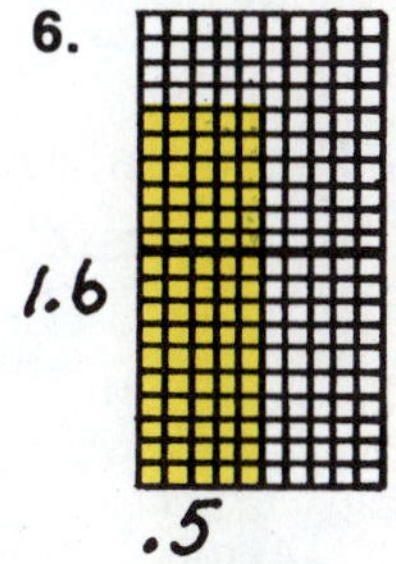

$$\begin{array}{r} 1.6 \\ \times\ .5 \\ \hline \end{array}$$

7. 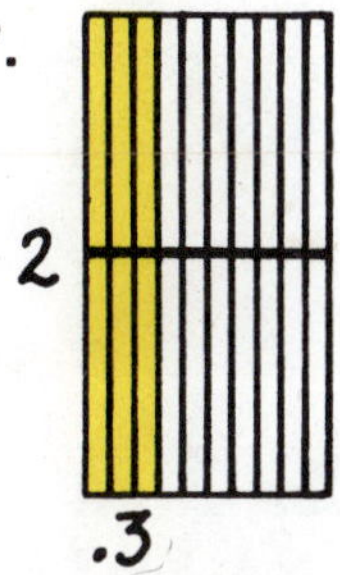

$$\begin{array}{r} 2 \\ \times .3 \\ \hline \end{array}$$

8. 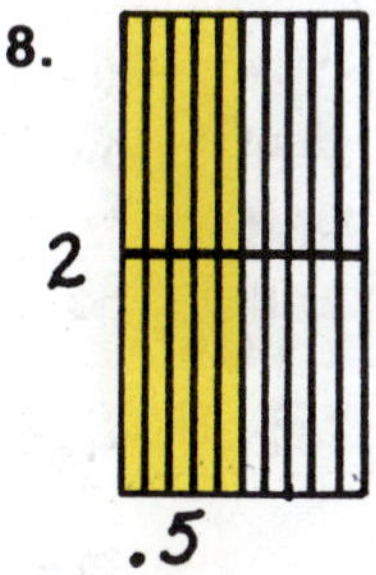

$$\begin{array}{r} 2 \\ \times .5 \\ \hline \end{array}$$

9. 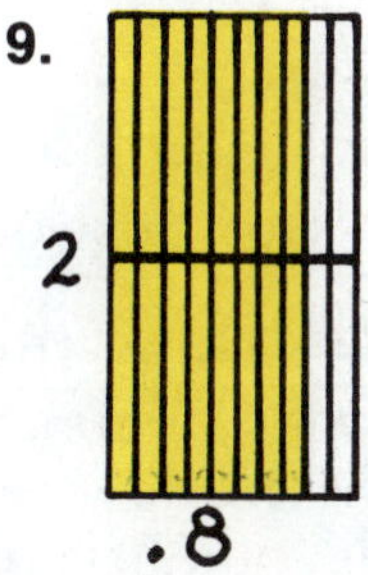

$$\begin{array}{r} 2 \\ \times .8 \\ \hline \end{array}$$

10. 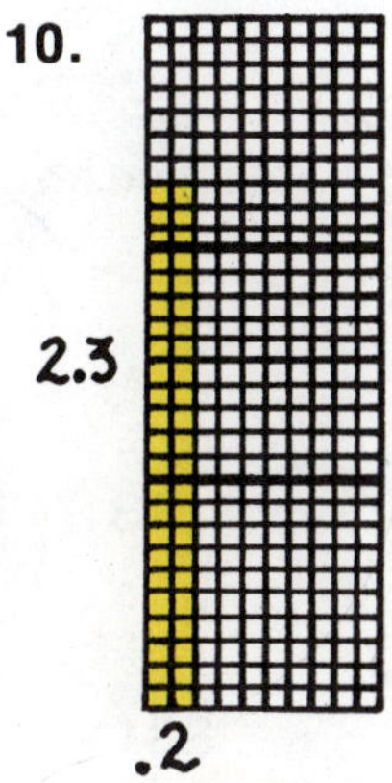

$$\begin{array}{r} 2.3 \\ \times .2 \\ \hline \end{array}$$

11. 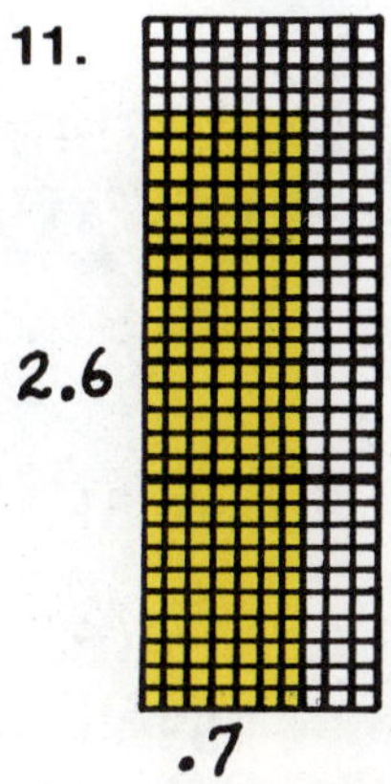

$$\begin{array}{r} 2.6 \\ \times .7 \\ \hline \end{array}$$

12. 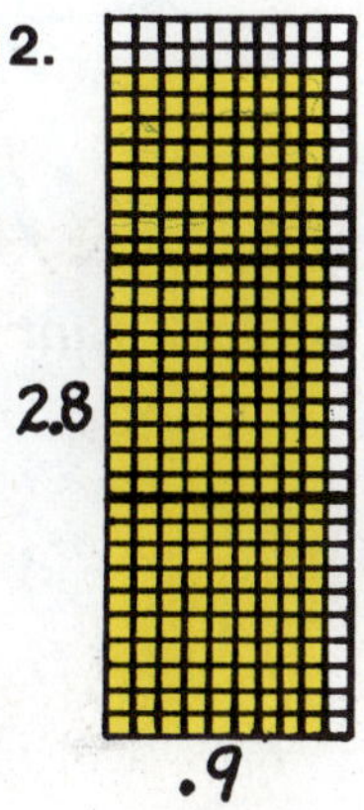

$$\begin{array}{r} 2.8 \\ \times .9 \\ \hline \end{array}$$

More about multiplying decimals

Here is a shortcut for placing the decimal point in the product.

.8 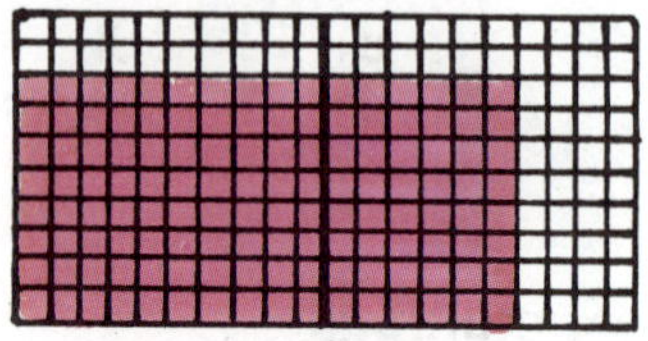 1.6

Step 1.
Multiply the numbers as whole numbers.

$$\begin{array}{r} \overset{4}{1.6} \\ \times\ .8 \\ \hline 128 \end{array}$$

Step 2.
Count the digits to the right of the decimal points.

$$\begin{array}{rl} \overset{4}{1.6} & 1 \text{ digit} \\ \times\ .8 & +1 \text{ digit} \\ \hline 128 & 2 \text{ digits} \end{array}$$

Step 3.
Count off the same number of digits in the product.

$$\begin{array}{r} \overset{4}{1.6} \\ \times\ .8 \\ \hline 1.28 \end{array}$$

If you first estimate the product, you will know whether your answer is reasonable.

$$\begin{array}{r} 19.84 \\ \times\ 1.9 \\ \hline \end{array} \qquad \begin{array}{rl} 19.84 & 2 \text{ digits} \\ \times\ 1.9 & +1 \text{ digit} \\ \hline 17856 & \\ 1984 & \\ \hline 37.696 & 3 \text{ digits} \end{array}$$

EXERCISES

Multiple choice. ***Hint:*** **Use the shortcut to place the decimal point.**

1. 5.3 × 2.7 — **a.** 143.1 **b.** 14.31 **c.** 1.431 **d.** .1431
2. 7.2 × 5.1 — **a.** 367.2 **b.** 36.72 **c.** 3.672 **d.** .3672
3. .43 × 7.7 — **a.** 331.1 **b.** 33.11 **c.** 3.311 **d.** .3311
4. .68 × 3.42 — **a.** 2325.6 **b.** 232.56 **c.** 23.256 **d.** 2.3256
5. 7.3 × .41 — **a.** 299.3 **b.** 29.93 **c.** 2.993 **d.** .2993

Multiple choice. ***Hint:*** **If you estimate the product, you may not have to multiply.**

6. 4.9×2.1
   a. 7.89
   b. 10.29
   c. 14.69
   d. 16.49

7. $8.02 \times .95$
   a. 7.6190
   b. 4.3590
   c. 12.1390
   d. 15.2490

8. 19.872×7.02
   a. 96.40144
   b. 102.18344
   c. 117.63544
   d. 139.50144

First estimate the product. Then multiply.

9. $3.74 \times .8$
10. $8.29 \times .9$
11. $65.3 \times .7$
12. 4.28×1.2
13. 93.1×2.9
14. 72.9×3.8
15. 8.34×4.7
16. 67.5×5.3
17. $5.83 \times .93$
18. $96.2 \times .89$
19. 75.31×4.68
20. 9.612×3.45
21. 53.84×8.32
22. 20.61×72.3
23. 73.85×1.03

Solve.

24. What is the price of 3 kilograms of apples?
25. What is the price of .85 kilogram of apples? Round your answer *up* to the nearest cent.
26. What is the price of 4 kilograms of oranges?
27. What is the price of .5 kilogram of oranges?
28. Can you buy 1.65 kilograms of oranges if you have $3?

Practice exercises

Sometimes you will need to write one or more zeros in the product before you can place the decimal point.

$$\begin{array}{r} 1.13 \\ \times\ .06 \\ \hline .0678 \end{array} \quad \begin{array}{r} 2 \text{ digits} \\ +2 \text{ digits} \\ \hline 4 \text{ digits} \end{array}$$

$$\begin{array}{r} .253 \\ \times .02 \\ \hline .00506 \end{array}$$

EXERCISES

Multiply.

1. $\begin{array}{r} .52 \\ \times .04 \\ \hline \end{array}$ **2.** $\begin{array}{r} 3.8 \\ \times .02 \\ \hline \end{array}$ **3.** $\begin{array}{r} .12 \\ \times .03 \\ \hline \end{array}$ **4.** $\begin{array}{r} 1.32 \\ \times .05 \\ \hline \end{array}$ **5.** $\begin{array}{r} .352 \\ \times .006 \\ \hline \end{array}$

6. $\begin{array}{r} 23.6 \\ \times .25 \\ \hline \end{array}$ **7.** $\begin{array}{r} 5.19 \\ \times .18 \\ \hline \end{array}$ **8.** $\begin{array}{r} 78.4 \\ \times 3.9 \\ \hline \end{array}$ **9.** $\begin{array}{r} 33.6 \\ \times 7.4 \\ \hline \end{array}$ **10.** $\begin{array}{r} 70.4 \\ \times .93 \\ \hline \end{array}$

Multiply. Then round the product to the nearest cent.

11. $\begin{array}{r} \$1.59 \\ \times 2.5 \\ \hline \end{array}$ **12.** $\begin{array}{r} \$2.63 \\ \times 86 \\ \hline \end{array}$ **13.** $\begin{array}{r} \$7.81 \\ \times 4.2 \\ \hline \end{array}$ **14.** $\begin{array}{r} \$5.73 \\ \times .73 \\ \hline \end{array}$ **15.** $\begin{array}{r} \$7.16 \\ \times .91 \\ \hline \end{array}$

16. $\begin{array}{r} \$2.19 \\ \times 12.5 \\ \hline \end{array}$ **17.** $\begin{array}{r} \$5.34 \\ \times 2.06 \\ \hline \end{array}$ **18.** $\begin{array}{r} \$3.28 \\ \times 32.1 \\ \hline \end{array}$ **19.** $\begin{array}{r} \$5.64 \\ \times 2.14 \\ \hline \end{array}$ **20.** $\begin{array}{r} \$1.97 \\ \times 1.66 \\ \hline \end{array}$

Find the price. Round your answer to the nearest cent.

21.

22.

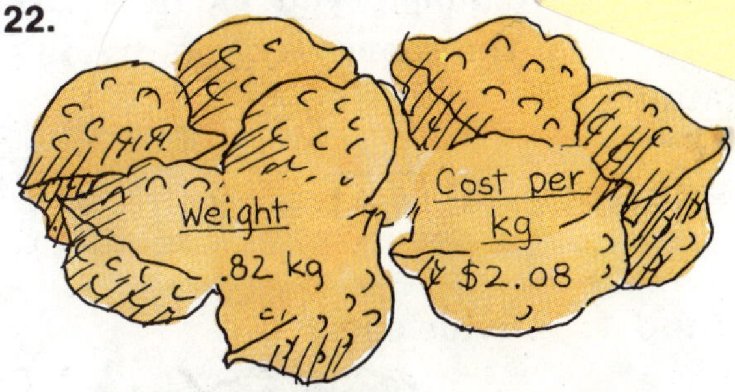

23.

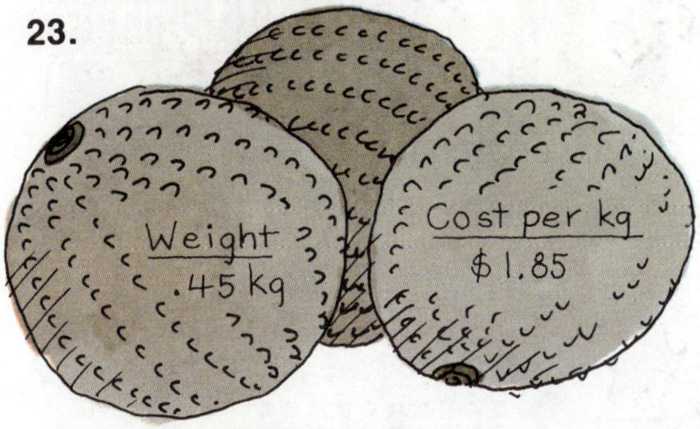

24.

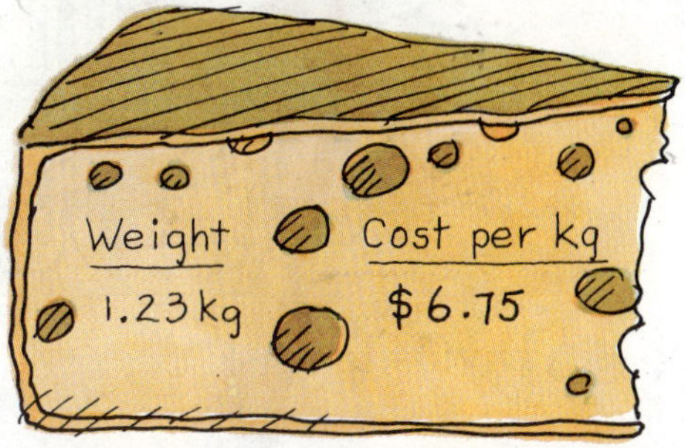

25.

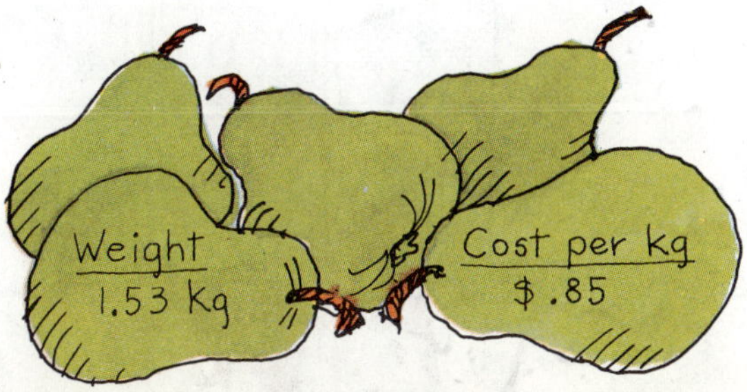

26.

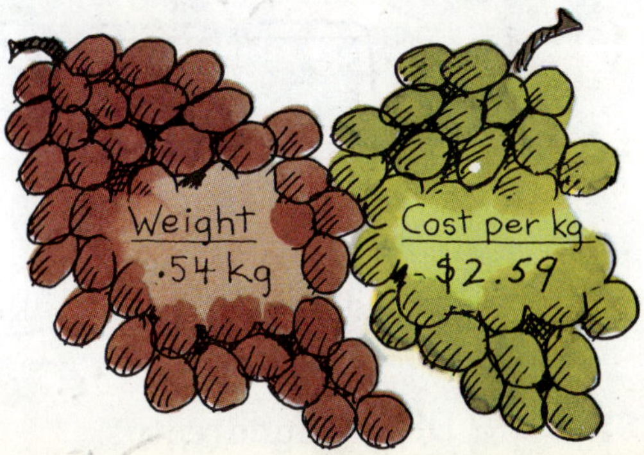

Keeping Skills Sharp

1. $2\frac{1}{3} + 1\frac{1}{3}$

2. $3\frac{2}{5} + 4\frac{1}{5}$

3. $5\frac{3}{8} + 1\frac{3}{8}$

4. $6\frac{1}{2} + 2\frac{1}{4}$

5. $5 + 3\frac{5}{6}$

6. $7\frac{1}{2} + 8\frac{1}{3}$

7. $9\frac{3}{4} + 3$

8. $6\frac{2}{3} + 9\frac{1}{2}$

9. $7\frac{5}{8} + 4\frac{3}{4}$

10. $8\frac{3}{4} + 8\frac{2}{3}$

11. $5\frac{5}{9} + 6\frac{5}{6}$

12. $9\frac{7}{8} + 9\frac{3}{4}$

Dividing a decimal

From the example, you should see that you divide a decimal the same way you divide a whole number.

EXAMPLE. 3)4.41

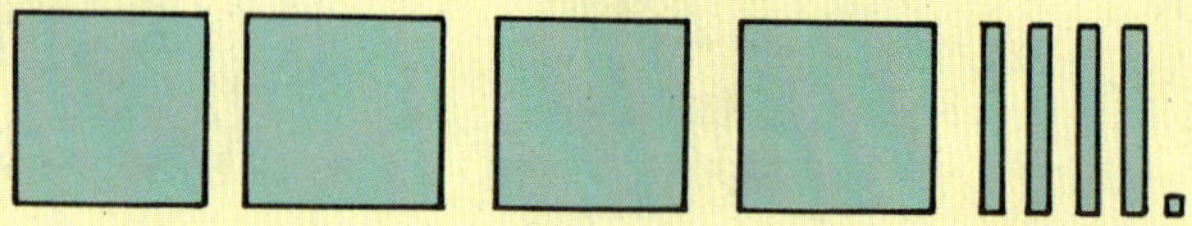

Step 1. Divide ones.

```
    1
3 )4.41
  -3
   1
```

Step 2. Regroup 1 one for 10 tenths.

```
    1
3 )4.41
  -3
   14
```

Step 3. Divide tenths.

```
    1.4
3 )4.41
  -3
   14
  -12
    2
```

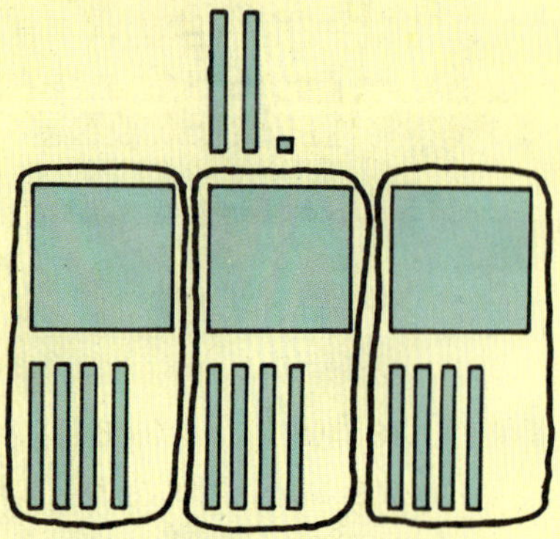

Step 4. Regroup 2 tenths for 20 hundredths.

```
    1.4
3 )4.41
  -3
   14
  -12
    21
```

Step 5. Divide hundredths.

```
    1.47
3 )4.41
  -3
   14
  -12
    21
   -21
     0
```

Another example.

```
       .954
36 )34.344
   -324
     194
    -180
     144
    -144
       0
```

EXERCISES

Divide.

1. $6\overline{)8.58}$	**2.** $8\overline{)99.2}$	**3.** $5\overline{)3.265}$	**4.** $7\overline{)17.71}$
5. $9\overline{)3.546}$	**6.** $7\overline{)39.06}$	**7.** $6\overline{)2.958}$	**8.** $9\overline{)558.9}$
9. $4\overline{)3.708}$	**10.** $8\overline{)51.92}$	**11.** $35\overline{)\$127.05}$	**12.** $64\overline{)\$231.04}$
13. $75\overline{)\$189.75}$	**14.** $92\overline{)\$124.20}$	**15.** $81\overline{)\$343.44}$	**16.** $72\overline{)198.00}$
17. $58\overline{)101.50}$	**18.** $26\overline{)22.334}$	**19.** $124\overline{)1.5748}$	**20.** $215\overline{)7.8475}$

Solve.

21.

How much for 1 kilogram?

22.

What is the price of 1 liter?

23. Jerry earned $29.40 for working 12 hours. How much was he paid per hour?

24. One hundred twenty-eight tickets were sold for a school play. The total receipts were $108.80. What was the price of each ticket?

Who am I?

25. If you divide me by 32, you get 3.65.

26. I am 5.63 less than the product of 42.5 and 7.4.

More about dividing decimals

Before regrouping, you sometimes have to write one or more zeros in the dividend.

EXAMPLE. $8\overline{)2.5}$

Step 1.
Divide and regroup.

$$\begin{array}{r} .3 \\ 8\overline{)2.50} \\ -2\,4 \\ \hline 10 \end{array}$$

Step 2.
Divide and regroup.

$$\begin{array}{r} .31 \\ 8\overline{)2.500} \\ -2\,4 \\ \hline 10 \\ -8 \\ \hline 20 \end{array}$$

Step 3.
Divide and regroup.

$$\begin{array}{r} .312 \\ 8\overline{)2.5000} \\ -2\,4 \\ \hline 10 \\ -8 \\ \hline 20 \\ -16 \\ \hline 40 \end{array}$$

Step 4.
Divide.

$$\begin{array}{r} .3125 \\ 8\overline{)2.5000} \\ -2\,4 \\ \hline 10 \\ -8 \\ \hline 20 \\ -16 \\ \hline 40 \\ -40 \\ \hline 0 \end{array}$$

EXERCISES
Divide.

1. $5\overline{)2.6}$
2. $4\overline{)4.6}$
3. $5\overline{)4.18}$
4. $6\overline{)15}$
5. $4\overline{)7.3}$
6. $8\overline{)25}$
7. $5\overline{)3.74}$
8. $4\overline{).31}$
9. $2\overline{).53}$
10. $5\overline{)9.27}$
11. $10\overline{).67}$
12. $12\overline{)9.3}$
13. $10\overline{).924}$
14. $16\overline{)27}$
15. $18\overline{)4.5}$
16. $24\overline{)25.2}$
17. $40\overline{)53.6}$
18. $16\overline{)5.2}$
19. $32\overline{)6.80}$
20. $56\overline{)9.8}$

The Gonzalez family kept this record of their car expenses while traveling in Canada. The gasoline tank was full at the beginning of their trip on July 7 and at the end of their trip on July 11.

Date	odometer reading in kilometers		Expenses			
			Gasoline		Other	
	Beginning	Ending	Liters	Cost	Item	Cost
July 7	28352	28643	79.2	$21.78	oil	$.95
July 8	28643	28930	80.4	$22.99		
July 9	28930	29135	54	$15.23	tire	$42.75
July 10	29135	29573	118.4	$33.03		
July 11	29573	30058	139	$40.80		

21. How many kilometers did they drive on July 7? On July 10?

22. How many kilometers did they drive during the first 3 days of their trip? During the entire trip?

23. How many kilometers did they average each day?

24. What was the cost per liter of the gasoline bought on July 9? Give your answer to the nearest tenth of a cent.

25. What was the total cost of gasoline? What were their total expenses?

26. During their trip, how many liters of gasoline were used? How many kilometers per liter did they average? Give your answer to the nearest tenth of a kilometer.

Decimals and fractions

These examples show how to change a decimal to a fraction in lowest terms.

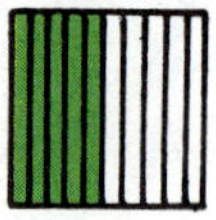 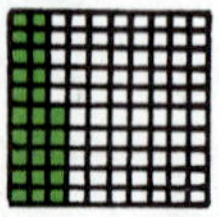 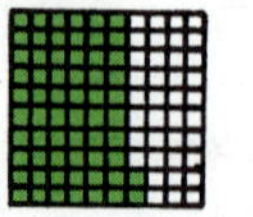 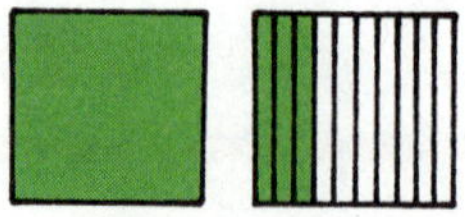

$.5 = \frac{5}{10} = \frac{1}{2}$

$.25 = \frac{25}{100} = \frac{1}{4}$

$.62 = \frac{62}{100} = \frac{31}{50}$

$1.3 = 1\frac{3}{10} = \frac{13}{10}$

To change a fraction to a decimal, divide the numerator by the denominator.

$$2\overset{.5}{\overline{)1.0}}$$

$\frac{1}{2} = .5$

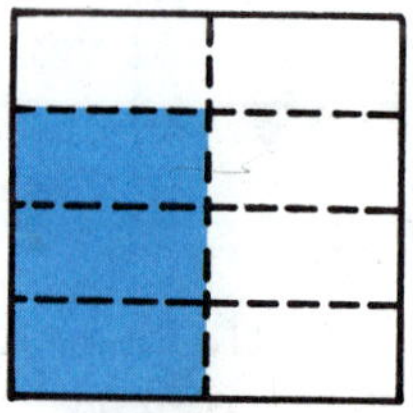

$$8\overset{.375}{\overline{)3.000}}$$

$$\begin{array}{r} -24 \\ \hline 60 \\ -56 \\ \hline 40 \\ -40 \\ \hline 0 \end{array}$$

$\frac{3}{8} = .375$

EXERCISES

Change to a fraction in lowest terms.

1. .1	**2.** .3	**3.** .6	**4.** .8	**5.** .9
6. .75	**7.** .45	**8.** .80	**9.** .06	**10.** .08
11. 1.5	**12.** 2.3	**13.** 1.4	**14.** 2.6	**15.** 3.5
16. 4.25	**17.** 2.75	**18.** 1.48	**19.** 2.96	**20.** 4.36

Change to a decimal.

21. $\frac{1}{4}$ 22. $\frac{1}{2}$ 23. $\frac{1}{8}$ 24. $\frac{1}{5}$ 25. $\frac{1}{10}$ 26. $\frac{3}{4}$

27. $\frac{5}{2}$ 28. $\frac{5}{8}$ 29. $\frac{7}{5}$ 30. $\frac{3}{8}$ 31. $\frac{11}{8}$ 32. $\frac{5}{4}$

33. $\frac{15}{8}$ 34. $\frac{7}{4}$ 35. $\frac{12}{5}$ 36. $\frac{3}{25}$ 37. $\frac{35}{20}$ 38. $\frac{15}{50}$

The batting average of a baseball player is the ratio of the number of "hits" to the number of "at bats." It is generally given as a decimal rounded to the nearest thousandth.

Here is how to find Carew's batting average.

```
          .3909
  330 )129.0000  ← Hits
   ↑    -990
At bats  3000
        -2970
          3000
         -2970
            30
```

Rounded to the nearest thousandth, his batting average is .391.

39. What is Singleton's average?
40. What is Fisk's average?
41. What is the average of the player in fifth place?
42. How much greater is Carew's average than Munson's?
43. Ty Cobb holds the record for greatest lifetime batting average. He had 4191 hits out of 11,429 times at bat. What was his lifetime batting average?

American League		
	At bat	*Hits*
Carew *Min*	330	129
Singleton *Bal*	281	95
Bostock *Min*	314	106
Bailor *Tor*	303	102
Dade *Cle*	229	76
Fisk *Bos*	285	91
Rivers *NY*	303	96
Rice *Bos*	349	109
Bannister *Chi*	347	108
Munson *NY*	333	100
Fairly *Tor*	269	83
Yastrzemski *Bos*	309	95

Multiplying or dividing by 10, 100, or 1000

Multiplying or dividing a decimal by 10 moves the decimal point 1 place.

3.46 × 10 = 34.6   3.46 ÷ 10 = .346

Multiplying or dividing a decimal by 100 moves the decimal point 2 places.

829.63 × 100 = 82963.

829.63 ÷ 100 = 8.2963

Multiplying or dividing a decimal by 1000 moves the decimal point 3 places.

792.46 × 1000 = 792460.

792.46 ÷ 1000 = .79246

Compare the number of zeros in 10, 100, or 1000 to the number of places that you moved the decimal point in each product or quotient. Can you find a rule?

EXERCISES

Use the shortcut to find each product or quotient.

1. a. 5.62×10
   b. $5.62 \div 10$
2. a. 83.5×100
   b. $83.5 \div 100$
3. a. 9.462×1000
   b. $9.462 \div 1000$
4. a. 74.2×100
   b. $74.2 \div 100$
5. a. 89.16×10
   b. $89.16 \div 10$
6. a. 71.42×1000
   b. $71.42 \div 1000$
7. a. 5643×10
   b. $5643 \div 10$
8. a. 81924×1000
   b. $81924 \div 1000$
9. a. 63057×100
   b. $63057 \div 100$

Complete each table.

10. 100 centimeters = 1 meter

	cm	m
a.	325	?
b.	26.5	?
c.	?	4.75
d.	?	7.4

11. 1000 meters = 1 kilometer

	m	km
a.	5826	?
b.	741.9	?
c.	?	1.6
d.	?	.745

Solve.

12. Longest Hot Dog
    Length: 4511 cm
    How many meters?

13. Largest Popsicle
    Weight: 1270 kg
    How many grams?

Keeping Skills Sharp

1. $5\frac{2}{3} - 1\frac{1}{3}$
2. $7\frac{5}{8} - 3\frac{1}{8}$
3. $2\frac{5}{9} - 1\frac{1}{3}$
4. $7\frac{3}{4} - 4\frac{5}{8}$
5. $9\frac{1}{2} - 3\frac{1}{4}$
6. $8\frac{1}{2} - 2\frac{1}{3}$
7. $9\frac{1}{2} - 3\frac{3}{4}$
8. $8\frac{1}{8} - 3\frac{1}{2}$
9. $5 - 3\frac{5}{8}$
10. $8 - 2\frac{1}{3}$
11. $7\frac{1}{4} - 2\frac{1}{3}$
12. $9\frac{1}{3} - 2\frac{3}{5}$

Dividing by a decimal

Dividing by a whole number is easy. Dividing by a decimal is harder. So whenever we have to divide by a decimal, we change the problem so that we can divide by a whole number instead. We change a hard job to an easier job.

Look at these examples. Notice that if we multiply both the divisor and the dividend by the same number, the quotient is not changed.

$2\overline{)8} = 4 \rightarrow 20\overline{)80} = 4 \rightarrow 200\overline{)800} = 4$

same quotient — same quotient

Multiply divisor and dividend by 10. — Multiply divisor and dividend by 10.

We use this fact to change all decimal divisors to whole numbers.

EXAMPLE 1.

Hard problem: $2.6\overline{)8.06}$

Easier problem: $26.\overline{)80.6}$

Multiply both divisor and dividend by 10.

```
      3.1
26. ) 80.6
     -78
       26
      -26
        0
```

EXAMPLE 2.

```
       125
.34. ) 42.50.
      -34
        85
       -68
        170
       -170
          0
```

EXAMPLE 3.

```
          870
.003. ) 2.610.
       -24
         21
        -21
          00
          -0
           0
```

EXERCISES

Divide.

1. $.3\overline{)1.44}$ 2. $.5\overline{).385}$ 3. $.2\overline{)1.32}$ 4. $.6\overline{)2.04}$ 5. $.4\overline{)1.84}$

6. $8\overline{)1.12}$ 7. $.5\overline{)16.5}$ 8. $.9\overline{)4.86}$ 9. $.2\overline{).130}$ 10. $.3\overline{)17.7}$

11. $.07\overline{)4.62}$ 12. $.08\overline{).272}$ 13. $.02\overline{)2.12}$ 14. $.005\overline{).515}$ 15. $.007\overline{)7.28}$

16. $.21\overline{)5.46}$ 17. $3.5\overline{)17.50}$ 18. $6.1\overline{)32.33}$ 19. $.36\overline{)1.872}$ 20. $.47\overline{)3.055}$

21. $.039\overline{).2496}$ 22. $.047\overline{)2.632}$ 23. $.252\overline{)47.88}$ 24. $1.24\overline{).4712}$

Solve.

25.

What is the price of 1 kilogram?

26.

How many kilograms can you buy with $9.24?

27. A bicycle club rode 38.7 kilometers in 4.5 hours. How many kilometers did they average each hour?

28. David worked 2.5 hours one evening and 1.75 hours the next. He was paid a total of $10. How much did he earn an hour? Give your answer to the nearest cent.

Copy and complete these multiplication boxes.

29.

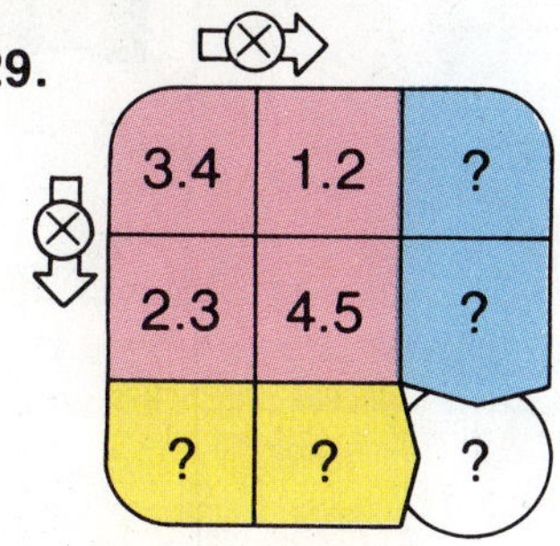

30.

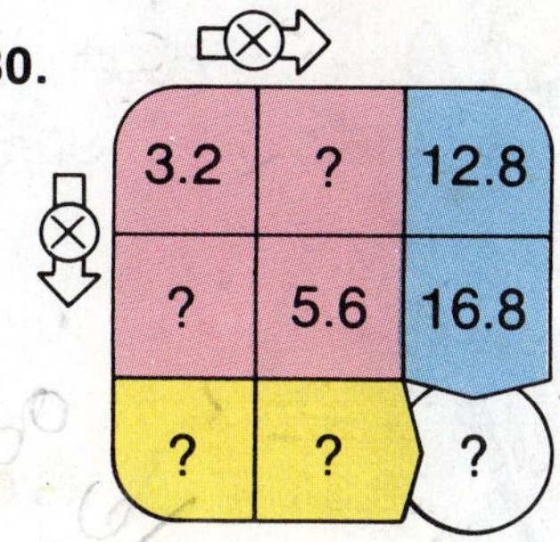

31.

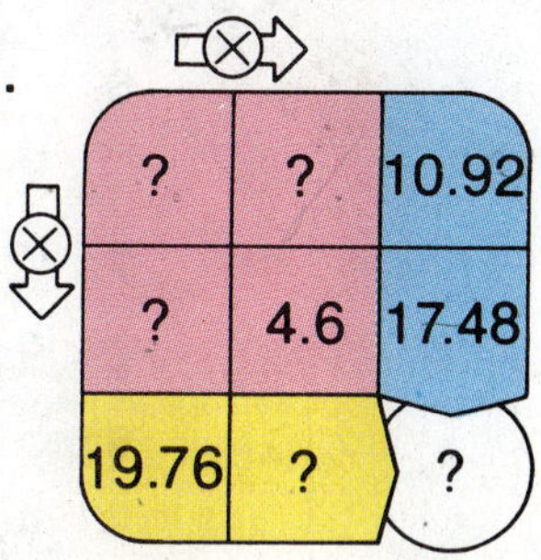

Estimating

If you estimate first, you will know whether your answer is reasonable.

To estimate the answer, I rounded each number to the nearest whole number. My estimate is 27.

$(4 \times 5) + 7 = 27$

$(3.9 \times 5.1) + 6.83 = ?$

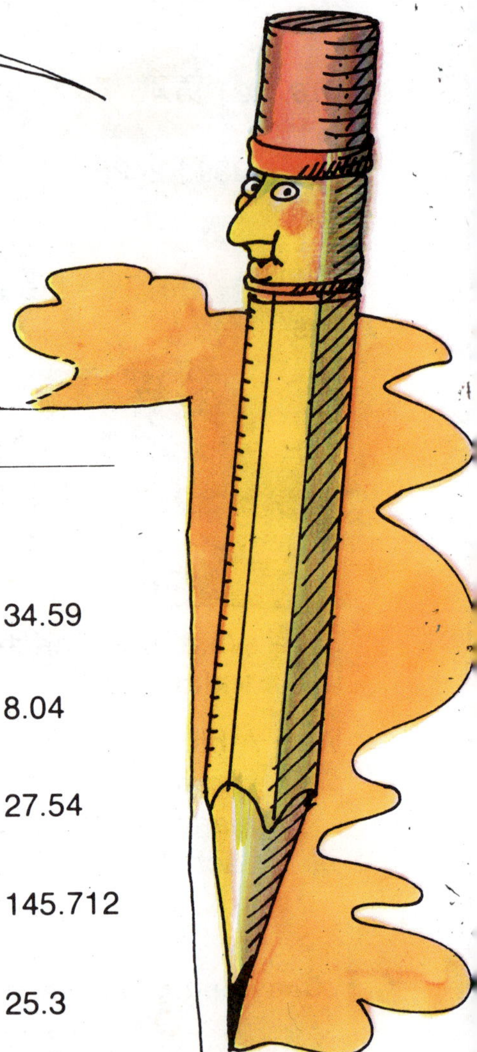

Work the problem. Is your answer "close" to the estimate?

EXERCISES

Here is a multiple choice "test." First see if you can pick out the correct answers by estimating. Then work out each problem.

TEST Name ____________________

Multiple choice.

1. $(8.8 - 2.7) \times 4.9 = ?$
   a. 24.09 **b.** 37.29 **c.** 29.89 **d.** 34.59
2. $(5.92 + 4.18) \div 2.5 = ?$
   a. 4.04 **b.** 6.24 **c.** 1.74 **d.** 8.04
3. $(3.6 \times 5.4) + 8.1 = ?$
   a. 20.24 **b.** 35.14 **c.** 34.34 **d.** 27.54
4. $6.9 \times (8.8 \times 2.1) = ?$
   a. 98.412 **b.** 127.512 **c.** 105.312 **d.** 145.712
5. $(25.42 \div 3.1) + 9.1 = ?$
   a. 17.3 **b.** 10.3 **c.** 13.3 **d.** 25.3
6. $(43.6 + 6.4) \times 2.01 = ?$
   a. 84.5 **b.** 116.5 **c.** 90.5 **d.** 100.5
7. $(33.75 - 4.25) \times 3.8 = ?$
   a. 122.1 **b.** 112.1 **c.** 102.1 **d.** 132.1
8. $5.78 + (228.92 \div 9.7) = ?$
   a. 19.38 **b.** 29.38 **c.** 36.08 **d.** 15.18

Skill Game

1. The target number was 30. How close did Jane come to the target number?
2. How close did David come to the target number?
3. Who came closer, Jane or David?

Play the game.

1. Have a game leader make these cards:

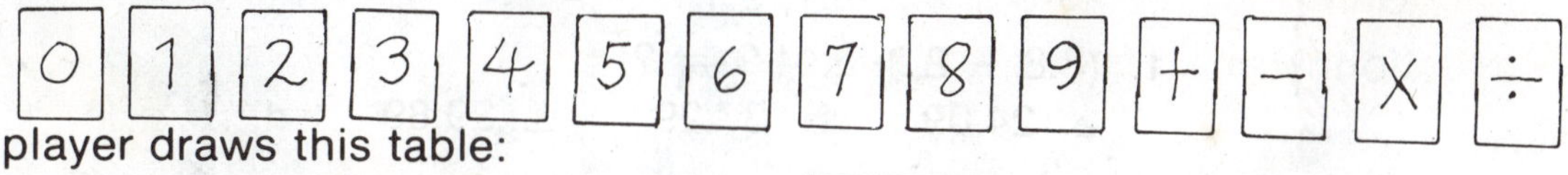

2. Each player draws this table:

3. The game leader picks two sign cards. The players write the two signs in their tables in either order.
4. The players fill in their tables as the six digits are picked.
5. The players place three decimal points in their tables.
6. The player getting closest to the target number wins the game.
7. Later, you may wish to change the target number. Note: When dividing, round your answer to the nearest hundredth.

Problem solving

In this gymnastic meet, four judges rate each contestant. The highest and lowest ratings are discarded. The two middle ratings are averaged for the contestant's score. The highest score wins.

1. Find each contestant's score.

	TEAM A
Adams:	8.7, 9.2, 8.8, 9.3
Davis:	8.5, 8.7, 8.7, 8.7
Langley:	9.1, 8.8, 9.0, 9.1
	TEAM B
Carter:	8.4, 8.6, 8.9, 8.7
Garcia:	9.3, 9.2, 9.1, 9.1
Johnson:	8.7, 8.3, 8.9, 9.0

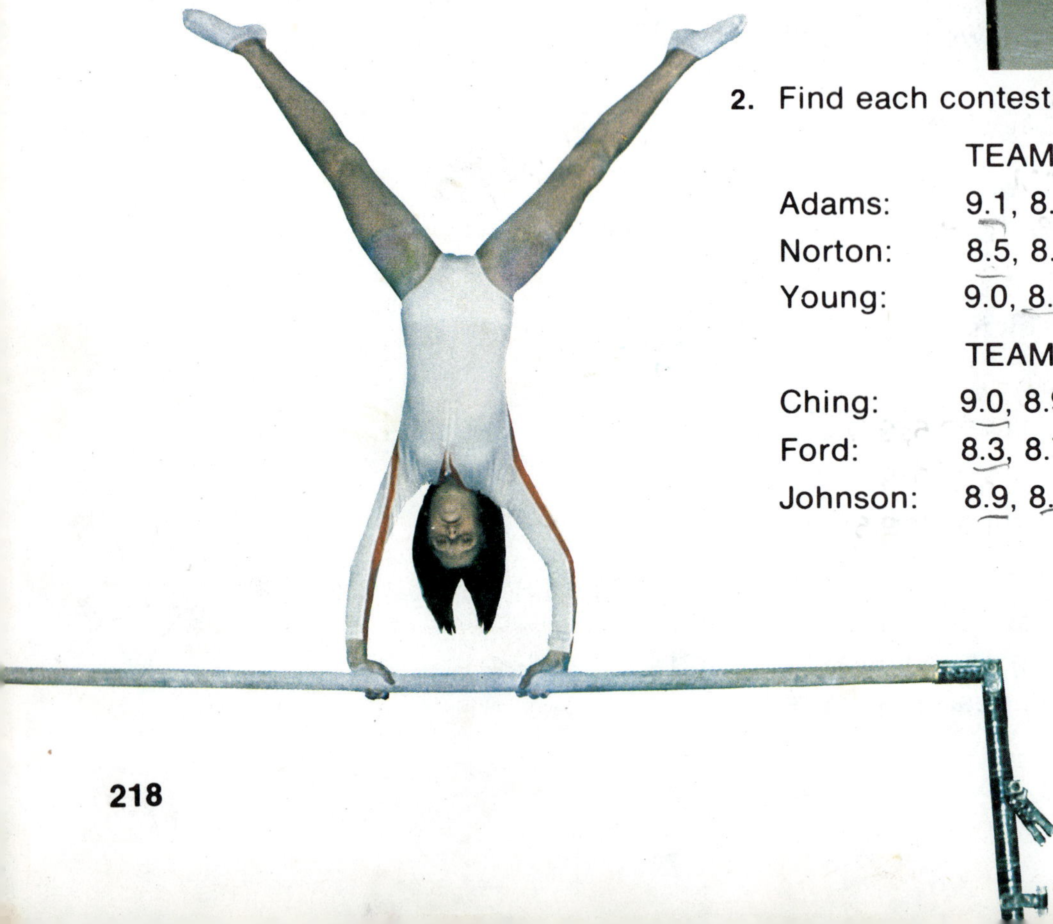

2. Find each contestant's score.

	TEAM A
Adams:	9.1, 8.9, 8.8, 8.9
Norton:	8.5, 8.6, 8.9, 8.6
Young:	9.0, 8.8, 9.2, 8.7
	TEAM B
Ching:	9.0, 8.9, 8.5, 8.8
Ford:	8.3, 8.7, 8.9, 8.8
Johnson:	8.9, 8.4, 8.6, 8.7

3. Find each contestant's score.

	TEAM A
Davis:	9.3, 9.5, 9.7, 9.4
Ross:	9.2, 9.5, 9.2, 8.7
Young:	8.5, 8.7, 8.3, 8.4

	TEAM B
Carter:	7.6, 7.3, 7.5, 7.8
Howey:	9.1, 9.0, 8.8, 8.9
Rogers:	9.6, 9.7, 9.8, 9.4

4. Find each contestant's score.

	TEAM A
Langley:	9.3, 9.7, 9.4, 9.7
Norton:	9.1, 9.0, 9.3, 8.6
Ross:	8.7, 8.8, 8.5, 8.5

	TEAM B
Ching:	9.8, 9.8, 9.5, 9.5
Ford:	9.2, 9.2, 9.0, 9.3
Rogers:	8.5, 8.4, 8.6, 8.6

5. To find a team's score in each event, add the scores of the three team members. High total wins. Which team won each event?

6. To find a team's total score, add its scores for the four events. High total wins. Which team, A or B, won the meet?

CHAPTER CHECKUP

Multiply. [pages 200–204, 212–213, 216–217]

1. $8 \times .2$
2. $.3 \times 5$
3. $6.4 \times .3$
4. 5.02×9
5. $6.59 \times .7$
6. $.3 \times .2$
7. $1.4 \times .3$
8. 3.46×10
9. 38.21×9.3
10. 456.4×5.18

Complete. [pages 210–211]

	11.	12.	13.	14.	15.	16.
Fraction in lowest terms	$\frac{3}{8}$	$\frac{3}{4}$	$\frac{9}{2}$	?	?	?
Decimal	?	?	?	.8	.48	1.16

Divide. [pages 206–208, 212–216]

17. $6\overline{)3.6}$
18. $9\overline{)2.313}$
19. $7\overline{).595}$
20. $8\overline{)56.48}$
21. $12\overline{)4.14}$
22. $100\overline{)435.8}$
23. $.25\overline{)5.75}$
24. $6.7\overline{)25.795}$
25. $2.24\overline{)4.0768}$

Solve. [pages 205, 209, 218–219]

26. 

What is the price of 12 packages?

27. 

How many candles can be bought for $1.04?

28. Barbara hiked 11.9 kilometers in 3.5 hours. How many kilometers did she average each hour?

29. To hike one trail, she took 2.5 hours and averaged 3.2 kilometers per hour. How long was the trail?

Project

Follow these steps to find the average reaction time of the students in your class.

1. Form a circle by grasping hands.
2. Your teacher or project leader will need a stopwatch (or a clock with a sweep-second hand) to time how long it takes a hand squeeze to be passed around the circle. The leader will start the squeeze around the circle. "Pass on" the squeeze as quickly as possible. *Do not look* to see when the squeeze is coming.
3. The project leader writes the time that it took for the squeeze to go completely around the circle.
4. Count the number of people in the circle and compute the average reaction time.
5. You may want to repeat the experiment to see if you can get a quicker reaction time.

CHAPTER REVIEW

Multiply.

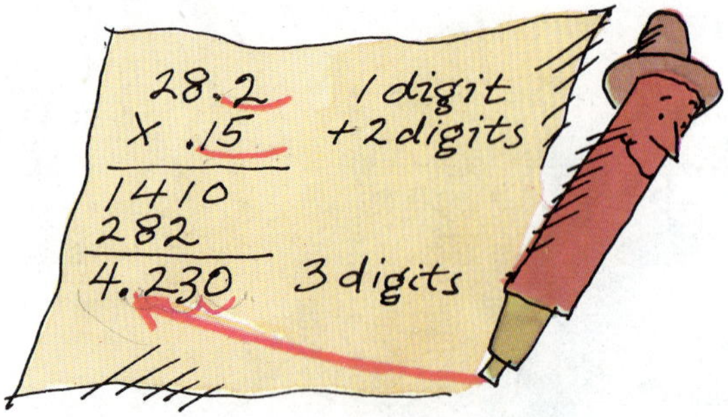

1. 3.2 × .6

2. 8.12 × .3

3. 4.15 × .08

4. 3.29 × .24

5. 53.9 × .33

6. 7.06 × 1.16

7. 57.8 × 43.8

Change to a fraction in lowest terms.

8. .8

9. .9

10. .25

11. .64

12. .96

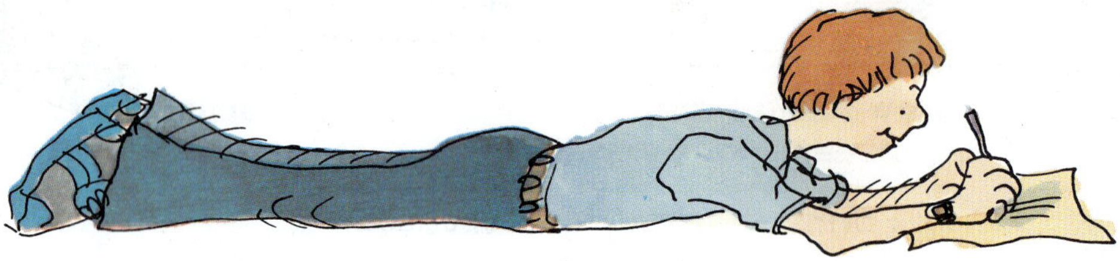

Change to a decimal.

13. $\frac{1}{2}$

14. $\frac{3}{8}$

15. $\frac{5}{8}$

16. $\frac{7}{4}$

17. $\frac{9}{2}$

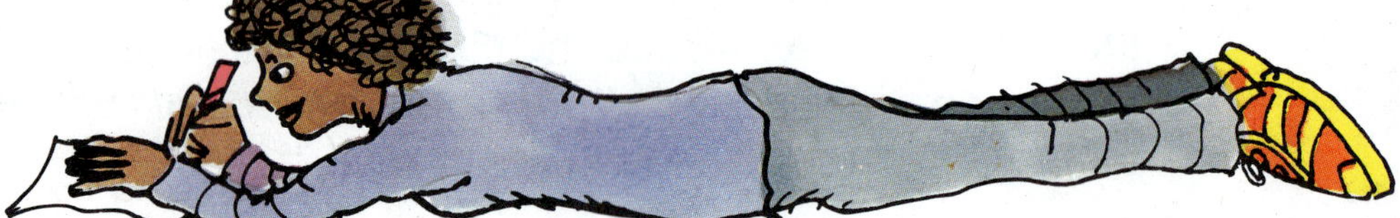

Divide.

18. .2)6.34

19. .3).234

20. .04)3.872

21. 1.8)12.06

22. .47)1.833

23. .082).3772

CHAPTER CHALLENGE

If you try to change $\frac{1}{3}$ to a decimal, the digit 3 keeps repeating.

$$
\begin{array}{r}
.33 \\
3\,\overline{)1.000} \\
-9 \\
\hline
10 \\
-9 \\
\hline
1
\end{array}
$$

We can write this: $\frac{1}{3} = .333\ldots$

The dots tell us that the 3 keeps repeating. Such a decimal is called a **repeating decimal.**

$\frac{1}{3} = .\overline{3}$

Now let's change $\frac{5}{11}$ to a decimal.

$$
\begin{array}{r}
.4545 \\
11\,\overline{)5.0000} \\
-44 \\
\hline
60 \\
-55 \\
\hline
50 \\
-44 \\
\hline
60 \\
-55 \\
\hline
5
\end{array}
$$

$\frac{5}{11} = .4545\ldots$

$\frac{5}{11} = .\overline{45}$

Change these fractions to repeating decimals.

1. $\frac{2}{3}$	**2.** $\frac{1}{9}$	**3.** $\frac{4}{3}$	**4.** $\frac{5}{9}$	**5.** $\frac{1}{6}$	**6.** $\frac{5}{6}$
7. $\frac{7}{6}$	**8.** $\frac{8}{11}$	**9.** $\frac{7}{11}$	**10.** $\frac{1}{7}$	**11.** $\frac{5}{7}$	**12.** $\frac{3}{13}$

MAJOR CHECKUP
Standardized Format

Choose the correct letter.

1. In 382,961,705, the digit in the hundred thousands place is

- **a.** 7
- **b.** 3
- **c.** 9
- **d.** none of these

2. Add.

$$\begin{array}{r} 26 \\ 539 \\ 6421 \\ +763 \\ \hline \end{array}$$

- **a.** 7749
- **b.** 6639
- **c.** 6749
- **d.** none of these

3. Subtract.

$$\begin{array}{r} 52006 \\ -3978 \\ \hline \end{array}$$

- **a.** 51,972
- **b.** 48,028
- **c.** 58,028
- **d.** none of these

4. Multiply.

$$\begin{array}{r} 8436 \\ \times 506 \\ \hline \end{array}$$

- **a.** 4,266,486
- **b.** 472,416
- **c.** 471,416
- **d.** none of these

5. Divide.

$146 \overline{)15079}$

- **a.** 13 R41
- **b.** 103 R41
- **c.** 13
- **d.** none of these

6. $\overleftrightarrow{AD}$ is perpendicular to

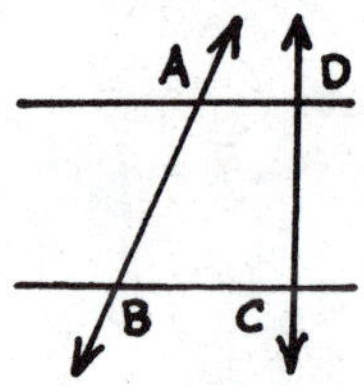

- **a.** $\overleftrightarrow{AB}$
- **b.** $\overleftrightarrow{BC}$
- **c.** $\overleftrightarrow{CD}$
- **d.** none of these

7. $3 + \frac{2}{3} =$

- **a.** $\frac{11}{3}$
- **b.** $\frac{5}{3}$
- **c.** $\frac{5}{9}$
- **d.** none of these

8. $\frac{7}{8} - \frac{3}{4} =$

- **a.** $\frac{13}{8}$
- **b.** $\frac{1}{8}$
- **c.** $\frac{1}{2}$
- **d.** none of these

9. $\frac{5}{6} \times \frac{3}{4} =$

- **a.** $\frac{15}{12}$
- **b.** $\frac{3}{2}$
- **c.** $\frac{5}{8}$
- **d.** none of these

10. $\frac{7}{4} \div \frac{3}{2} =$

- **a.** $\frac{6}{7}$
- **b.** $\frac{7}{6}$
- **c.** $\frac{21}{8}$
- **d.** none of these

11. $53.8 + 6.45 =$

- **a.** 11.83
- **b.** 118.3
- **c.** 60.25
- **d.** none of these

12. $82.3 - 3.56 =$

- **a.** 4.67
- **b.** 78.74
- **c.** 46.7
- **d.** none of these

9 Measurement

Time

A. M. is used for times after 12:00 midnight and before 12:00 noon.

P. M. is used for times after 12:00 noon and before 12:00 midnight.

The time shown on the clock is in the afternoon. It is 2:45 P. M.

EXERCISES

Daylight or dark?

1. 11:25 A.M.
2. 11:25 P.M.
3. 12:00 noon
4. 12:00 midnight

Complete.

60 seconds = 1 minute

5. To change minutes to seconds, ? by 60.
6. 8 minutes = ? seconds
7. 15 minutes = ? seconds

8. To change seconds to minutes, ? by 60.
9. 840 seconds = ? minutes
10. 232 seconds = ? minutes ? seconds

60 minutes = 1 hour (h)

11. To change hours to minutes, _?_ by 60.

12. 9 hours = _?_ minutes

13. 12 hours = _?_ minutes

14. To change minutes to hours, _?_ by 60.

15. 420 minutes = _?_ hours

16. 378 minutes = _?_ hours _?_ minutes

24 hours = 1 day

17. To change days to hours, _?_ by 24.

18. 10 days = _?_ hours

19. 21 days = _?_ hours

20. To change hours to days, _?_ by 24.

21. 432 hours = _?_ days

22. 320 hours = _?_ days _?_ hours

Add.

Regroup 60 minutes for 1 hour.

23. 5 h 40 min + 2 h 28 min

24. 4 h 15 min + 6 h 32 min

25. 8 h 30 min + 3 h 30 min

26. 15 h 42 min + 6 h 36 min

Seconds

27. 22 min 14 s + 18 min 19 s

28. 15 min 25 s + 7 min 43 s

29. 42 min 53 s + 19 min 20 s

Subtract.

Regroup 1 hour for 60 minutes

30. 7 h 15 min − 2 h 40 min

31. 8 h 38 min − 5 h 21 min

32. 8 h 15 min − 2 h 30 min

33. 16 h 10 min − 9 h

34. 10 min 15 s − 5 min 50 s

35. 24 min 10 s − 15 min 36 s

36. 35 min − 14 min 25 s

More about time

Classes at Willow Elementary School begin at 8:30. One hour and 45 minutes later, recess begins. There are several ways to find the recess time. Here are two.

Can you think of still another way to find recess time?

EXERCISES

What time is

1. 30 minutes later than 6:30?
2. 45 minutes later than 8:45?
3. 1 hour and 10 minutes later than 9:36?
4. 1 hour earlier than 5:15?
5. 1 hour and 15 minutes later than 4:55?
6. 1 hour and 35 minutes earlier than 4:30?

How much time do you have

7. from the beginning of school to recess?
8. from the end of recess to lunch time?
9. from the end of lunch time to the next recess?
10. from the end of recess to the end of the school day?
11. How much time do you spend in the classroom during one day? During a week?

Solve.

12. Jan arrived at the bus stop at 7:55. The bus was due at 8:13. How much time did she have to wait?
13. Terry started on a bike tour at 8:35. Five hours and 42 minutes later he completed the tour. What time was that?
14. Marcia worked 2 hours and 45 minutes one morning and 3 hours and 35 minutes in the afternoon. How long did she work that day?
15. One morning Alex got up at 7:30 after sleeping $9\frac{1}{2}$ hours. What time did he go to bed?
16. The car pictured was the first car to finish the Indianapolis 500 race with an average speed over 160 mph. Its recorded time was

    3:04:05.54
    (h = 3, min = 04, s = 05.54)

    The year before, Al Unser won in 3:10:11.56. How much slower was Al Unser's time?

Driver: Mark Donohue

Time zones

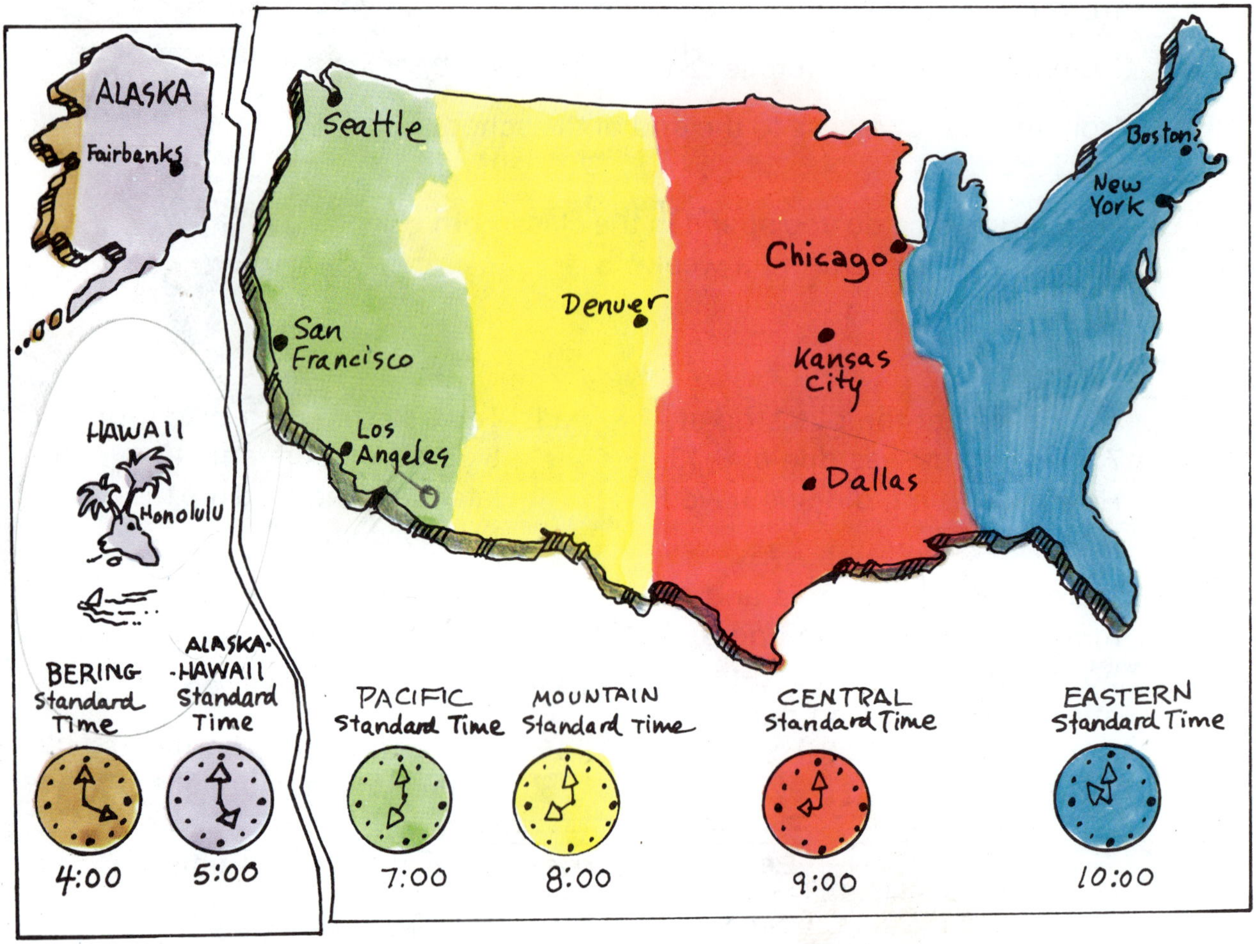

Have you ever heard a statement like this?

The world is divided into 24 different time zones. The map above shows the time zones of the United States. Why do you think we need time zones?

EXERCISES

Give the time zone of

1. Los Angeles
2. New York
3. Chicago
4. Denver
5. Honolulu
6. your school

When it's 9:00 A.M. in Chicago, what time is it in

7. Dallas?
8. Denver?
9. Seattle?
10. Fairbanks?
11. Honolulu?
12. New York?

When it's 7:30 in Denver, what time is it in

13. Boston?
14. Kansas City?
15. San Francisco?

When it's 12:00 noon in Kansas City, what time is it in

16. New York?
17. Los Angeles?
18. Denver?

Flight	Leaves	Arrives
Chicago–Dallas	9:00 A. M.	11:15 A. M.
Dallas–Los Angeles	5:00 P. M.	6:01 P. M.
Los Angeles–Denver	1:50 P. M.	4:48 P. M.
Denver–New York	10:00 P. M.	3:30 A. M.

The table lists the departure and arrival times of some flights. Arrival and departure times are *local* times. Find the time taken for each flight.

19. Chicago–Dallas
20. Dallas–Los Angeles
21. Los Angeles–Denver
22. Denver–New York

Length

The basic unit for measuring length in the metric system is the **meter.**

10 millimeters (mm) = 1 centimeter (cm)
10 centimeters (cm) = 1 decimeter (dm)
10 decimeters (dm) = 1 meter (m)

Notice that 10 of one unit is the same as 1 of the next larger unit. The decimeter (dm) is used less frequently than the other units.

EXERCISES

Measure

1. the length of your pencil to the nearest millimeter.
2. the width of a piece of paper to the nearest centimeter.
3. the height of a bulletin board to the nearest decimeter.
4. the length of your classroom to the nearest meter.

Here are some other units used for measuring length.

10 meters (m) = 1 dekameter (dam)
10 dekameters (dam) = 1 hectometer (hm)
10 hectometers (hm) = 1 kilometer (km)

The dekameter (dam) and the hectometer (hm) are seldom used.

5. Measure off a 10-meter distance in your classroom or hall.

6. If possible, measure off a 100-meter distance on the school grounds.

7. One of your "giant steps" is about a meter. If possible, see how many giant steps it takes you to go around your school grounds. Is the distance less than a kilometer?

Forgetful Frannie forgot to write the units. Think about the size of each object and then tell whether the unit should be millimeter, centimeter, meter, or kilometer.

8. My height is 148.
9. My little finger is 39 long.
10. I can throw a ball 30.
11. I live 4 from my school.
12. My record high jump is 86.

Complete.

13. 1 m = ? cm
14. 1 m = ? mm
15. 1 km = ? m
16. 2 cm = ? mm
17. 14 m = ? cm
18. 25 km = ? m

Changing units

Our place-value system and the metric system are both based on 10. This means that the two systems "fit together."

For example, in the measurement

24.38 dm

the 2 stands for 20 dm, or 2 m;

the 4 stands for 4 dm;

the 3 stands for .3 dm, or 3 cm;

the 8 stands for .08 dm, or 8 mm;

This means that we can label the places with metric units.

m	dm	cm	mm
2	**4**	**3**	**8**

Notice these things:

1. The unit used in the measurement goes in the ones place.
2. The other units go in order with the digits.

This gives us a short way to change from one unit to another.

EXAMPLE 1. Change 24.38 dm to m.

Step 1. Think about the metric units with the digits. dm goes in the ones place.

(m)(dm)(cm)(mm)
2 4. 3 8 dm

Step 2. Move the decimal point so that m is in the ones place. Change the unit label.

(m)(dm)(cm)(mm)
2.4. 3 8 ~~dm~~ m

EXAMPLE 2. Change 52.4 m to cm.

Step 1. Think about the metric units with the digits. m goes in the ones place.

(dam)(m)(dm)
5 2 .4 m

Step 2. Move the decimal point so that cm is in the ones place. Change the unit label.

(dam)(m)(dm)(cm)
5 2 .4 0. ~~m~~ cm

EXERCISES

Complete.

1. 5.73 dm = _?_ cm
2. 53.4 dm = _?_ m
3. 48.6 cm = _?_ dm
4. 528.4 cm = _?_ m
5. 42.6 cm = _?_ mm
6. 4328 m = _?_ km
7. 426.1 cm = _?_ dm
8. 426.1 cm = _?_ mm
9. 437 mm = _?_ dm
10. 437 mm = _?_ cm
11. 2.4 m = _?_ cm
12. 3.2 m = _?_ mm
13. 73.2 mm = _?_ m
14. 6.4 km = _?_ m
15. 38 m = _?_ km
16. 84 cm = _?_ m
17. 75.3 mm = _?_ cm
18. 763 mm = _?_ m
19. 5 km = _?_ mm
20. 47 mm = _?_ km
21. 420 cm = _?_ m

Answer.

22. One race in the Olympic Games is 5000 m long. How many kilometers is that?
23. At one time the world high-jump record was 2.32 m. How many centimeters is that?
24. Randy Matson once threw the shot 214.76 dm. How many meters is that?
25. In 1968, Bob Beamon set a long-jump record that may never be broken. His record was 890.21 cm. How many meters is that?

Perimeter

The distance around a figure is called the **perimeter** of the figure.

We can find the perimeter of the mosaic by adding the lengths of the sides.

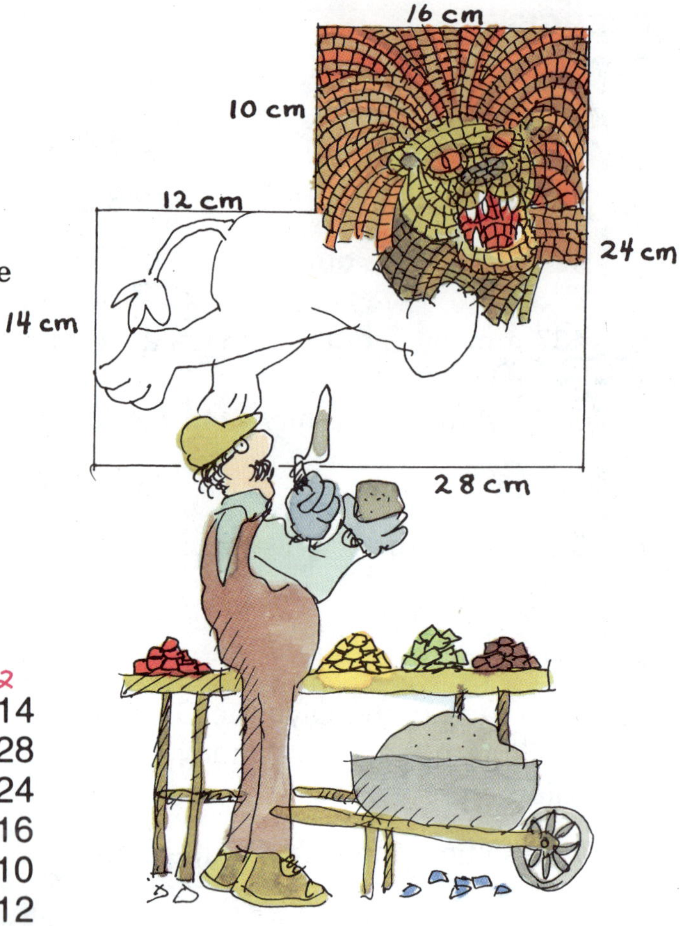

$$\begin{array}{r} 14 \\ 28 \\ 24 \\ 16 \\ 10 \\ +12 \\ \hline 104 \end{array}$$

The perimeter is 104 cm.

EXERCISES

Find the perimeter of each figure.

1. 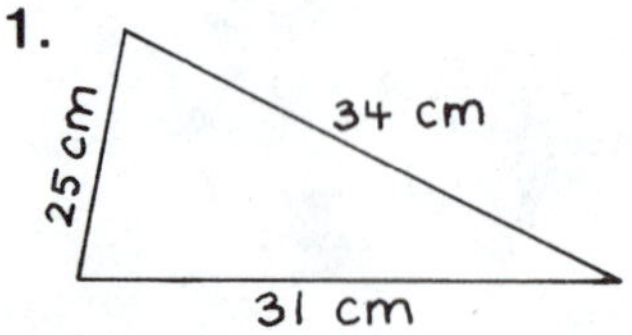

2. 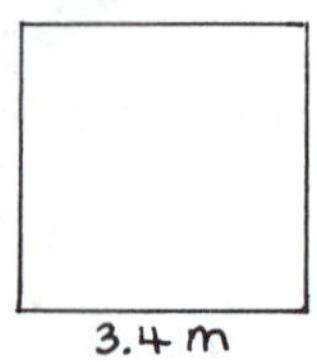

3. 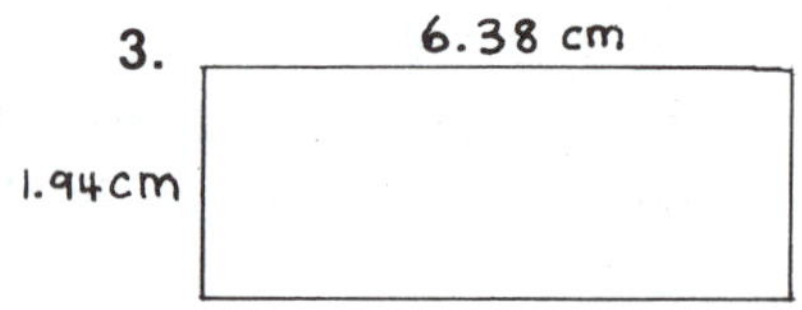

4. 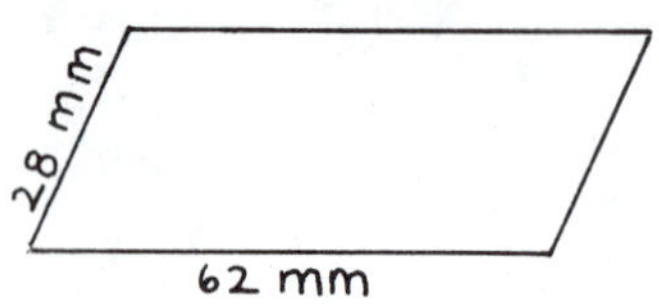

5. 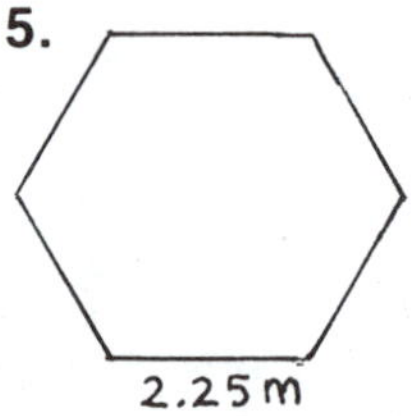

6. 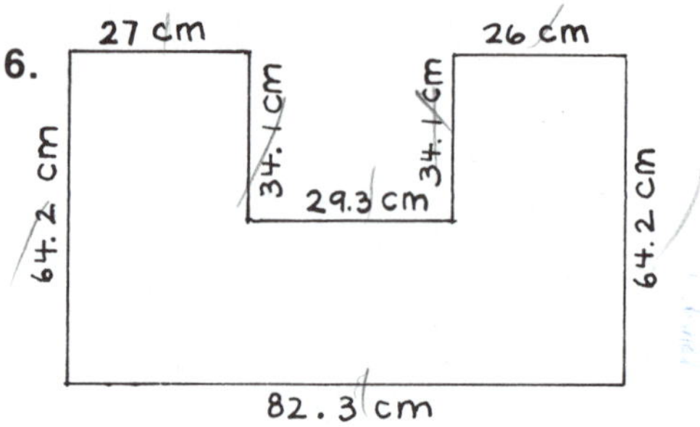

True or false?

7. The perimeter of an equilateral triangle is 3 times the length of one side.

8. The perimeter of a square is 4 times the length of one side.

9. The perimeter of a rectangle is 2 times the sum of the length and the width.

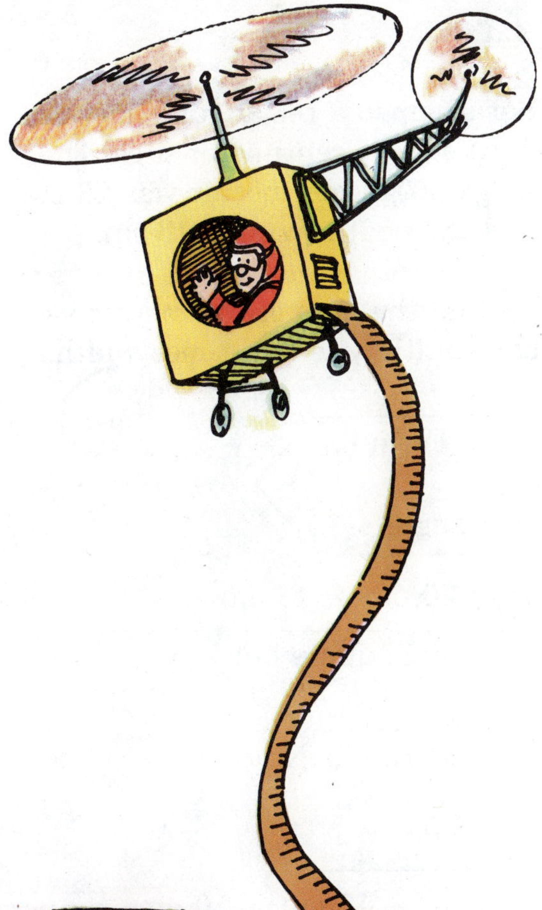

10. Measure the length and the width of a rectangular desk top to the nearest centimeter. Compute the perimeter.

11. Measure the width and the height of a bulletin board to the nearest hundredth of a meter. Compute the perimeter.

12. If possible, find the perimeter of your school grounds to the nearest meter.

Solve.

13. Mr. Smith has a rectangular yard that is 56 meters wide and 32 meters long. How much fencing will he need to enclose the yard? If fencing costs $5.85 a meter, how much will the whole fence cost?

14. Ms. Davis jogs around a square park that measures 75 meters on a side. One morning she jogged around the park 7 times. Did she jog more or less than 2 kilometers? How much more or less?

Area

Sarah made a rectangular mosaic from square-centimeter tiles. The length (l) of her mosaic was 20 cm and the width (w) was 18 cm.

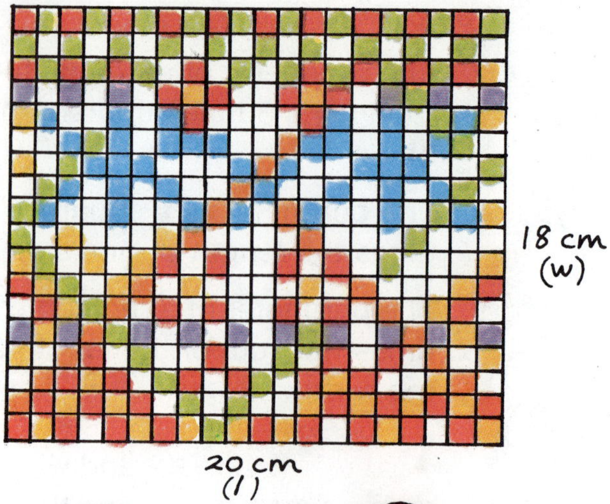

To find the area of the mosaic we can multiply length times width.

Area Formula for a Rectangle $A = l \times w$

$A = 20 \text{ cm} \times 18 \text{ cm}$

$A = 360$ square cm

Area Formula for a Triangle $A = \frac{1}{2} \times b \times h$

$A = \frac{1}{2} \times 16 \text{ cm} \times 10 \text{ cm}$

$A = \frac{1}{2} \times 160$ square cm

$A = 80$ square cm

EXERCISES

Give each area.

1. 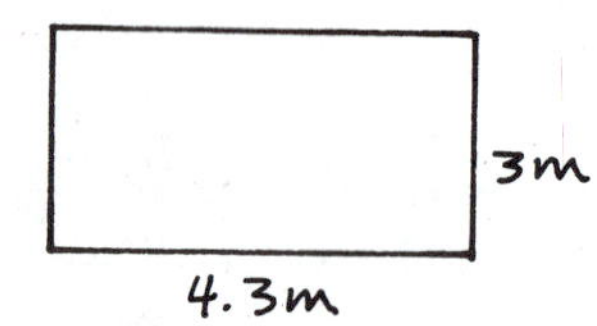

2. 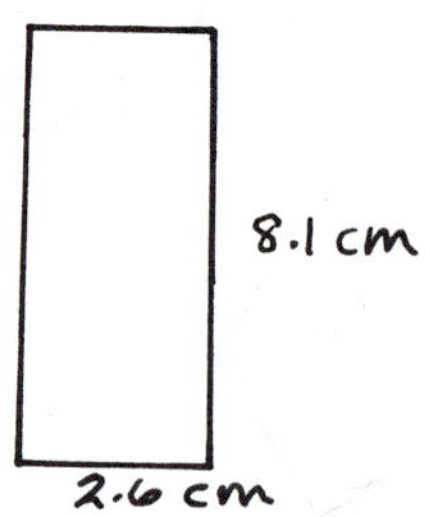

3. 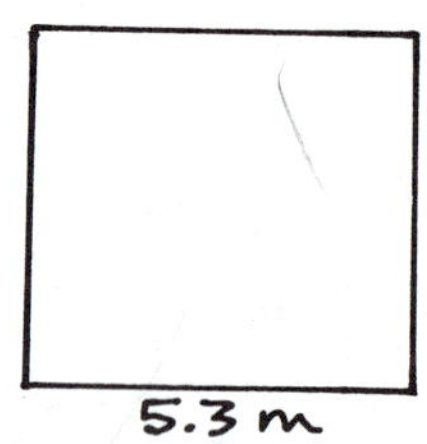

4. 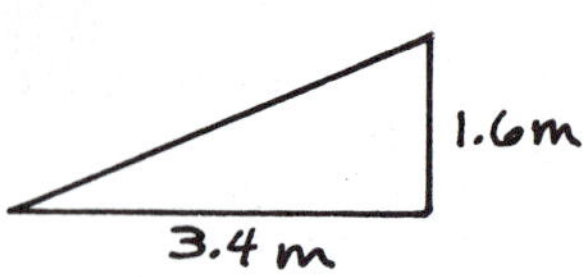

5. 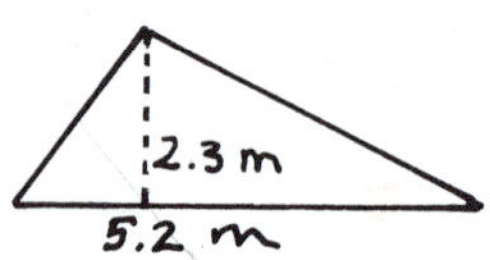

6. 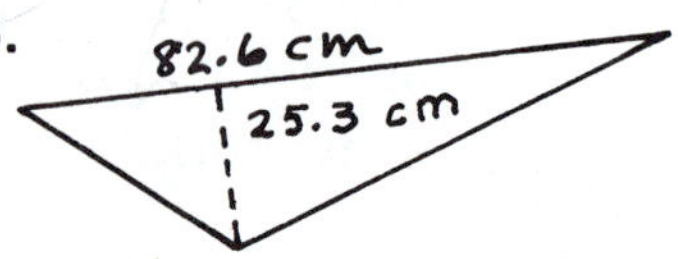

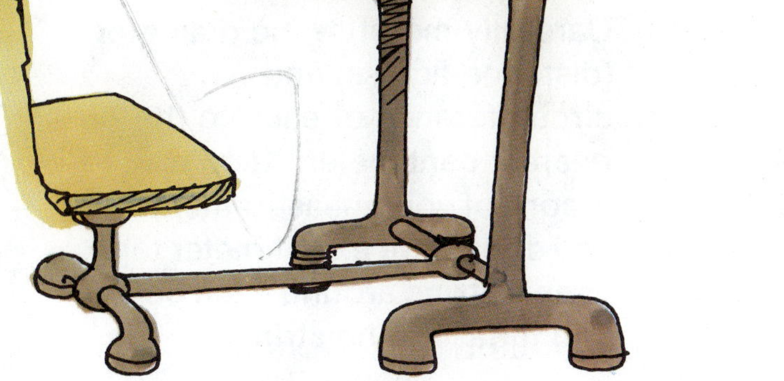

7. Measure the length and width of your desk to the nearest centimeter. Compute the area.

8. Measure the length and width of your classroom to the nearest hundredth of a meter. Compute the area.

Solve.

9. A rectangular room is 8.2 m long and 5.6 m wide. How many square meters of carpeting are needed to cover the floor?

10. A square kitchen, 4.5 m on a side, is to be covered with vinyl that cost $10.35 a square meter. How much will the vinyl cost?

11. A flower bed in the shape of a right triangle is to be covered with plastic. The two shorter sides measure 3.6 m and 4.2 m. How much plastic is needed to cover the flower bed?

12. Mr. Kline has 34 square meters of carpeting. He wants to carpet two rectangular rooms. One is 4 m by 4.6 m and the other is 5.2 m by 4.8 m. How much more carpeting does he need?

13. Miss Ross's backyard is a 32.5 m by 46.2 m rectangle. She wants to plant grass seed. If 1 box of seed covers 50 square meters, how many boxes should she buy? Round your answer to the nearest whole number.

14. Below is the floor plan of a living room. If 1 square meter of carpeting costs $11.75, how much will the carpeting cost?

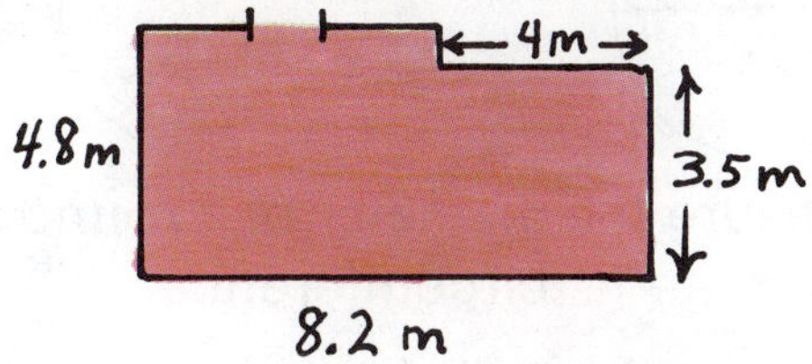

Circumference and area of a circle

The distance around a circle is called the **circumference** of the circle.

EXERCISES

1. Find some round objects. Carefully measure the diameter (distance across) and circumference of each to the nearest centimeter. Keep a record of your measurements. (If you don't have a centimeter tape, wrap a string around each object and measure the string.)

Object	diameter (d)	circum-ference (c)	$\frac{c}{d}$

2. Divide the circumference *(C)* by the diameter *(d)* and round each quotient to the nearest hundredth. Enter your results in the last column. Did you get close to the same quotient each time? If not, check your work.

3. If you multiply $\frac{C}{d}$ and the diameter, what should you get?

> To find the approximate circumference of a circle, multiply 3.14 by the diameter.

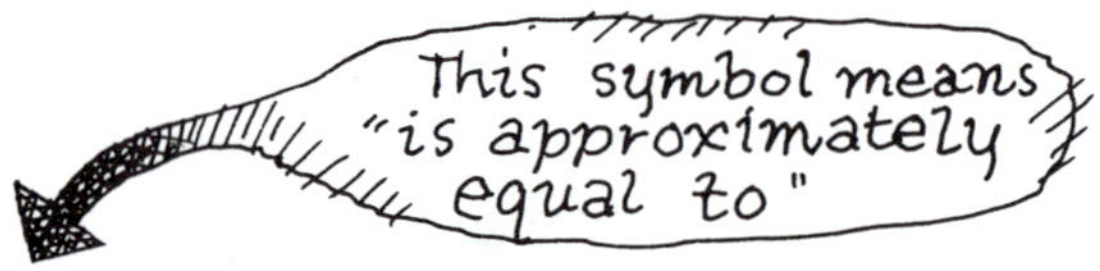

$C \approx 3.14 \times d$

Find each circumference.

4. 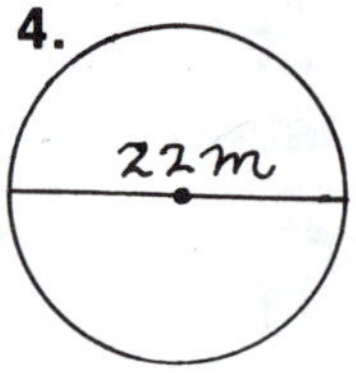

5. 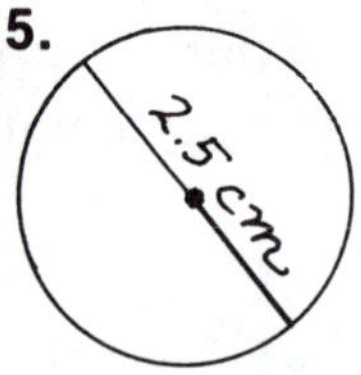

6. 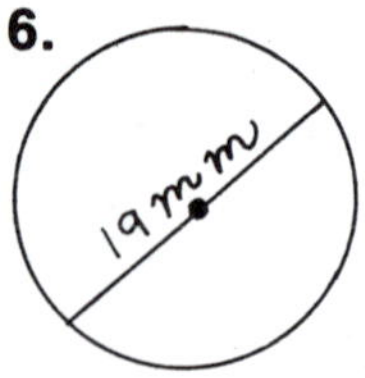

7. 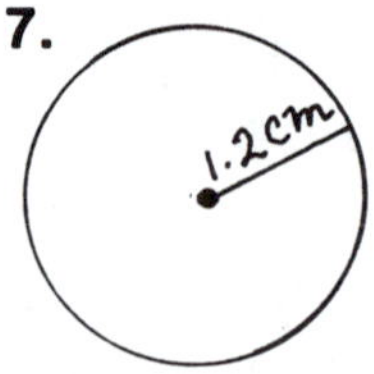

8. Measure the diameter of a round wastebasket. Compute its circumference.

If we cut the circle into pieces as shown, the pieces can be fitted together to make a figure that looks like a rectangle.

The length would be $\frac{1}{2}$ the circumference ($3.14 \times r$) and the width would be r.

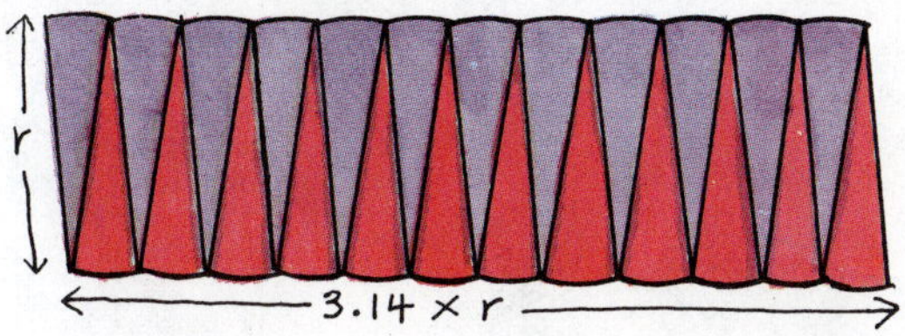

> To find the approximate area of a circle, use this formula:
>
> $$A \approx 3.14 \times r \times r$$

Find each area.

9. 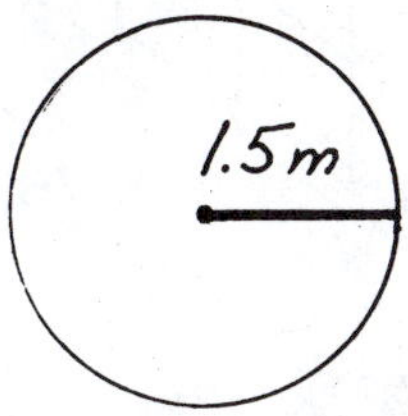

10. 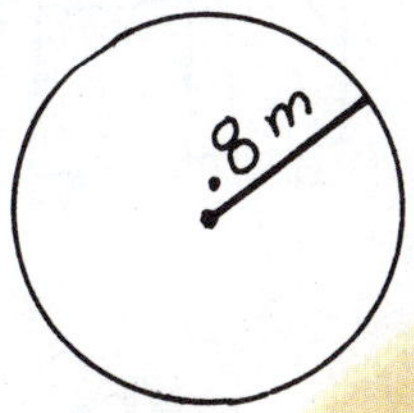

11. 

12. 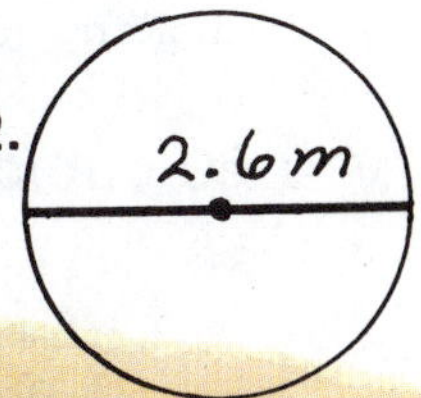

13. Find a circular object in your classroom. Measure its diameter. Determine the radius and compute the area.

Solve.

14. What is the area of a pizza that measures 28 cm across? (The diameter is 28 cm.)

15. Sam the pizza man sells a pizza 25 cm in diameter for \$2.20. A 36-cm pizza costs twice as much. Which size is the better buy?

Surface area and volume

The **surface area** of a rectangular solid is the sum of the areas of the six faces.

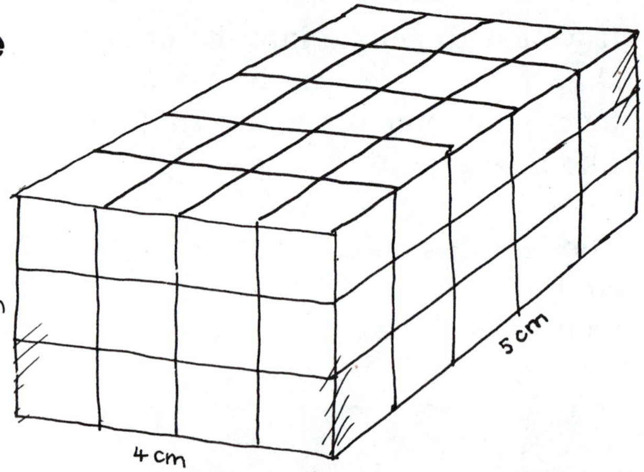

To find the **volume**, we find how many cubic centimeters it takes to fill the box. We can find the volume by multiplying length (l) times width (w) times height (h).

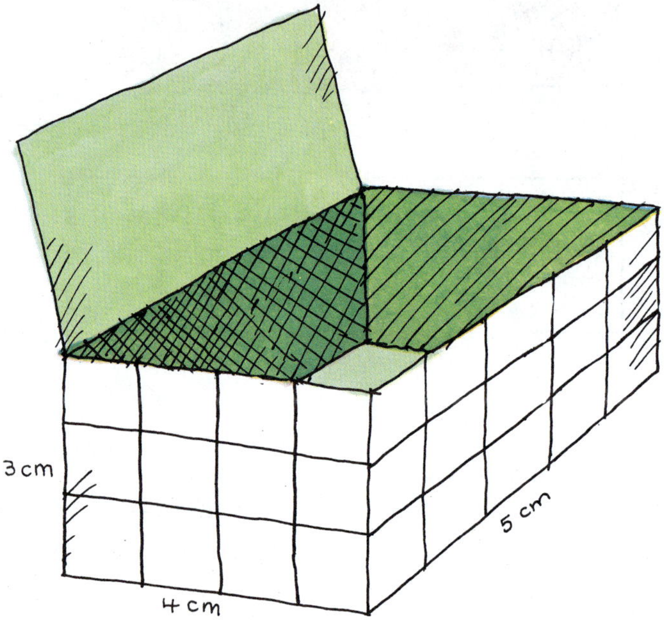

Volume Formula for a Rectangular Solid

$V = l \times w \times h$

$V = 5 \text{ cm} \times 4 \text{ cm} \times 3 \text{ cm}$

Number of cubes in bottom layer — Number of layers

$V = 60$ cubic centimeters

EXERCISES

What is the area of the

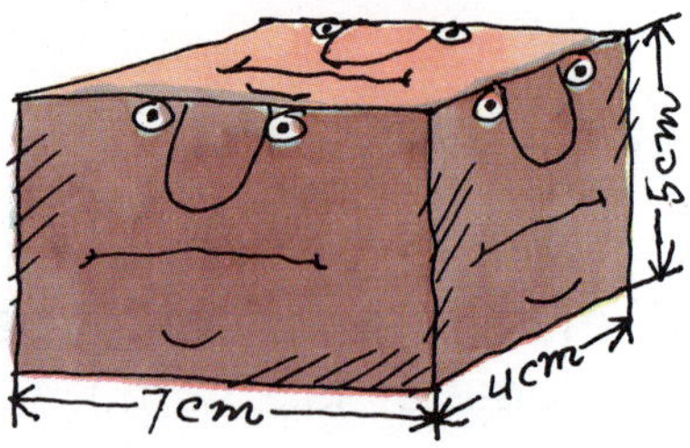

1. front face?
2. back face?
3. top face?
4. bottom face?
5. right face?
6. left face?
7. What is the surface area?

Find the surface area and volume of each solid.

8.

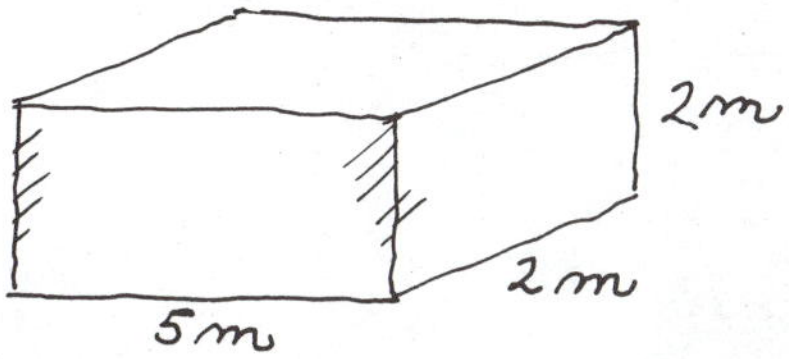

9.

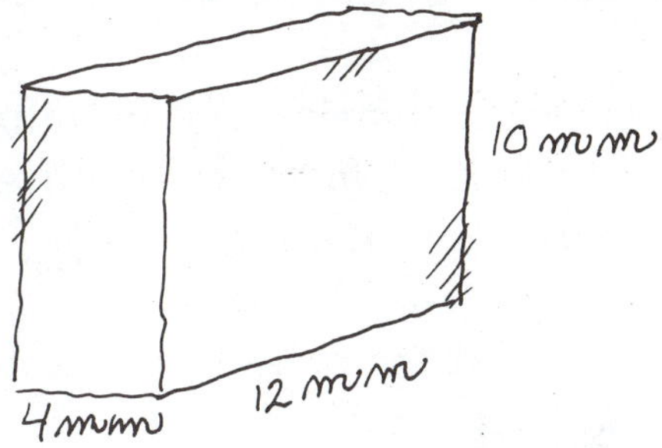

10.

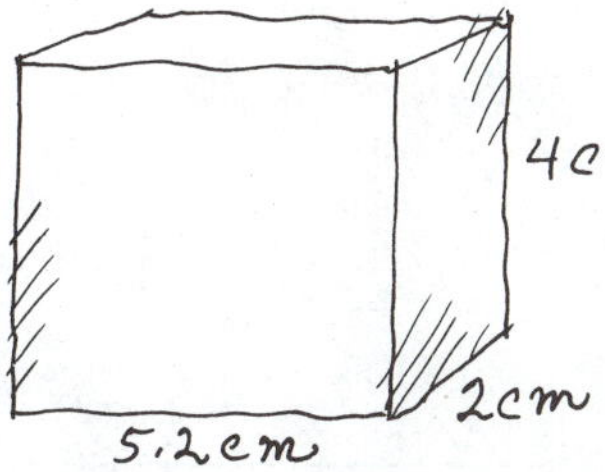

11.

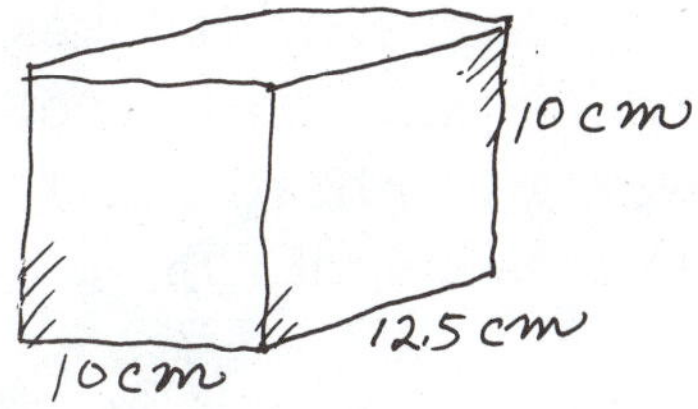

12. Get a rectangular box. Measure its length, width, and height to the nearest centimeter. Compute its surface area and volume.

13. Determine the surface area and volume of your classroom.

Solve.

14. A rectangular package that is 10.5 cm by 6.5 cm by 4 cm is to be wrapped with paper. How much paper is needed?

15. An aquarium that is 20.6 cm wide and 32 cm long is to be filled to a depth of 18 cm with water. How many cubic centimeters of water are needed?

16.

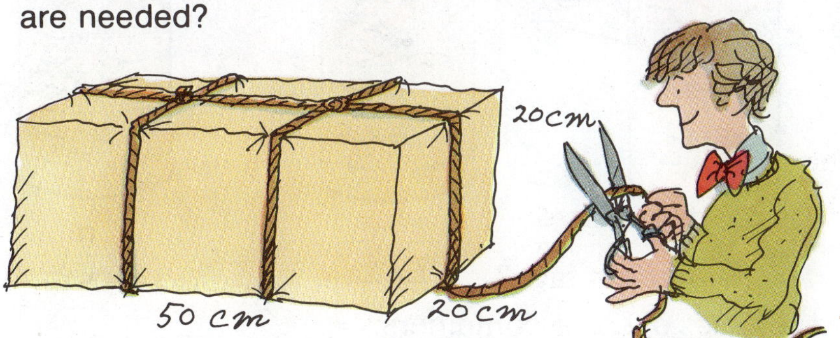

How much string was used to tie this package? Ignore the knots.

Liquid volume and weight

A unit for measuring liquid volume is the **liter** (L). Notice that a liter is 1000 cubic centimeters.

Another unit for measuring liquid volume is the **milliliter** (mL).

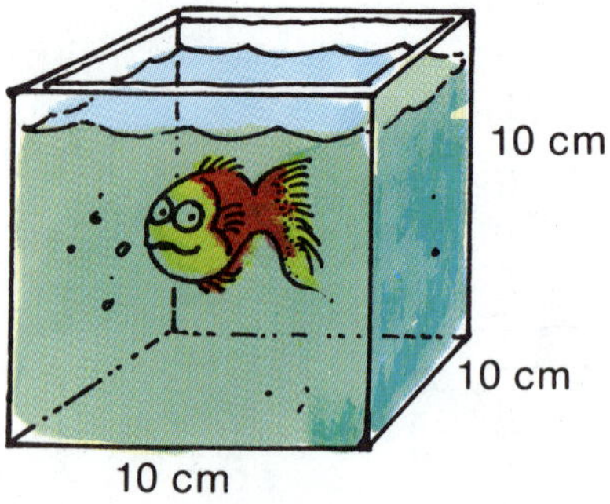

1 liter (L)

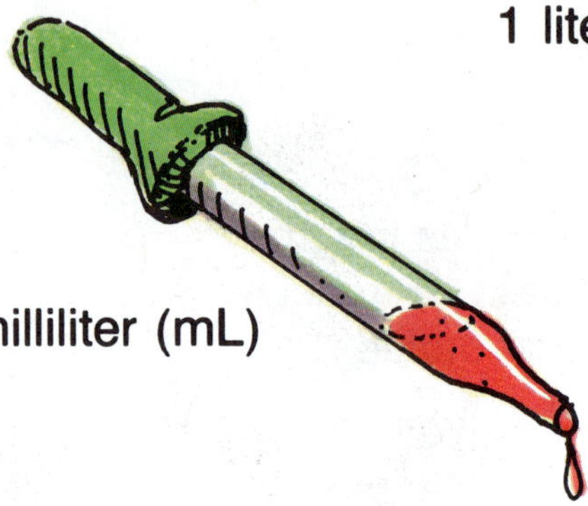

1 milliliter (mL)

1000 mL = 1 L
or .001 L = 1 mL

EXERCISES

Liter or milliliter? Which unit would be used to measure the volume of

1. a small can of juice? 2. an aquarium?

3. a wading pool? 4. a drinking glass?

How much liquid is in each container?

5. 

_?_ mL
_?_ L

6. 

_?_ mL
_?_ L

7. 

_?_ mL
_?_ L

8. 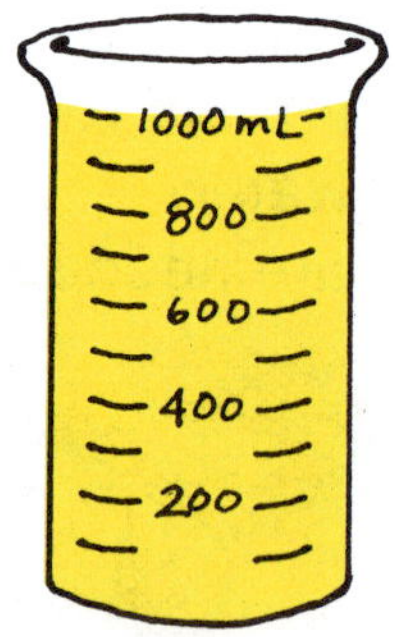

_?_ mL
_?_ L

9. Estimate the liquid volumes of some containers. Fill each container to see how close your estimate is to the actual liquid volume.

A unit for measuring weight is the **gram** (g).

A smaller unit for measuring weight is the **milligram** (mg).

1000 mg = 1 g

.001 g = 1 mg

A larger unit for measuring weight is the **kilogram** (kg).

1000 g = 1 kg

.001kg = 1 g

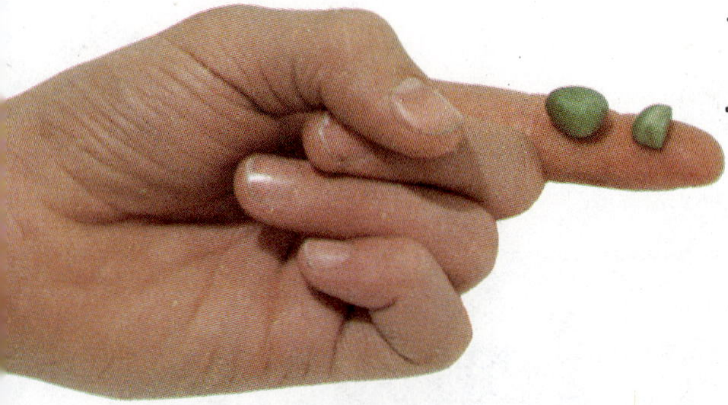

about 1 gram (g)

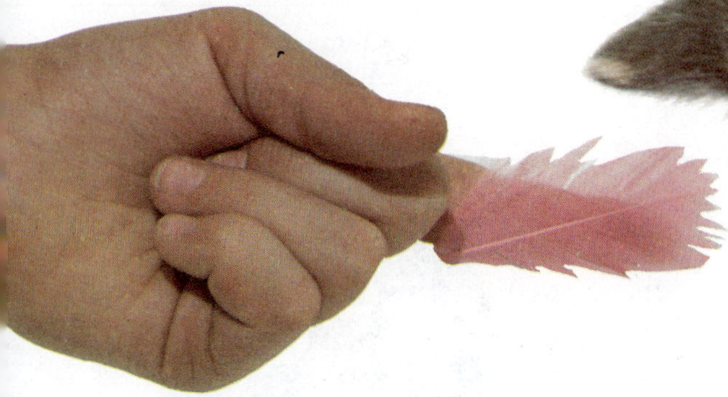

about 1 milligram (mg)

about 1 kilogram (kg)

EXERCISES

Milligram, gram, or kilogram? Which unit would be used to measure the weight of

10. a bicycle? **11.** a candy bar? **12.** a small needle? **13.** a bag of potatoes?

Which is heavier

14. 900 g or 1 kg? **15.** 350 mg or 3 g? **16.** 250 g or .24 kg?

17. Estimate the weights of some objects. Weigh each object to see how close your estimate is to the actual weight.

Length in the customary system

There are several units of length that are used in the customary system.

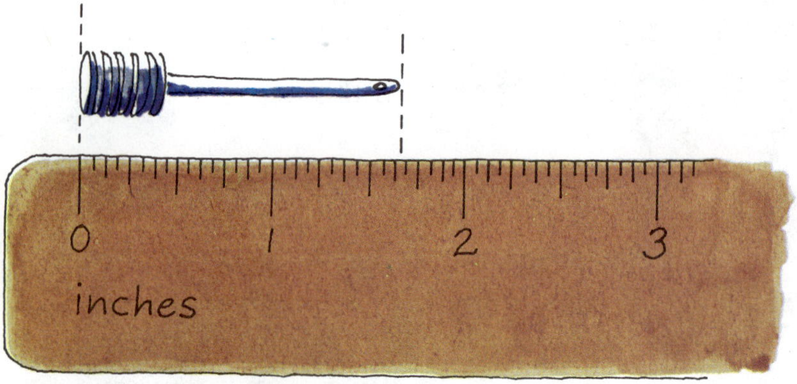

The length of the needle measured to the nearest

inch is 2 in.(") $\frac{1}{2}$ in. is $1\frac{1}{2}$".

$\frac{1}{4}$ in. is $1\frac{3}{4}$". $\frac{1}{8}$ in. is $1\frac{5}{8}$".

What is the length measured to the nearest $\frac{1}{16}$ in.?

Here are some other units that are used for measuring length in the customary system.

12 in. = 1 foot (ft)

3 ft(') = 1 yard (yd)

5280 ft = 1 mile (mi)

How many yards are in 1 mile?

EXERCISES

Draw segments of these lengths.

1. $3\frac{1}{2}$" **2.** $2\frac{1}{8}$" **3.** $1\frac{15}{16}$" **4.** $3\frac{3}{4}$" **5.** $1\frac{1}{16}$"

6. $4\frac{1}{4}$" **7.** $2\frac{7}{8}$" **8.** $2\frac{1}{2}$" **9.** $2\frac{13}{16}$" **10.** $1\frac{3}{8}$"

Measure

11. the width of your desk to the nearest $\frac{1}{4}$ inch.
12. the length of the chalkboard to the nearest foot.
13. the length of your classroom to the nearest yard.
14. the width of your classroom to the nearest $\frac{1}{2}$ yard.

Copy and complete.

	inches	feet	yards
15.	36	3	1
16.		9	
17.	144		
18.		15	

Complete.

19. $\frac{1}{2}$ ft = ? in.
20. $\frac{3}{4}$ ft = ? in.
21. $\frac{2}{3}$ ft = ? in.
22. $\frac{1}{2}$ yd = ? in.
23. $\frac{2}{3}$ yd = ? in.
24. $\frac{2}{3}$ yd = ? ft
25. 7 ft 8 in. = ? in.
26. 4 yd 1 ft = ? ft
27. 8 ft 5 in. = ? in.
28. 75 in. = ? ft ? in.
29. 158 in. = ? ft ? in.
Hint: Divide by 12.
30. 50 ft = ? yd ? ft

Keeping Skills Sharp

1. $6\overline{)9.30}$
2. $8\overline{)50.4}$
3. $5\overline{)68.4}$
4. $4\overline{)9.24}$
5. $8\overline{)5.52}$
6. $.5\overline{)4.25}$
7. $.3\overline{)7.32}$
8. $.6\overline{)29.4}$
9. $.08\overline{)6.72}$
10. $.09\overline{).7353}$
11. $2.6\overline{)66.82}$
12. $2.9\overline{).3045}$
13. $.84\overline{)18.9}$
14. $.46\overline{)16.008}$
15. $6.1\overline{)23.119}$
16. $6.1\overline{)1.5677}$
17. $.49\overline{).6615}$

Perimeter and area

Perimeter:

$$\begin{array}{r} 8 \text{ in.} \\ 15 \text{ in.} \\ 8 \text{ in.} \\ +15 \text{ in.} \\ \hline 46 \text{ in.} \end{array}$$

IMPORTANT NOTICE
Before computing, make sure that the units are the same.

Area:

$A = l \times w$

$A = 15 \text{ in.} \times 8 \text{ in.}$

$A = 120$ square in.

Perimeter:

$$\begin{array}{r} {}^{2} \\ 4'8'' \\ 3'6'' \\ +5'10'' \\ \hline 14' \end{array}$$

Regrouped 24 in. for 2 ft

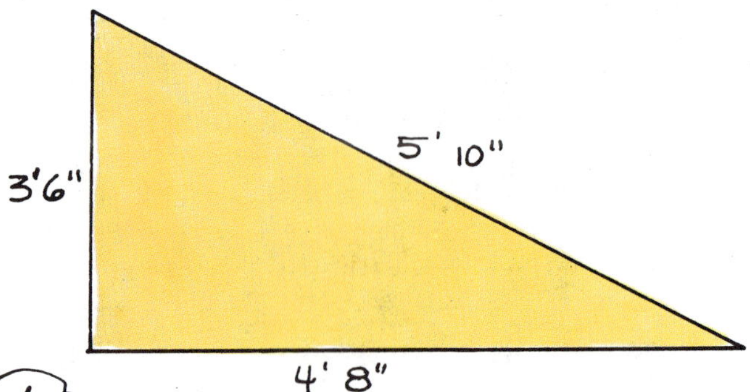

Area:

$A = \frac{1}{2} \times b \times h$

$A = \frac{1}{2} \times 56'' \times 42''$

$A = 1176$ square in.

EXERCISES

Add. Regroup whenever possible.

1. 5 ft 3 in. + 4 ft 7 in.

2. 9 ft 8 in. + 4 ft 7 in.

3. 4 ft 9 in. + 3 ft 11 in.

4. 6 ft 8 in. + 2 ft 5 in.

5. 4 yd 1 ft + 3 yd 2 ft

6. 3 yd 2 ft + 8 yd 2 ft

7. 4 yd 1 ft 5 in. + 1 yd 1 ft 8 in.

8. 9 yd 2 ft 7 in. + 5 yd 2 ft 9 in.

Multiply. Remember to work with a single unit.

9. 9 in. × 5 in.
10. 18 ft × 12 ft
11. 16 ft × 5 yd
12. 2 ft × 1 ft 3 in.
13. 4 yd × 3 yd 1 ft
14. 5 ft 4 in. × 3 ft 2 in.

Give the perimeter and area of each figure.

15. 

16. 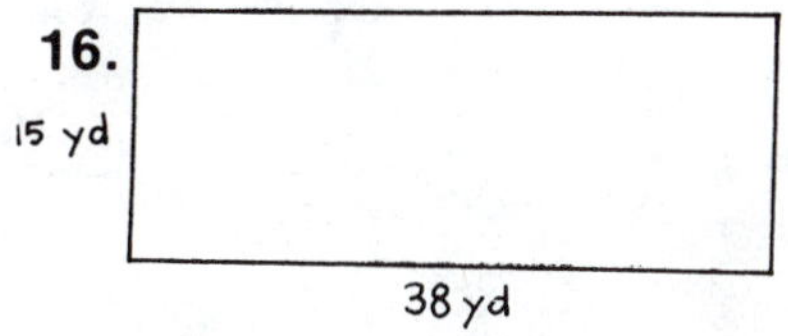

17. 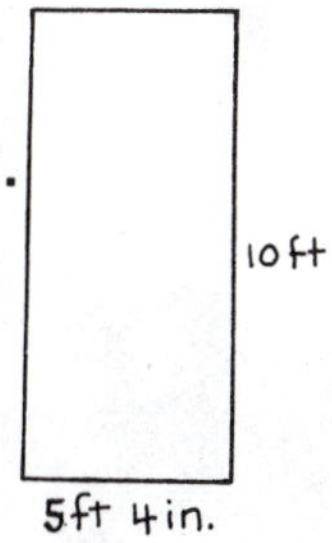

18. 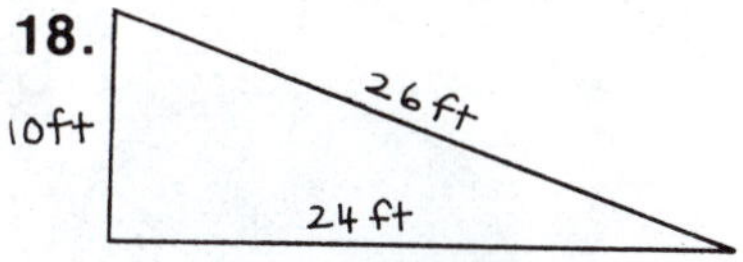

19. 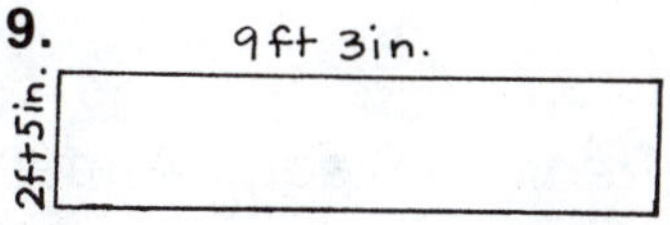

20. 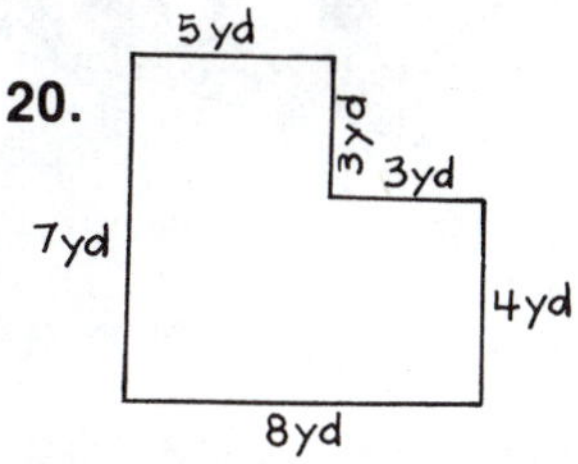

21. How many square inches are in 1 square foot?
22. How many square feet are in 1 square yard?

Solve.

23. A square garden measures $8\frac{1}{2}$ ft on a side. How much fencing is needed to fence in the garden?

24. A rectangular room is 13 feet wide and 23 feet long. What is its floor area?

25. A rectangular room is 18 feet long and 12 feet wide. How many 9-inch by 9-inch tiles are needed to cover the floor?

26. A roll of carpet is 96 feet long and 15 feet wide. How many square feet of carpeting are in the roll? How many square yards?

27. A rectangular porch 29 feet long and 16 feet wide is to be painted. Suppose that a gallon of paint covers 475 square feet. Will one gallon be enough?

28. A rectangular table top measures 3' 2" by 4' 6". How many 1" by 1" tiles are needed to cover the table top?

Surface area and volume

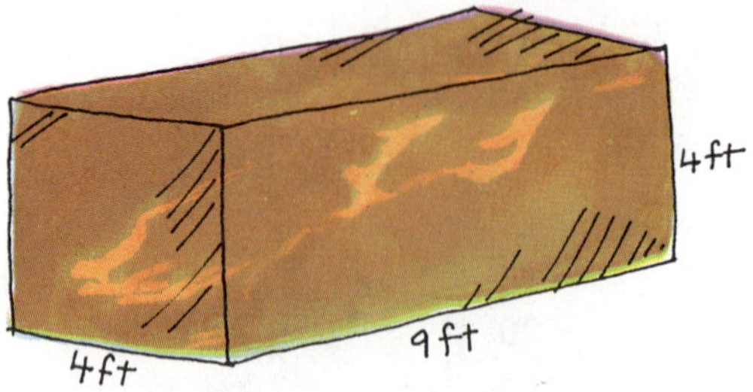

$$\begin{array}{r} 16 \\ 16 \\ 36 \\ 36 \\ 36 \\ +36 \\ \hline 176 \end{array}$$

Surface area: 176 square feet

Volume:

$V = l \times w \times h$

$V = 9 \text{ ft} \times 4 \text{ ft} \times 4 \text{ ft}$

$V = 144$ cubic feet

EXERCISES

Is the question about length (perimeter), area (surface area), or volume?

1.

How much sand will the truck hold?

2.

How much fence is needed to fence the yard?

3.

How high is the balloon?

4.

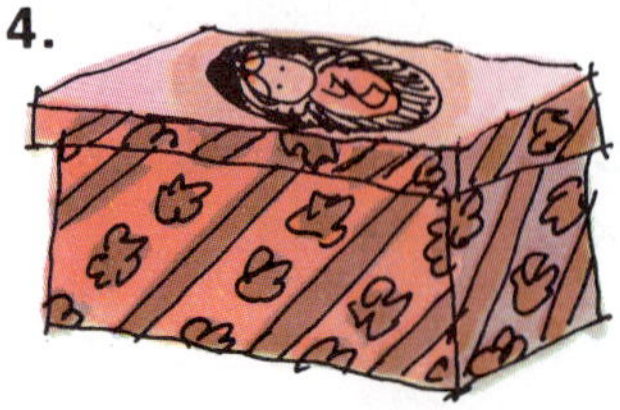

How much paper is needed to wrap the package?

5.

How deep is the well?

6.

How much carpeting is needed to cover the floor?

Give the surface area and volume of each rectangular solid.

7.

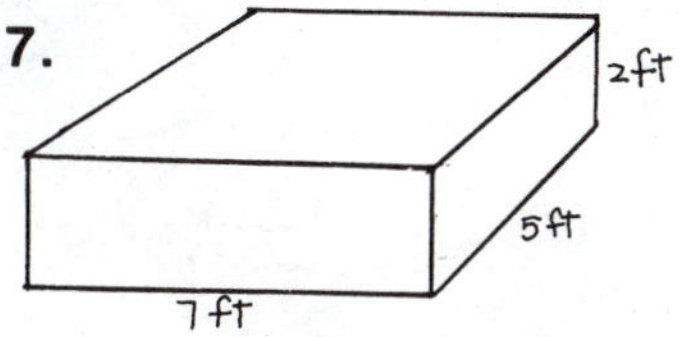

8.

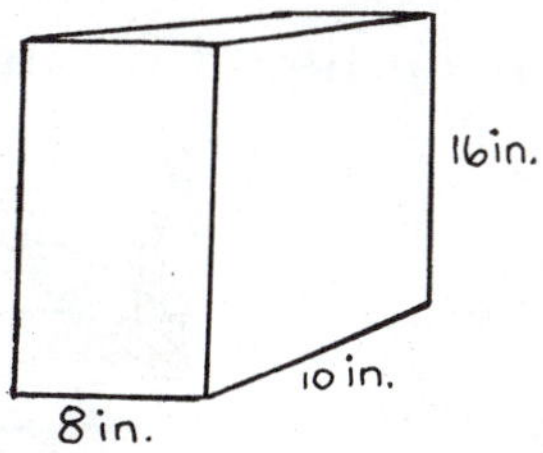

9.

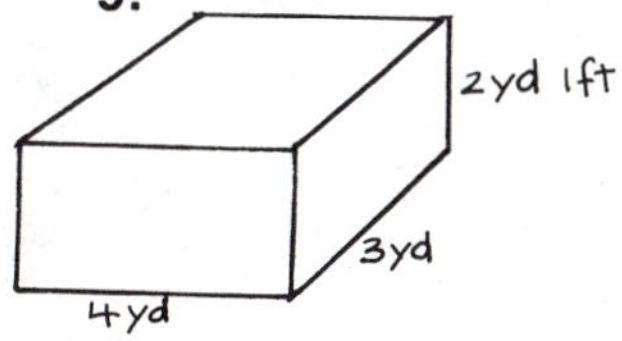

Solve.

10.

a. What is the volume of the package?

b. How much wrapping paper did it take to cover the package?

c. What is the volume of a package that is twice as wide, twice as long and twice as high?

d. The volume is how many times as great as the volume of the package shown?

e. If only the height of the package were doubled, what would happen to the volume?

11.

a. The cereal company wishes to sell a box that holds twice as much cereal as the one shown. Could they just make the new box twice as wide?

b. What is the surface area of the box shown?

c. If the larger box were twice as wide, would the surface area be twice as much?

d. If the larger box were twice as long, would the surface area be the same as that of the box that is twice as wide?

12. How many cubic inches in a cubic foot?

13. How many cubic feet in a cubic yard?

14. A rectangular concrete form is 2 feet wide, 18 feet long, and 6 feet high. How many cubic yards of concrete are needed to fill the form?

15. A candy box is 2" by 8" by 4". Give the dimensions of a box that holds the same amount of candy but has the smallest possible surface area.

Liquid volume and weight

Here are some units for measuring liquid volume in the customary system.

2 cups (c) = 1 pint (pt)

2 pints = 1 quart (qt)

2 quarts = 1 half-gallon

2 half-gallons = 1 gallon (gal)

EXERCISES

Complete.

1. 2 pt = ? c **2.** 6 qt = ? pt **3.** 2 qt = ? c

4. 1 gal = ? qt **5.** 1 gal = ? c **6.** 3 half-gallons = ? pt

7. $1\frac{1}{2}$ pt = ? c **8.** $2\frac{1}{2}$ qt = ? pt **9.** $2\frac{1}{2}$ qt = ? c

10. $1\frac{1}{2}$ gal = ? qt **11.** $1\frac{1}{2}$ gal = ? pt **12.** $1\frac{3}{4}$ gal = ? qt

13. 8 pt = ? qt **14.** 12 c = ? pt **15.** 12 c = ? qt

16. 16 qt = ? gal **17.** 14 pt = ? qt **18.** 10 half-gallons = ? gal

19. Estimate the liquid volumes of some containers. Use a measuring cup to see how close your estimates are to the actual volumes.

These units are used to measure weight in the customary system.

16 ounces (oz) = 1 pound (lb)

2000 pounds (lb) = 1 ton

Which unit would be used to weigh a

20. baseball?

21. bicycle?

22. large jet airliner?

23. wrist watch?

Complete.

24. 2 lb = ? oz

25. $2\frac{1}{2}$ lb = ? oz

26. 3 tons = ? lb

27. $4\frac{3}{4}$ tons = ? lb

28. 54 oz = ? lb ? oz

29. 78 oz = ? lb ? oz

30. First estimate the weights of some objects. Then weigh each object to see how close your estimate was to the actual weight.

CHAPTER CHECKUP

Complete. [pages 232–235]

1. 2 m = ? cm 2. 425 cm = ? m 3. 3.5 km = ? m 4. 758 m = ? km

Give the perimeter (or circumference) and area. [pages 236–241, 248–249]

5. 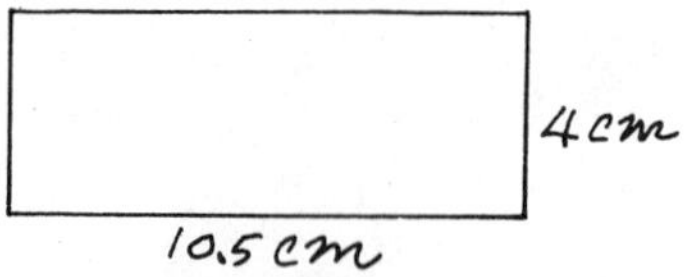

6. 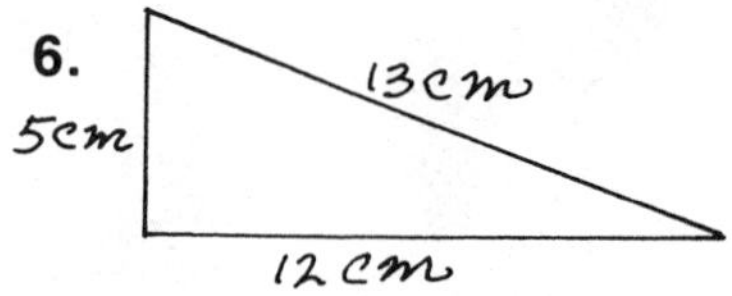

7. 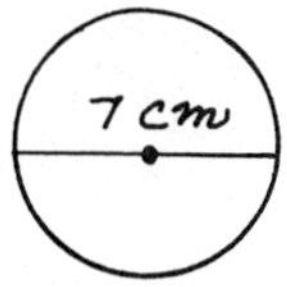

Give the surface area and volume.
[pages 242-243, 250–251]

8. 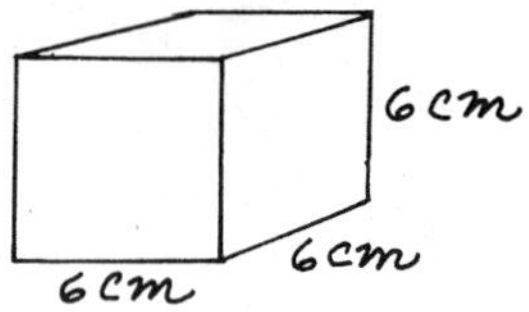

9. 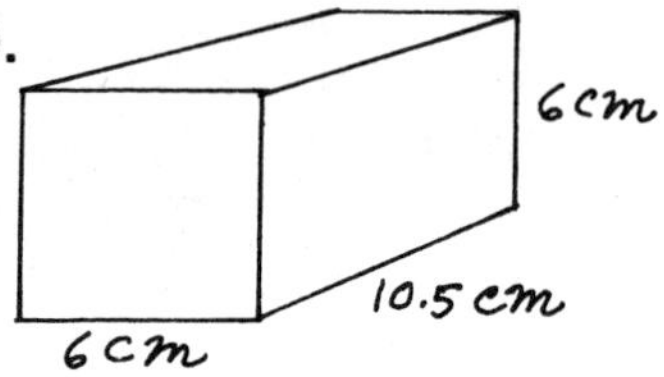

Complete. [pages 244-245]

10. 3 L = ? mL 11. 450 mL = ? L 12. 8 g = ? mg 13. 84 kg = ? g

Draw segments of these lengths. [pages 246–247]

14. $2\frac{1}{4}$ inches 15. $1\frac{1}{8}$ inches 16. $2\frac{5}{16}$ inches

Complete. [pages 246–247]

17. 2 feet = ? inches
18. 3 yards = ? feet
19. 15 feet = ? yards
20. 1 mile = ? yards
21. 5 miles = ? yards
22. 1 mile = ? feet
23. $2\frac{1}{2}$ feet = ? inches
24. $4\frac{1}{3}$ yards = ? feet
25. 15 inches = ? foot ? inches
26. 20 inches = ? foot ? inches
27. 11 feet = ? yards ? feet
28. 75 inches = ? feet ? inches

Complete. [pages 252–253]

29. 8 cups = ? pints
30. 12 quarts = ? gallons
31. $2\frac{1}{2}$ gallons = ? quarts
32. 6 pounds = ? ounces
33. 8000 pounds = ? tons
34. 140 ounces = ? pounds ? ounces

Project

Estimate how long it would take you to walk from New York City to Los Angeles.

1. Measure off 50 or 100 meters on the school grounds.
2. See how long it takes you to walk 100 meters.
3. Compute how long it would take you to walk a kilometer. Give your answer to the nearest tenth of a minute.
4. Next compute how many minutes it would take you to walk 4800 kilometers from New York to Los Angeles. (Round to the nearest minute.)
5. Convert to hours.
6. If you walked 8 hours each day, how many days would it take you to walk from New York City to Los Angeles?

CHAPTER REVIEW

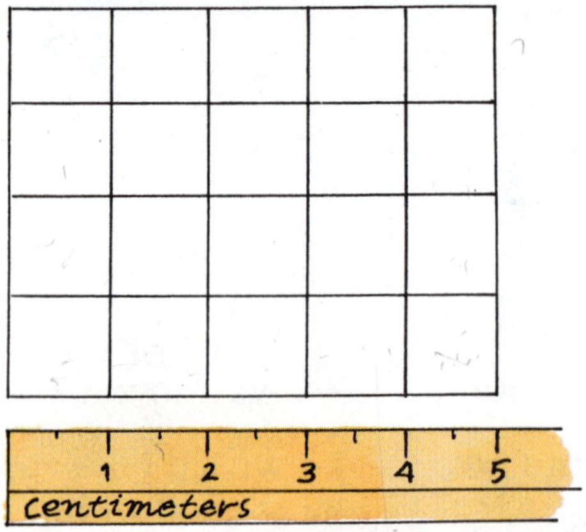

1. The length is ? cm.
2. The width is ? cm.
3. The perimeter is ? cm.
4. The area is ? square cm.

Give the area of the

5. front.
6. back.
7. top.
8. bottom.
9. right side.
10. left side.
11. What is the surface area?
12. What is the volume?

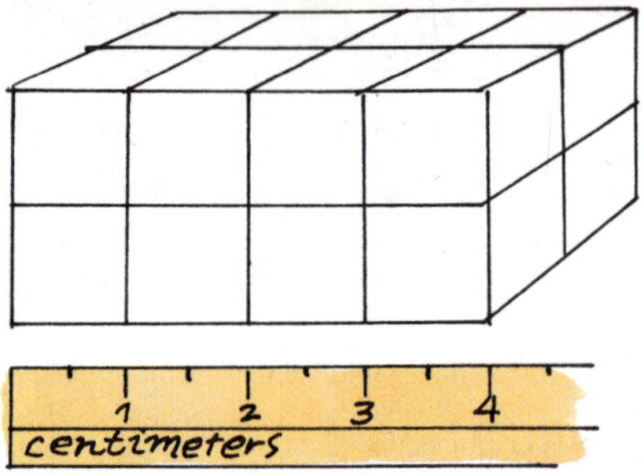

Complete.

13. 1 liter = ? milliliters
14. 1000 grams = ? kilogram

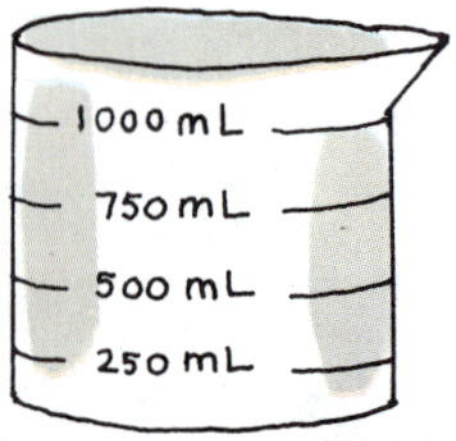

Give the letter.

15. $2\frac{7}{8}$ inches
16. $1\frac{1}{2}$ inches
17. $2\frac{7}{16}$ inches
18. $2\frac{1}{4}$ inches

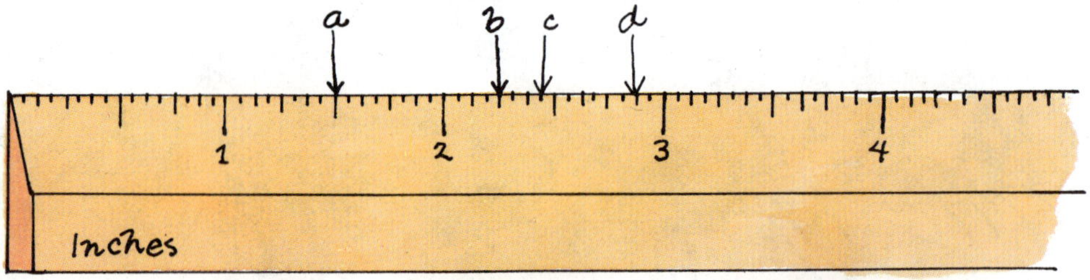

Complete.

19. 1 foot = ? inches
20. 1 yard = ? feet
21. 1 yard = ? inches

CHAPTER CHALLENGE

The metric unit for measuring land area is the **hectare.** A square whose sides are 100 meters long has an area of 1 hectare.

Complete.

1. The **square meter** is used for measuring smaller areas. How many square meters are in a hectare?
2. 120 square meters = ? hectare
3. 3478 square meters = ? hectare
4. 6.7 hectares = ? square meters
5. .34 hectare = ? square meters
6. The hectare is used to measure areas of such regions as farms and parks. To measure very large areas such as the area of a state or country the unit used is the square kilometer. How many hectares are in a square kilometer?
7. 2.6 square kilometers = ? hectares
8. 8973 hectares = ? square kilometers
9. Find the area of your school grounds in square meters. In hectares.

MAJOR CHECKUP

Standardized Format

Choose the correct letter.

1. 59,643,205 rounded to the nearest million is

- **a.** 59,000,000
- **b.** 59,600,000
- **c.** 60,000,000
- **d.** none of these

2. Multiply.

$$\begin{array}{r} 394 \\ \times 165 \\ \hline \end{array}$$

- **a.** 63,210
- **b.** 51,110
- **c.** 59,790
- **d.** none of these

3. Divide.

$118\overline{)6352}$

- **a.** 54
- **b.** 53 R98
- **c.** 54 R98
- **d.** none of these

4. How many lines of symmetry does a square have?

- **a.** 2
- **b.** 4
- **c.** 3
- **d.** none of these

5. $5\frac{4}{5} + 9\frac{1}{2} =$

- **a.** $14\frac{5}{7}$
- **b.** $15\frac{3}{10}$
- **c.** $14\frac{3}{10}$
- **d.** none of these

6. $15 - 8\frac{5}{8} =$

- **a.** $6\frac{3}{8}$
- **b.** $7\frac{5}{8}$
- **c.** $7\frac{3}{8}$
- **d.** none of these

7. $3\frac{1}{2} \times 4 =$

- **a.** $12\frac{1}{2}$
- **b.** 9
- **c.** 14
- **d.** none of these

8. $2\frac{1}{2} \div 1\frac{1}{4} =$

- **a.** $3\frac{1}{8}$
- **b.** 2
- **c.** $\frac{1}{2}$
- **d.** none of these

9. $14.36 + 9.053 =$

- **a.** 10.498
- **b.** 104.98
- **c.** 1049.8
- **d.** none of these

10. $57.3 - 2.86 =$

- **a.** 28.7
- **b.** 54.44
- **c.** 2.87
- **d.** none of these

11. $78.2 \times .15 =$

- **a.** 1.173
- **b.** 117.3
- **c.** 11.73
- **d.** none of these

12. $7.605 \div 4.5 =$

- **a.** 1.69
- **b.** 16.9
- **c.** 169
- **d.** none of these

10
Consumer Mathematics

Computing prices

The apples cost 3 for 58¢. Alison is buying 1 apple. Here is how to compute the price.

Step 1. Compute the exact price of 1 apple.

$$\begin{array}{r} 19\tfrac{1}{3}¢ \\ 3\overline{)58¢} \\ -3 \\ \hline 28 \\ -27 \\ \hline 1 \end{array}$$

Step 2. *Round up* to the next cent. The price of 1 apple is 20¢.

Oranges are 5 for 69¢. Tom is buying 2 oranges. Here is how to compute the price.

Step 1. Compute the exact price of 1 orange.

$$\begin{array}{r} 13\tfrac{4}{5}¢ \\ 5\overline{)69¢} \\ -5 \\ \hline 19 \\ -15 \\ \hline 4 \end{array}$$

Step 2. Multiply the exact price by 2.

$$13\frac{4}{5}¢ \times 2 = 27\frac{3}{5}¢$$

Step 3. Round up to the next cent. The price of two oranges is 28¢.

To find the price of 7 oranges, add the price of 5 oranges (69¢) and the price of 2 oranges (28¢).

EXERCISES

Compute the price.

1. a. 1 melon
   b. 3 melons
2. a. 1 ear of corn
   b. 8 ears of corn
3. a. 1 tomato
   b. 2 tomatoes
4. a. 1 pear
   b. 2 pears
5. a. 1 plum
   b. 8 plums
   Hint: Add the price of 6 and the price of 2.
6. a. 1 tangerine
   b. 11 tangerines

Solve.

7. You have: $1.75

   You buy: 

   How much money do you have left?

8. You have: $2

   You buy: 

   How much money do you have left?

9. What is the price of 2 melons and 9 ears of corn?

10. Which costs more, 2 pears or 3 tangerines? How much more?

11. How many plums can you buy with 72¢? How much money would be left?

12. Suppose that you have $3 and want to buy $1\frac{1}{2}$ dozen ears of corn and as many tomatoes as possible. How many tomatoes can you buy? How much money would you have left?

Computing discounts

Merchants often reduce prices and have a sale. You have probably seen sales advertised in store windows, in newspapers, and on television. If you need an item, you can save money by buying it on sale.

The following example shows you how to compute the sale price of the record.

Step 1. To find $\frac{1}{3}$ of \$3.79, divide by 3.

$$
\begin{array}{r}
\$1.26\frac{1}{3} \\
3\overline{)\$3.79} \\
-3 \\
\hline
7 \\
-6 \\
\hline
19 \\
-18 \\
\hline
1
\end{array}
$$

Step 2. Round $\$1.26\frac{1}{3}$ to the nearest cent for the discount.

\$1.26

Step 3. Subtract the discount from the regular price to find the sale price.

$$
\begin{array}{r}
\$3.79 \\
-1.26 \\
\hline
\$2.53
\end{array}
$$

The sale price is \$2.53.

EXERCISES

Compute the discount and the sale price.

1. 

2. 

3. 

4. 

5. 

6. 

Solve.

7. Jan got \$10 for her birthday. She bought a \$4.19 record for $\frac{1}{3}$ off. How much money did she have left?

8. David has \$2.25. He wants to buy a \$4.69 record that is on sale for $\frac{1}{4}$ off. How much more money does he need?

9. What is the total discount on a \$3.79 record marked $\frac{2}{3}$ off and a \$4.19 record marked $\frac{1}{2}$ off? What is the total price?

★10. On the first day of the sale, a \$4.95 record was marked $\frac{1}{3}$ off. The next day it was marked $\frac{3}{4}$ off. By how much was the discount increased?

Keeping Skills Sharp

1. $2 \times 1\frac{1}{2}$
2. $3 \times 1\frac{1}{4}$
3. $4 \times 2\frac{1}{3}$
4. $3 \times 2\frac{1}{2}$
5. $1\frac{1}{2} \times 1\frac{1}{2}$
6. $2\frac{1}{2} \times 1\frac{1}{3}$
7. $3\frac{1}{4} \times 1\frac{2}{3}$
8. $2\frac{3}{4} \times 3\frac{1}{3}$
9. $2\frac{2}{3} \times 3\frac{1}{4}$
10. $5\frac{3}{8} \times 4\frac{3}{4}$
11. $3\frac{2}{5} \times 5\frac{5}{8}$
12. $4\frac{1}{3} \times 4\frac{3}{5}$

Unit pricing

Have you ever had to decide which of two items is the better buy? To decide, you should ask yourself questions like these:

> Are the two items of the same quality? (An item of poor quality is never a good buy.)
> Can I use the larger amount (quantity)?
> If not, you should buy the smaller quantity.

When two items are of the same quality and you could use either quantity, you may have to compute the **unit price** to decide which is the better buy.

EXAMPLE. Which is the better buy?

The unit price is the cost per ounce of ketchup. By comparing the unit prices, you can see that the 12-ounce bottle is a slightly better buy.

EXERCISES

Compute the unit price of each item to three decimal places.

1. 

2. 

3. 

4. 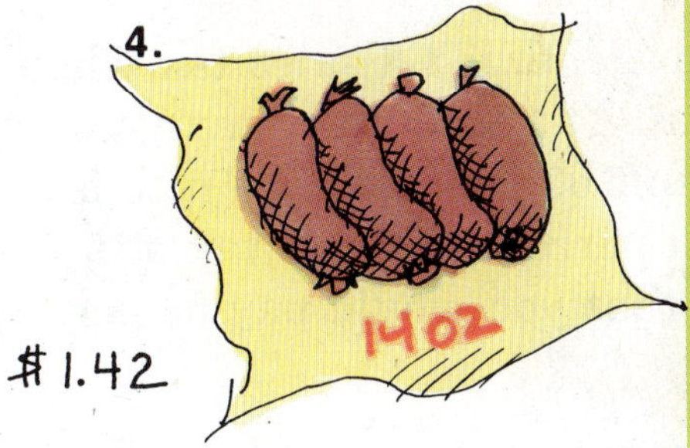

5. 

Tell which is the better buy.

6. MILK
   1 quart for 58¢ or 1 gallon for \$2.29

7. ICE CREAM
   1 quart for 99¢ or 1 half-gallon for \$1.65

8. KITE STRING
   200 yards for \$1.29 or 240 yards for \$1.52

9. PEANUTS
   3 pounds for \$2.55 or 5 pounds for \$4.35

10. Many grocery stores display unit prices so that shoppers can tell which is the better buy. See if your store displays unit prices.

Keeping Skills Sharp

1. $2\frac{1}{2} \div 2$
2. $2\frac{5}{8} \div 3$
3. $1\frac{3}{4} \div 2$
4. $2\frac{2}{3} \div 4$
5. $5 \times 1\frac{1}{4}$
6. $3 \times 2\frac{5}{8}$
7. $3\frac{1}{2} \times 1\frac{1}{2}$
8. $4\frac{1}{2} \times 2\frac{1}{3}$
9. $3\frac{1}{2} \times 2\frac{2}{3}$
10. $4\frac{3}{4} \times 4\frac{2}{3}$
11. $5\frac{3}{8} \times 3\frac{3}{4}$
12. $4\frac{2}{5} \times 3\frac{1}{8}$
13. $6 \div 1\frac{1}{3}$
14. $2\frac{1}{2} \div 3\frac{3}{5}$
15. $6\frac{1}{4} \div 1\frac{1}{4}$
16. $9\frac{3}{7} \div 2\frac{1}{5}$

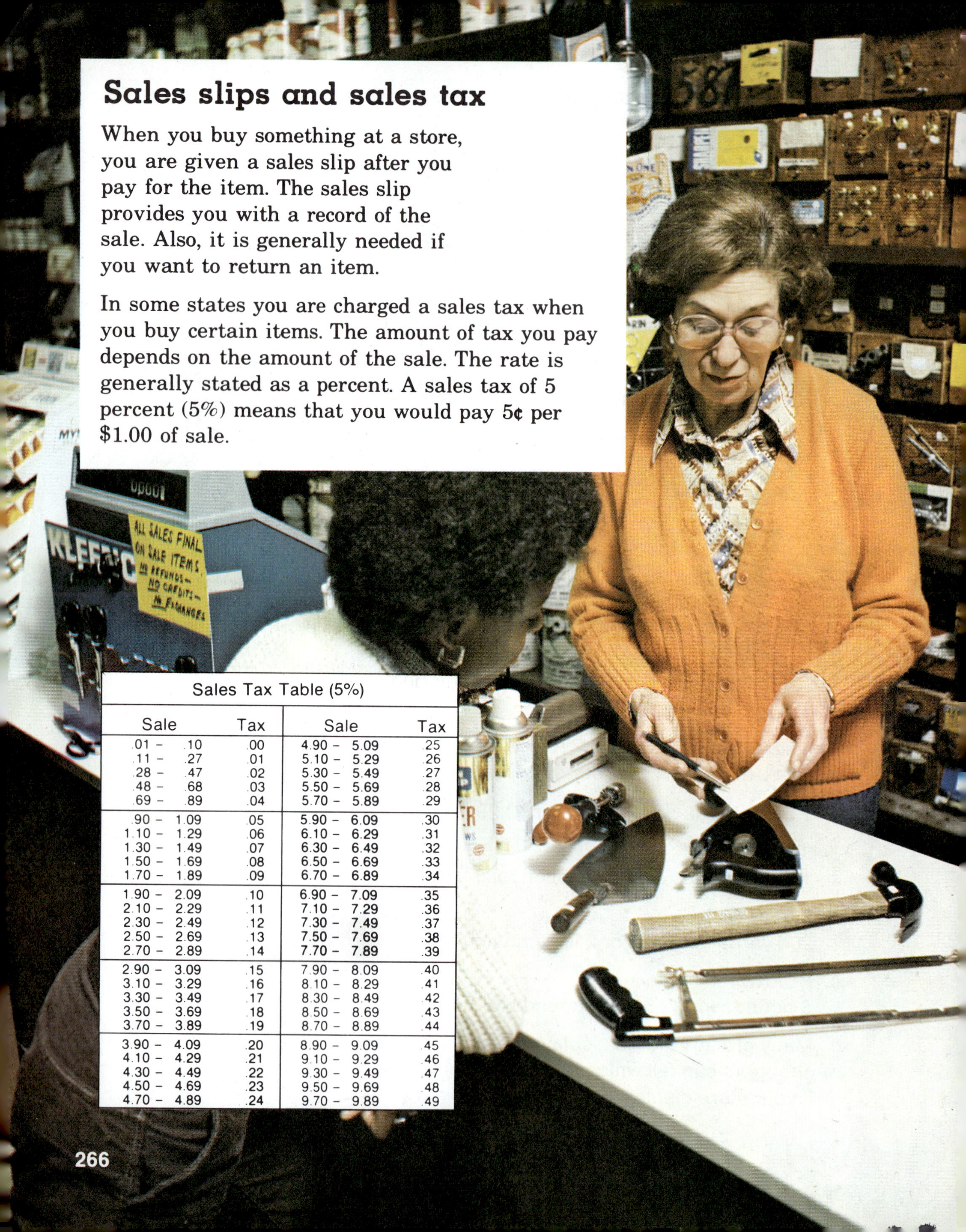

Sales slips and sales tax

When you buy something at a store, you are given a sales slip after you pay for the item. The sales slip provides you with a record of the sale. Also, it is generally needed if you want to return an item.

In some states you are charged a sales tax when you buy certain items. The amount of tax you pay depends on the amount of the sale. The rate is generally stated as a percent. A sales tax of 5 percent (5%) means that you would pay 5¢ per $1.00 of sale.

Sales Tax Table (5%)

Sale	Tax	Sale	Tax
.01 – .10	.00	4.90 – 5.09	.25
.11 – .27	.01	5.10 – 5.29	.26
.28 – .47	.02	5.30 – 5.49	.27
.48 – .68	.03	5.50 – 5.69	.28
.69 – .89	.04	5.70 – 5.89	.29
.90 – 1.09	.05	5.90 – 6.09	.30
1.10 – 1.29	.06	6.10 – 6.29	.31
1.30 – 1.49	.07	6.30 – 6.49	.32
1.50 – 1.69	.08	6.50 – 6.69	.33
1.70 – 1.89	.09	6.70 – 6.89	.34
1.90 – 2.09	.10	6.90 – 7.09	.35
2.10 – 2.29	.11	7.10 – 7.29	.36
2.30 – 2.49	.12	7.30 – 7.49	.37
2.50 – 2.69	.13	7.50 – 7.69	.38
2.70 – 2.89	.14	7.70 – 7.89	.39
2.90 – 3.09	.15	7.90 – 8.09	.40
3.10 – 3.29	.16	8.10 – 8.29	.41
3.30 – 3.49	.17	8.30 – 8.49	.42
3.50 – 3.69	.18	8.50 – 8.69	.43
3.70 – 3.89	.19	8.70 – 8.89	.44
3.90 – 4.09	.20	8.90 – 9.09	.45
4.10 – 4.29	.21	9.10 – 9.29	.46
4.30 – 4.49	.22	9.30 – 9.49	.47
4.50 – 4.69	.23	9.50 – 9.69	.48
4.70 – 4.89	.24	9.70 – 9.89	.49

EXERCISES

Suppose that you had to pay a 5% sales tax on the following amounts. Give both the sales tax and the total cost.

1. \$7.42	**2.** \$3.18	**3.** \$6.26	**4.** \$5.23	**5.** \$1.09
6. \$9.28	**7.** \$4.32	**8.** \$3.45	**9.** \$6.79	**10.** \$5.26

Use the sales slip to answer the following questions.

11. At what store was the purchase made?

12. On what date was the purchase made?

13. How many grocery items were bought?

14. What was the total price of the items? Do not include the sales tax.

15. How much was the sales tax?

16. How much money was the clerk given?

17. How much change did the clerk give back?

18. Suppose that the item priced at \$2.98 was returned. How much should be refunded? Don't forget the sales tax.

19. Suppose that the item priced at \$1.23 was returned and you bought another taxed item for \$1.69. How much would you owe the clerk?

20. Collect some sales slips. See what information is included on each slip.

```
B & J Market
4-16-81

$    .91  GR
$   1.68  MT
$    .93  GR TX
$    .59  PR
$    .79  PR
$   2.98  GR TX
$    .48  PR
$   3.72  MT
$   1.23  GR TX
$    .46  GR
$    .66  GR TX

$  14.43  SUB
     .29  TX
$  14.72  TOTL
$  15.00  CASH
$    .28  CHANGE
```

Installment buying

Many expensive items can be bought on an installment plan. For example, you could buy the bicycle by paying $36.72 when you received the bicycle (the down payment) and then $11.50 a month for 12 months (the monthly installments). Let's compare the cost of buying the bicycle on the installment plan with the cost of buying the bicycle for cash.

Cash

$156

Installment Plan

```
  $11.50 ← monthly installment
  ×  12
  -----
   2300
  1150
  -----
$138.00
+ 36.72 ← down payment
-------
$174.72
```

The installment plan costs $18.72 more. The $18.72 is charged by the merchant for lending the money. This charge is the interest or finance charge.

EXERCISES

Compute the total installment cost.

1.

$38 down

$17.50 a month for 12 months

2.

$125 down

$21.75 a month for 18 months

3.

$315 down

$32.40 a month for 2 years

Complete.

	Cash Price	Down Payment	Monthly Payment	Number of Payments	Installment Price	Interest or Finance Charge
4.	$98	$21.84	$7.00	12		
5.	$170	$56.40	$10.60	12		
6.	$256	$60.35	$14.75	15		
7.	$394	$56.98	$21.35	18		
8.	$628	$53.12	$28.14	24		

Solve.

9. A $169 lawn mower can be bought for $41.52 down and $11.75 a month for 12 months. How much could be saved by paying cash?

10. If the installment price is $330 and the down payment is $37.80, what is the monthly payment for 12 monthly installments?

11. See if you can find some examples of installment buying at some of your local stores.

A checking account

You have probably seen many people pay for things by writing checks. In order to write a check you have to have a checking account at a bank. To open a checking account, you give the bank some money (a **deposit**). When some of your money is needed, you write the bank a note (a check) telling them to give the money to you or to someone else.

EXAMPLE.
Mr. Weaver bought a used power saw from Mr. Cooper. He paid for the saw by giving Mr. Cooper this check. The check instructs the Alcancía National Bank to pay Harold Cooper $42.75 from the checking account of John K. Weaver.

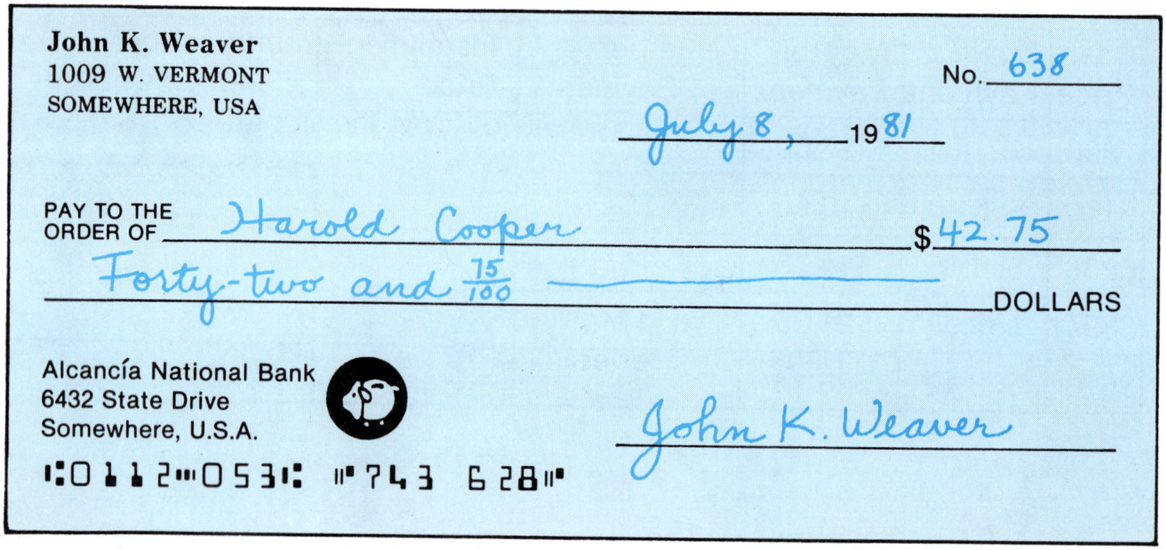
John K. Weaver
1009 W. VERMONT
SOMEWHERE, USA

No. 638

July 8, 1981

PAY TO THE ORDER OF Harold Cooper $42.75

Forty-two and 75/100 DOLLARS

Alcancía National Bank
6432 State Drive
Somewhere, U.S.A.

John K. Weaver

⑆0112⑈0531⑆ ⑈743 628⑈

Before Mr. Cooper can receive the money, he must **endorse** the check. That means that he signs his name on the back of the check. The endorsed check is proof that Harold Cooper received the money (cashed the check).

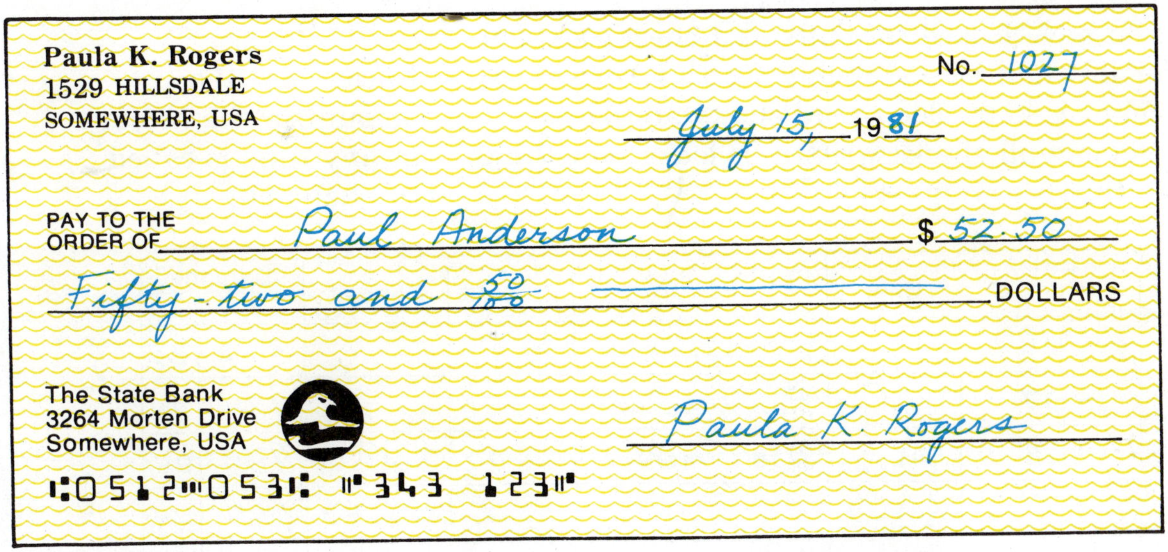
Paula K. Rogers
1529 HILLSDALE
SOMEWHERE, USA

No. 1027

July 15, 1981

PAY TO THE ORDER OF Paul Anderson $ 52.50

Fifty-two and 50/100 DOLLARS

The State Bank
3264 Morten Drive
Somewhere, USA

Paula K. Rogers

⑆0512⑈0531⑆ ⑈343 123⑈

EXERCISES

1. On what date was the check written?
2. To whom was the check written?
3. Who wrote the check?
4. What is the amount of the check?
5. Where does Ms. Rogers have her checking account?
6. On July 16, 1981, Ms. Rogers bought $43.69 worth of groceries at Herb's Food Market. She wrote a check to pay for the groceries. What would the completed check look like?
7. **a.** After writing the $43.69 check for groceries, Ms. Rogers had $227.14 in her account. The next day she deposited $194.78. How much money did she have in her checking account then?

   b. The following morning Ms. Rogers wrote checks for $10.27, $24.32, and $42.05. How much money did she have left in her account then?
8. Try to obtain blank checks from some local banks. If you had a checking account, would you know how to write a check?

More about a checking account

If you have looked carefully at a checkbook you probably noticed a section where a record of the account is kept. The record shows all deposits made and all checks written. It looks something like this:

Notice that the amount in the account at any time is called the **balance.**

Check No.	Date	Pay to	Amount	Deposit	Balance	
					534	23
1506	Apr 5	Keith's Shoes	16.97		517	26
1507	Apr 8	Eastern Tel.	43.20		474	06
1508	Apr 8	Don Smith	15.00		459	06
	Apr 10			206.43	665	49
1509	Apr 15	Southern Electric	32.25		633	24
1510	Apr 17	Cash	40.00		593	24
	Apr 24			75.00	668	24
1511	Apr 30	Libby's Lumber	56.27		611	97

EXERCISES

Refer to the record above for questions 1 – 6.

1. What was the balance after the check was written on April 15?
2. Who received the check numbered 1508?
3. On what dates were deposits made?
4. What was the amount of the check written to Libby's Lumber?
5. What was the amount of the check written on April 15?
6. On April 17 a check was made payable to "Cash." What do you think that means?

Compute each balance.

	Check No.	Date	Pay to	Amount	Deposit	Balance	
						611	97
7.	1512	May 1	Davis Hardware	46.25			
8.	1513	May 14	Bob's Pharmacy	15.00			
9.	1514	May 18	Anchorage Oil Co	138.44			
10.		May 21			120.00		
11.	1515	May 21	Cash	65.00			
12.	1516	May 26	H. A. Thomas	83.75			
13.		May 31			142.58		
14.	1517	June 2	Bell Telephone	38.74			

15. Some banks charge a service charge for a checking account. For example, they might charge \$.75 a month plus \$.10 a check. At that rate, what would be the service charge for the account above during May?

16. Other banks do not charge a service charge if the account always has a balance of, say, \$300 or more. Would there be a service charge on the account above during the month of May?

17. Find out if your local banks have a service charge for checking accounts. If they do, find out how they compute the service charge.

CHAPTER CHECKUP

Compute the price. [pages 260–263, 266–267]

1. 1 pair of socks
2. 5 pairs of socks
3. 1 pair of tennis shoes
4. 2 sweatshirts
5. 3 pair of socks plus a 5% sales tax (Use the table on page 266.)

Solve. [pages 264–265, 268–273]

6. A store sells an 8-oz cola for 35¢ and a 10-oz cola for 42¢. Which is the better buy?

7. An FM radio is priced at $64.95. It can be bought on the installment plan for $18.50 down and 6 monthly payments of $9.20 each. How much cheaper is paying cash?

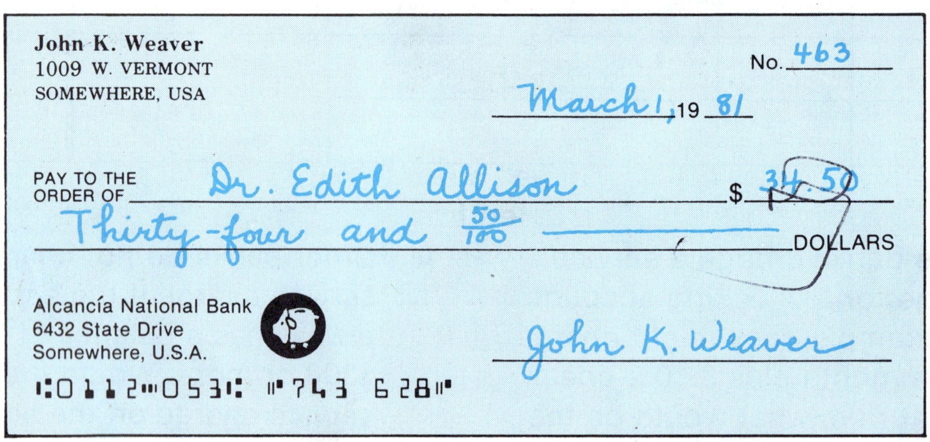
John K. Weaver
1009 W. VERMONT
SOMEWHERE, USA

No. 463

March 1, 19 81

PAY TO THE ORDER OF Dr. Edith Allison $ 34.50

Thirty-four and 50/100 DOLLARS

Alcancía National Bank
6432 State Drive
Somewhere, U.S.A.

John K. Weaver

⑆0112⑈0531⑆ ⑈743 628⑈

8. What is the check number?
9. To whom was the check written?
10. Who wrote the check?
11. What is the amount of the check?
12. Before writing the check, Mr. Weaver's balance was $275.19. What was his balance after writing the check?

Project

Stocks	Sales 100s	High	Low	Close	Net Chg.
AlldCh	349	43⅜	42⅝	43¼	+¼
AlldMnt	11	12½	12⅜	12½	+⅛
AlldPd	6	12⅞	12¾	12⅞	..
AlldStr	86	23⅛	22⅞	23⅛	–⅛
AlldSup	58	2½	2⅜	2½	..
AllisCh	366	25¼	24⅞	24 7⁄3	–¾
AllrAu	1	11	11	11	..
AlphPrt	65	16½	15⅞	16½	+⅝
→ Alcoa	626	44½	43⅝	44¼	+⅛
AmlSug	46	26	24½	26	+1½
Amax	304	37	35¾	35⅞	–⅞
AMBAC	109	35⅜	34⅛	35¼	+1⅛
Amcord	325	13⅞	13½	13⅞	..
Amrce	25	22½	22⅛	22⅛	–⅛
Amrc	1	38	38	38	+½
AHess	578	29½	29	29⅜	+¼
AHes	118	66½	65¼	66½	+½
AAirFilt	46	19¼	18⅞	18⅞	..
AmAir	593	10	9⅝	9⅞	–⅛
AmAir	137	3⅝	3⅜	3½	–⅛
AAir	95	21⅜	21¼	21⅜	+⅛
ABaker	16	16¼	16⅛	16⅛	..
ABrnds	92	43¾	43⅛	43½	+⅜
ABrd	14	24⅛	23⅞	24⅛	+⅛
ABdcst	1257	42¾	41⅝	42¼	+⅛

Refer to the stock quotations above to answer questions 1–3.

1. The day's high for 1 share of Alcoa (Aluminum Co. of America) was $44\frac{1}{2}$. What was the day's low? What was the price when the stock exchange closed?

2. The number in the "Net Change" column tells how today's closing price differed from yesterday's closing price. What was the net change of Alcoa? What was yesterday's closing price?

3. The number in the "Sales" column tells how many *hundred shares* were sold. How many shares of Alcoa were sold?

4. **a.** Pick out some companies and keep daily records of the closing prices of their stocks for a week.
   b. Make a line graph of the daily closing prices of your stocks.

CHAPTER REVIEW

1. a. What is the exact price of 1  pencil?
   b. What would you have to pay for 1 pencil? *Hint:* Round up to the next cent.

2. What would you have to pay for 4 pencils? *Hint:* Add the price of 3 pencils and 1 pencil.

3. To find the sale price of the book
   a. Find $\frac{1}{3}$ of \$2.35.
   b. Round to the nearest cent for the amount of discount.
   c. Subtract the discount from \$2.35.

4. Which is the better buy? *Hint:* Compute the price per ounce of the candy in each box.

5. What is the number of the check?
6. On what date was the check written?
7. Who wrote the check?
8. How much is the check for?

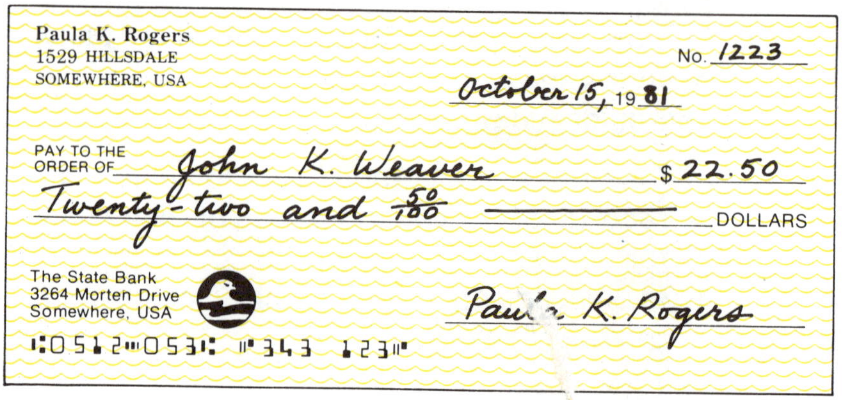

Paula K. Rogers
1529 HILLSDALE
SOMEWHERE, USA

No. 1223

October 15, 19 81

PAY TO THE ORDER OF John K. Weaver $ 22.50

Twenty-two and $\frac{50}{100}$ DOLLARS

The State Bank
3264 Morten Drive
Somewhere, USA

Paula K. Rogers

CHAPTER CHALLENGE

When people have extra money they can put it in a savings account and earn **interest.** The amount of interest earned depends on the amount of money in the account (the **principal**), the percent of interest paid (the **rate**), and the length of time that the money is in the account (the **time**). The amount of interest may be computed by using this formula.

$$I = p \times r \times t$$

INTEREST → I, PRINCIPAL → p, RATE → r, TIME → t

EXAMPLE. How much interest would be paid on \$120 in 3 months if the bank pays a yearly rate of 6% ($\frac{6}{100}$) and computes the interest every 3 months ($\frac{1}{4}$ year)?

$I = p \times r \times t$

$I = \$120 \times \frac{6}{100} \times \frac{1}{4}$

$I = \$1.80$

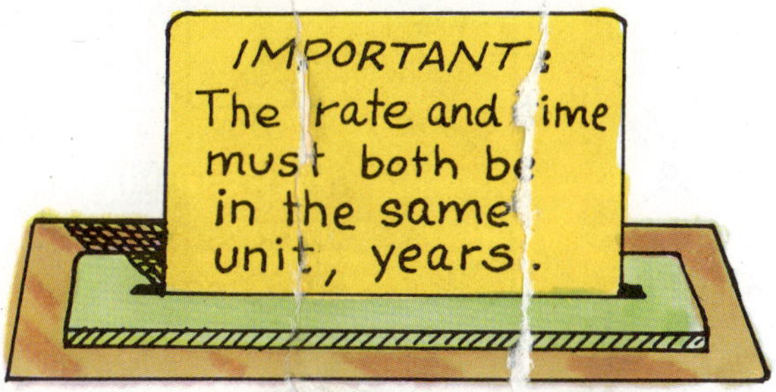

Compute the interest. Assume that the savings bank computes the interest every 3 months.

1. Principal: \$300
   Rate: 6%
   Time: 3 months
2. Principal: \$240
   Rate: 5%
   Time: 3 months
3. Principal: \$340
   Rate: 7%
   Time: 3 months
4. Principal: \$280
   Rate: 6%
   Time: 3 months
5. Add the principal and interest in exercise 4.
6. Using the same rate and time as exercise 4, compute the interest for the sum in exercise 5. Round your answer to the nearest cent.

MAJOR CHECKUP
Standardized Format

Choose the correct letter.

1. The 6 in 586,492,173 stands for
   a. 6 billion
   b. 6 thousand
   c. 6 million
   d. none of these

2. Which angle is a 90° angle?
   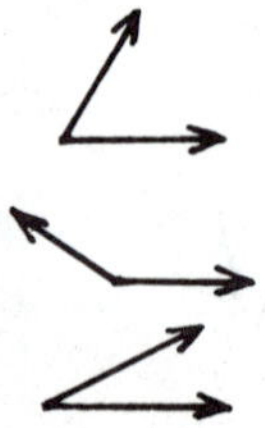
   a.
   b.
   c.
   d. none of these

3. Add.
$$\begin{array}{r} 9\frac{5}{6} \\ +6\frac{2}{3} \\ \hline \end{array}$$
   a. $15\frac{1}{2}$
   b. $16\frac{1}{2}$
   c. $15\frac{1}{6}$
   d. none of these

4. Subtract.
$$\begin{array}{r} 16\frac{1}{3} \\ -7\frac{3}{4} \\ \hline \end{array}$$
   a. $9\frac{5}{12}$
   b. $8\frac{7}{12}$
   c. $8\frac{5}{12}$
   d. none of these

5. $\frac{5}{9} \times \frac{3}{4} =$
   a. $\frac{12}{5}$
   b. $\frac{20}{27}$
   c. $\frac{5}{12}$
   d. none of these

6. $\frac{3}{8} \div \frac{5}{4} =$
   a. $\frac{3}{10}$
   b. $\frac{10}{3}$
   c. $\frac{15}{32}$
   d. none of these

7. .038 + 2.65 =
   a. 3.03
   b. .303
   c. 2.688
   d. none of these

8. 52.6 − 3.48 =
   a. 1.74
   b. 49.12
   c. 17.4
   d. none of these

9. Multiply.
$$\begin{array}{r} 32.8 \\ \times .75 \\ \hline \end{array}$$
   a. 2.46
   b. 24.6
   c. .246
   d. none of these

10. Divide.
$$.83\overline{)4.6563}$$
   a. 5.61
   b. 56.1
   c. .561
   d. none of these

11. 3.8 m = ? cm
   a. 38
   b. .038
   c. 380
   d. none of these

12. Find the surface area.
   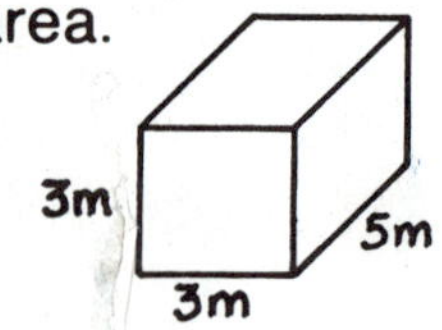

   a. 45 square m
   b. 48 square m
   c. 63 square m
   d. none of these

11 Ratio and Percent

READY OR NOT ?

What fraction of the stamps are blue?

1\.

2\.

3\.

4\.

5\.

Ratios

John and a friend collect returnable bottles. For each 3 bottles they return they are paid 7¢. The **ratio** of bottles to cents is 3 to 7. The ratio can be written as a fraction.

$\frac{3}{7}$ bottles to cents

Read as "three to seven."

The ratio of cents to bottles is:

$\frac{7}{3}$ cents to bottles

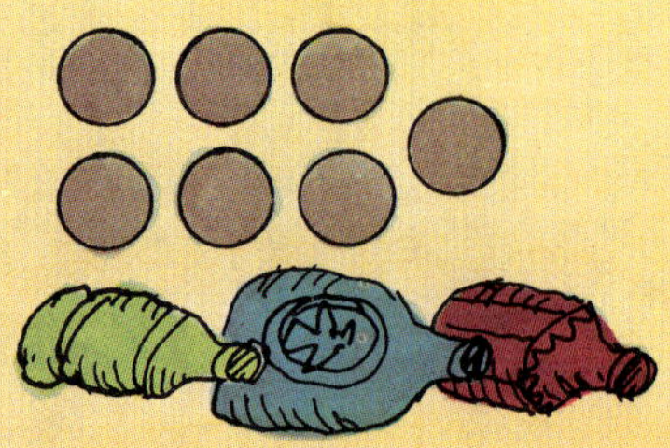

EXERCISES

1. A certain cookie recipe calls for 3 cups of brown sugar and 4 cups of flour.

What is the ratio of

a. sugar to flour?

b. flour to sugar?

2. Give the ratio.

   a. chocolates to grams
   b. grams to chocolates
   c. chocolates to cents
   d. cents to chocolates
   e. grams to cents
   f. cents to grams

3. Give the ratio.

   a. pears to kilograms
   b. cents to pears
   c. cents to grams
   d. pears to grams
   e. pears to cents
   f. kilograms to pears

4. Give each ratio.

   a. length of square B to length of square A.
   b. area of square B to area of square A.

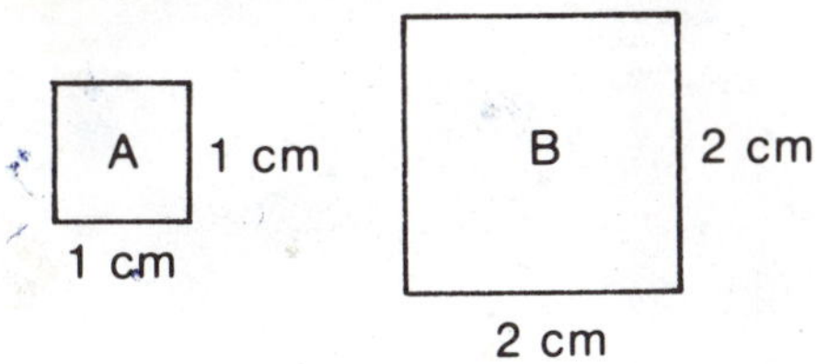

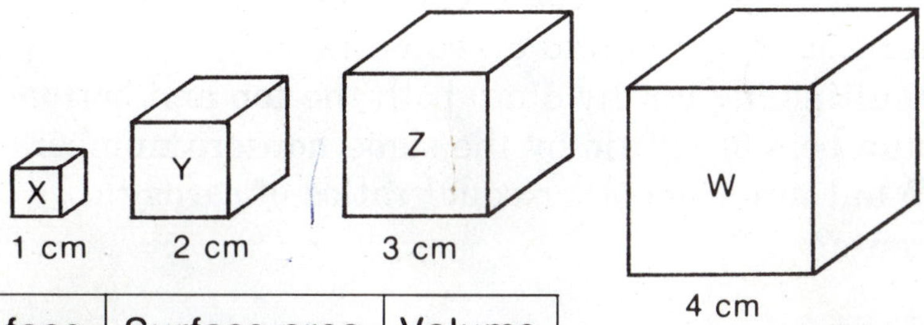

Copy and fill in the table.

	Ratio	Edge	Area of 1 face	Surface area	Volume
5.	Y to X				
6.	Z to X				
7.	W to X				

Equal ratios

A package of 3 baseball cards costs 5¢. Below are some ratios of cards to cents. They are **equal ratios.**

cards to cents $\frac{3}{5} = \frac{6}{10} = \frac{9}{15} = \frac{12}{20}$

The equal ratios tell you that you can buy 3 cards for 5¢, 6 cards for 10¢, 9 cards for 15¢, or 12 cards for 20¢. You can find an equal ratio by multiplying (or dividing) both the top and bottom numbers of a ratio by the same nonzero number. What are some other equal ratios of cards to cents?

EXERCISES

Copy and complete the equal ratios.

1. $\frac{3}{20} = \frac{?}{40} = \frac{?}{60} = \frac{?}{80} = \frac{?}{100}$

2. balls to cents $\frac{2}{15} = \frac{?}{30} = \frac{6}{?} = \frac{?}{60} = \frac{10}{?}$

Copy and complete.

3. $\frac{1}{2} \overset{\times 2}{\underset{\times 2}{=}} \frac{?}{4}$ **4.** $\frac{1}{3} \overset{\times 4}{\underset{\times 4}{=}} \frac{?}{12}$ **5.** $\frac{2}{3} \overset{\times 3}{\underset{\times 3}{=}} \frac{6}{?}$ **6.** $\frac{3}{8} \overset{\times 2}{\underset{\times 2}{=}} \frac{6}{?}$

7. $\frac{10}{12} \overset{\div 2}{\underset{\div 2}{=}} \frac{5}{?}$ **8.** $\frac{9}{6} \overset{\div 3}{\underset{\div 3}{=}} \frac{?}{2}$ **9.** $\frac{5}{20} \overset{\div 5}{\underset{\div 5}{=}} \frac{1}{?}$ **10.** $\frac{12}{16} \overset{\div 4}{\underset{\div 4}{=}} \frac{?}{4}$

11. $\frac{1}{5} = \frac{?}{15}$ **12.** $\frac{5}{8} = \frac{25}{?}$ **13.** $\frac{5}{9} = \frac{?}{18}$ **14.** $\frac{4}{3} = \frac{12}{?}$

15. $\frac{14}{4} = \frac{?}{2}$ **16.** $\frac{20}{12} = \frac{5}{?}$ **17.** $\frac{9}{4} = \frac{27}{?}$ **18.** $\frac{14}{16} = \frac{?}{8}$

Solve.

To get a certain shade of green, a paint store mixes 2 parts of blue paint with 3 parts of yellow paint.

19. How many liters of yellow are needed for 2 liters of blue?

20. How many liters of blue are needed for 6 liters of yellow?

21. How much of each color would you use to make 30 liters of green paint?

Problem solving

If you need to, list some equal ratios to solve the problem.

1. The ratio of adult tickets to student tickets sold by one class was 3 to 5. If they sold 30 adult tickets, how many student tickets did they sell?

2. The tables were placed in groups of two. Each group used 14 chairs. How many chairs were needed if there were 18 tables in all?

3. The art class made 3 center-pieces for each 2 tables. How many centerpieces were made for the 18 tables?

4. To decorate the edges of the tables, 2 packages of crepe ribbon were needed for 3 tables. How many tables would 8 packages of ribbon decorate?

5. The recipe for punch called for 5 parts grape juice to 2 parts ginger ale. How much ginger ale should they mix with 25 liters of grape juice?

6. The chili recipe called for 3 kilograms of ground beef to 4 kilograms of drained kidney beans. How much ground beef should be added to 12 kilograms of kidney beans?

7. Some volunteers baked brownies for the chili supper. They planned on 240 people attending the supper and baked 3 brownies for every 2 people. How many brownies did they bake?

8. For every 3 tickets sold, 25¢ was put in a door-prize fund. How much should have been in the fund if 252 tickets were sold?

Percent

Percent means per hundred. It is a ratio in which the second number (or denominator) is 100.

Alice bought 100 stamps for her collection. 23 of the stamps were Canadian.

The ratio of Canadian stamps to all stamps is $\frac{23}{100}$. We can also write the ratio as a percent:

$$23\%$$

Read as "23 percent."

EXERCISES

What percent of the square is shaded?

1. 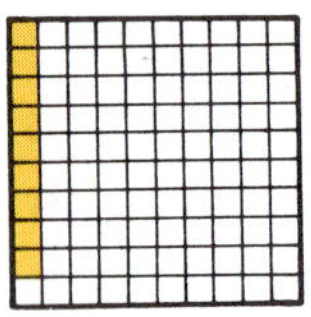

2. 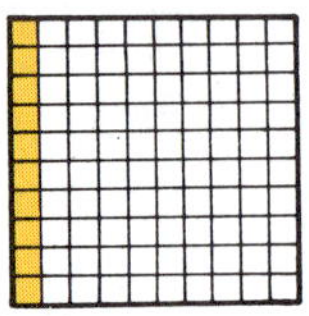

3. 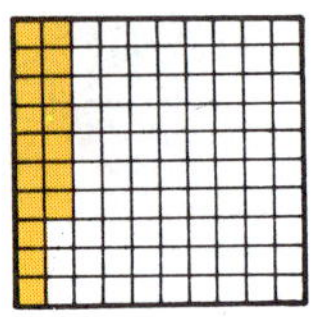

4. 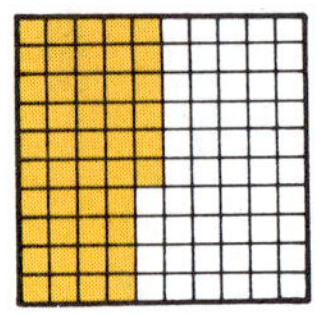

5. 

6. 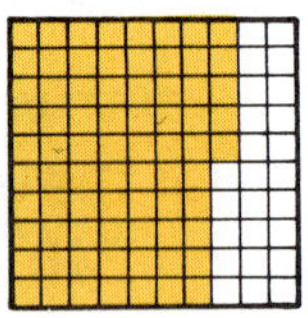

7. 

8. 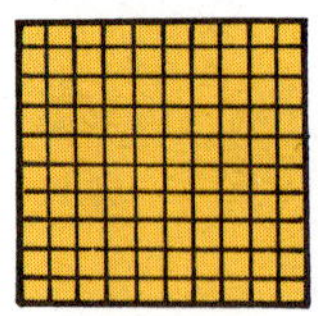

Write as percents.

9. $\frac{13}{100}$
10. $\frac{5}{100}$
11. $\frac{20}{100}$
12. $\frac{37}{100}$
13. $\frac{62}{100}$
14. $\frac{85}{100}$
15. $\frac{95}{100}$
16. $\frac{100}{100}$
17. $\frac{150}{100}$
18. $\frac{235}{100}$

Solve.

19. 100 students
42 ride bicycles to school.
What percent ride bicycles to school?

20. 100-question test
Got 82 correct.
What percent were answered correctly?

21. Had $1.
Spent 56¢.
What percent of the money was spent?

22. Had $1.
Spent 39¢.
What percent of the money was left?

A sixth-grade class kept a record of the number of people in each car that passed their school. Then they made this graph.

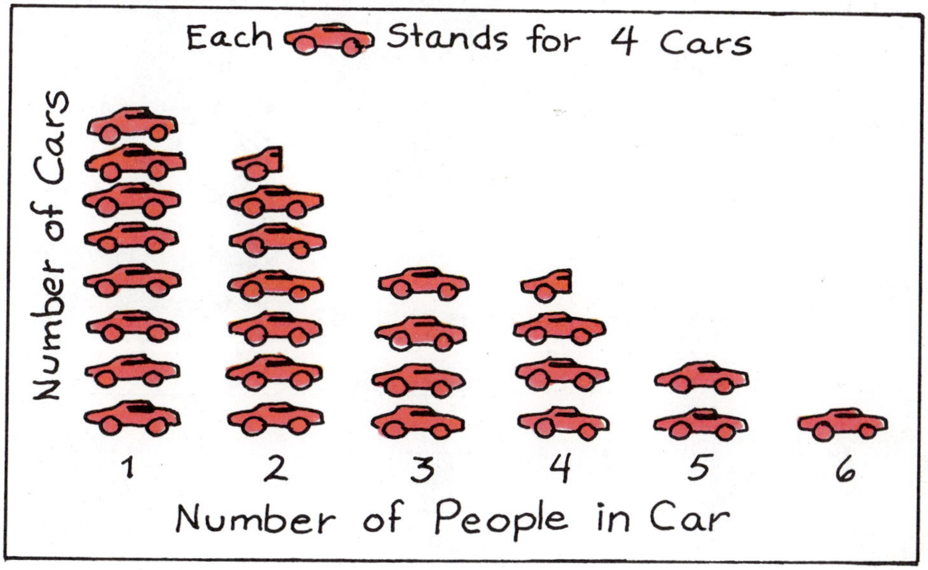

23. How many cars passed their school? Careful, the answer is not 25!

24. How many cars had only 1 person?

25. What percent of the cars had 6 people?

26. What percent of the cars had 4 or more people?

Keeping Skills Sharp

1. $\begin{array}{r} 242 \\ \times .05 \\ \hline \end{array}$

2. $\begin{array}{r} 321 \\ \times .06 \\ \hline \end{array}$

3. $\begin{array}{r} 532 \\ \times .08 \\ \hline \end{array}$

4. $\begin{array}{r} 21.6 \\ \times .24 \\ \hline \end{array}$

5. $\begin{array}{r} 39.7 \\ \times 3.6 \\ \hline \end{array}$

More about percent

Player	Number of hits	Number of times at bat
Clark	10	50
Davis	12	48
Dennis	9	36
Ford	12	48
Helms	18	50
Jacobson	11	55
Lopez	14	50
Lowry	18	60
Morris	16	40
Peterson	20	50

Is Clark a better hitter than Davis? We can use percents to compare the two hitters.

Clark's "hits" to "at bats" ratio is $\frac{10}{50}$. To change the ratio to a percent, we can think of an equal ratio having a denominator of 100.

$$\frac{10}{50} = \frac{20}{100} = 20\%$$

Clark got a hit 20% of the times that he went to bat.

To find Davis's hitting percentage we also work with equal ratios.

$$\frac{12}{48} = \frac{1}{4} = \frac{25}{100} = 25\%$$

Notice that we reduced $\frac{12}{48}$ before getting an equal ratio having a denominator of 100.

Who is the better hitter?

EXERCISES

Write as a percent. ***Hint:*** **Think of an equivalent fraction having a denominator of 100.**

1. $\frac{1}{2}$ 2. $\frac{1}{10}$ 3. $\frac{1}{4}$ 4. $\frac{3}{20}$ 5. $\frac{3}{10}$ 6. $\frac{7}{4}$

7. $\frac{7}{10}$ 8. $\frac{3}{2}$ 9. $\frac{3}{4}$ 10. $\frac{8}{25}$ 11. $\frac{5}{4}$ 12. $\frac{7}{20}$

13. $\frac{28}{200}$ 14. $\frac{42}{200}$ 15. $\frac{12}{300}$ 16. $\frac{3}{15}$ 17. $\frac{8}{40}$ 18. $\frac{9}{60}$

19. $\frac{16}{40}$ 20. $\frac{18}{60}$ 21. $\frac{48}{80}$ 22. $\frac{16}{50}$ 23. $\frac{6}{24}$ 24. $\frac{27}{45}$

Solve. Refer to the table on page 288.

25. What is Helms's hitting percent?
26. What is Lopez's hitting percent?
27. Who has the greater hitting percent, Ford or Lopez?

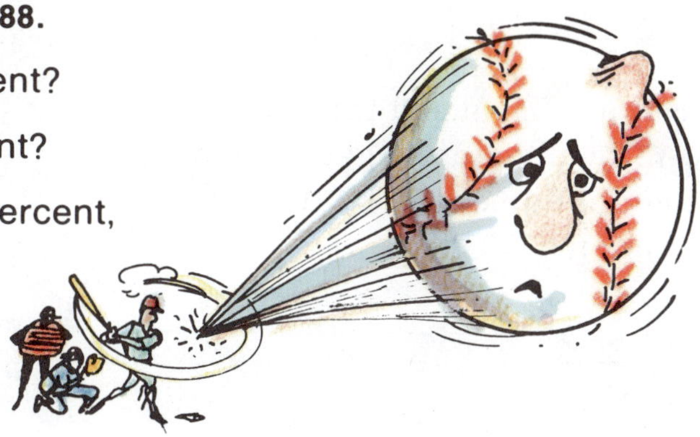

Solve.

28. 25-kilometer hike
16 kilometers hiked before lunch
What percent was hiked before lunch?

29. \$150 bicycle
\$60 saved
What percent has been saved?

Mary got \$20 for her birthday. She bought these items:

What percent of the money did she

30. spend for the record?
31. spend for the tennis racket?
32. not spend?

Decimals and percent

You have changed some ratios to percents by finding an equal ratio that has a denominator of 100.

$$\frac{3}{4} = \frac{75}{100} = 75\%$$

We can also change a ratio to a percent by dividing to get a decimal. Divide the numerator by the denominator.

$$4\overline{)3.00} = .75$$

$$\begin{array}{r} 3.00 \\ -2\,8 \\ \hline 20 \\ -20 \\ \hline 0 \end{array}$$

$$\frac{75}{100} \qquad 75\%$$

Sometimes we use a mixed number to write a percent.

$$\frac{3}{8} = ?$$

$$8\overline{)3.00} = .37\tfrac{1}{2}$$

$$\begin{array}{r} 3.00 \\ -2\,4 \\ \hline 60 \\ -56 \\ \hline 4 \end{array}$$

$$\frac{37\frac{1}{2}}{100} \qquad 37\tfrac{1}{2}\%$$

EXERCISES

Change each decimal to a percent.

1. .13	**2.** .24	**3.** .43	**4.** .75	**5.** .80
6. .92	**7.** .64	**8.** 1.25	**9.** 2.42	**10.** 3.00
11. .5 (.50)	**12.** .4	**13.** .9	**14.** .08	**15.** .01

Change to a percent. ***Hint:*** **Divide numerator by denominator.**

16. $\frac{1}{3}$	**17.** $\frac{2}{3}$	**18.** $\frac{1}{6}$	**19.** $\frac{1}{12}$	**20.** $\frac{5}{6}$
21. $\frac{5}{8}$	**22.** $\frac{9}{4}$	**23.** $\frac{1}{8}$	**24.** $\frac{5}{12}$	**25.** $\frac{5}{9}$
26. $\frac{7}{4}$	**27.** $\frac{7}{6}$	**28.** $\frac{6}{5}$	**29.** $\frac{5}{16}$	**30.** $\frac{1}{9}$
31. $\frac{11}{12}$	**32.** $\frac{7}{16}$	**33.** $\frac{7}{12}$	**34.** $\frac{7}{8}$	**35.** $\frac{2}{9}$

Copy and complete.

	36.	**37.**	**38.**	**39.**	**40.**
Fraction	$\frac{1}{2}$	$\frac{2}{3}$			
Decimal	.5		.16	.64	
Percent	50%				120%

41. "To change a decimal to a percent, I move the decimal point __?__ places to the right and write a percent sign."

42. "To change a percent to a decimal, remove the percent sign and move the decimal point 2 places to the __?__."

Four students took a 16-problem math quiz. The teacher wrote the number correct in red.

43. Compute the percent that each student got correct and round to the nearest whole-number percent.

Finding a percent of a number

Ann took 24 pictures. 75% of the pictures were of her new puppy.

Here are two ways to find how many pictures were of the puppy.

$$75\% \text{ of } 24 = \underline{\quad ? \quad}$$

1. Change the percent to a fraction and multiply.

$$75\% = \frac{75}{100} = \frac{3}{4}$$

$$75\% \text{ of } 24 = \frac{3}{4} \times 24 = 18$$

$$\frac{3}{\cancel{4}_1} \times \cancel{24}^{6} = 18$$

2. Change the percent to a decimal and multiply.

$$75\% = \frac{75}{100} = .75$$

$$75\% \text{ of } 24 = .75 \times 24 = 18$$

$$\begin{array}{r} 24 \\ \times .75 \\ \hline 120 \\ 168 \\ \hline 18.00 \end{array}$$

EXERCISES

Compute.

1. 20% of 75
2. 25% of 96
3. 50% of 72
4. 30% of 80
5. 35% of 60
6. 60% of 95
7. 8% of 92
8. 12% of 200
9. 18% of 56
10. 36% of 70
11. 100% of 36
12. 125% of 62

A **circle graph** can be used to show percents. The circle graph shows how a youth club raised $240 to attend a summer camp.

13. What is the sum of the percents shown on the circle graph?
14. Which activity raised the most money?
15. Which activity raised the least money?
16. How much money was raised by the car wash?
17. How much was raised by the chili supper?
18. How much more money came from the bake sale than from the candy sale?
19. Which fund-raising activity raised $60?
20. What percent of the money was not a gift?

Keeping Skills Sharp

1. 74)4.218
2. 43)36.12
3. 66)5.610
4. 91)409.5
5. 46)3.082
6. 5.5)37.40
7. 1.9)15.77
8. 7.1)29.82
9. .82)508.4
10. 5.4)3.888
11. 13.2)3.168
12. 14.6)48.18
13. 46.3)50.93
14. 2.78)13.066
15. 25.5)30.855
16. 38.7)15.093
17. 9.16)52.212

Practice exercises

Write as percents.

1. **a.** $\frac{1}{5}$ **b.** $\frac{2}{5}$ **c.** $\frac{3}{5}$ **d.** $\frac{4}{5}$

2. **a.** $\frac{1}{4}$ **b.** $\frac{2}{4}$ **c.** $\frac{3}{4}$ **d.** $\frac{4}{4}$

3. **a.** $\frac{1}{8}$ **b.** $\frac{2}{8}$ **c.** $\frac{3}{8}$ **d.** $\frac{4}{8}$

4. **a.** $\frac{1}{6}$ **b.** $\frac{2}{6}$ **c.** $\frac{3}{6}$ **d.** $\frac{4}{6}$

Compute.

5. 20% of 40

6. 50% of 38

7. 25% of 56

8. 18% of 100

9. 42% of 50

10. 64% of 200

11. 58% of 75

12. 76% of 93

13. 92% of 42

14. 100% of 96

15. 150% of 84

16. 175% of 64

Solve.

17. Out of a class of 32 students, 4 were absent. What percent of the class was absent?

18. Nine hundred sixty students attend Thomas Elementary School. One day 5% of the students were absent. How many were absent?

19. A sporting goods store reduced all prices 15% for a sale. What is the sale price of a $37 rod and reel?

20. Sarah's daily calorie intake was 2860. Her doctor told her to reduce her calorie intake by 20%. How many calories a day could she have then?

Skill Game

1. Divide the class into two teams, team A and team B.
2. A player from team A goes to the chalkboard and writes a fraction, a decimal, or a percent problem.
3. A volunteer from team B goes to the chalkboard and works the problem. If the work is correct, team B earns 1 point. If incorrect, the player who wrote the problem may earn 1 point by working the problem correctly.
4. Next, a player from team B goes to the chalkboard and makes up a problem for a volunteer from team A. Play continues until each player has worked on a problem. The team with the greater number of points wins the game.

CHAPTER CHECKUP

Give the ratio. [280–281]

1. cookies to cents
2. cookies to grams
3. cents to grams

Copy and complete. [pages 282–285]

4. $\frac{2}{3} = \frac{?}{9}$ **5.** $\frac{4}{5} = \frac{?}{10}$ **6.** $\frac{3}{4} = \frac{12}{?}$ **7.** $\frac{5}{8} = \frac{20}{?}$

8. $\frac{16}{12} = \frac{?}{3}$ **9.** $\frac{24}{18} = \frac{4}{?}$ **10.** $\frac{20}{4} = \frac{?}{1}$ **11.** $\frac{32}{24} = \frac{4}{?}$

Write as a percent. [pages 286–289, 294–295]

12. $\frac{17}{100}$ **13.** $\frac{6}{100}$ **14.** $\frac{60}{100}$ **15.** $\frac{135}{100}$ **16.** $\frac{249}{100}$

17. $\frac{1}{2}$ **18.** $\frac{3}{4}$ **19.** $\frac{18}{45}$ **20.** $\frac{18}{40}$ **21.** $\frac{13}{25}$

Change each decimal to a percent. [pages 290–291, 295]

22. .26 **23.** .82 **24.** .08 **25.** .01 **26.** .93

27. 1.54 **28.** 2.00 **29.** .4 **30.** .1 **31.** 2.6

Change to a percent. [pages 290–291, 294–295]

32. $\frac{1}{4}$ **33.** $\frac{1}{3}$ **34.** $\frac{2}{3}$ **35.** $\frac{5}{8}$ **36.** $\frac{7}{12}$ **37.** $\frac{9}{8}$ **38.** $\frac{11}{6}$ **39.** $\frac{2}{7}$

Solve. [pages 292–294]

40. What is the sale price?

41. Alex got 80% of the test problems correct. There were 65 problems in all. How many did he get correct?

42. Connie has 150 coins in her collection. Thirty-six percent of the coins are dimes. How many dimes does she have?

Project

Architects work with a ratio when they make a drawing. For example, in the scale drawing shown, $\frac{1}{8}$" on the drawing represents 1' of actual distance.

1. Measure the length and width of your classroom to the nearest meter.
2. Draw a floor plan of your classroom. Let 2 cm on your drawing represent 1 meter in your classroom.
3. Determine the size and location of some objects (desks, tables, etc.) in your classroom. Include them in your floor plan.

CHAPTER REVIEW

1. Copy and complete the equal ratios of cartons to bottles.

cartons to bottles $\frac{1}{6} = \frac{?}{12} = \frac{3}{?} = \frac{?}{24}$

Percent means per hundred

$\frac{12}{100} = 12\%$

Write as a percent.

2. $\frac{15}{100}$ 3. $\frac{28}{100}$ 4. $\frac{70}{100}$ 5. $\frac{95}{100}$

6. $\frac{6}{100}$ 7. $\frac{9}{100}$ 8. $\frac{150}{100}$ 9. $\frac{275}{100}$

Complete.

	10.	11.	12.	13.	14.	15.
Fraction	$\frac{1}{2}$	$\frac{1}{4}$	$\frac{3}{4}$	$\frac{2}{5}$	$\frac{3}{5}$	$\frac{7}{10}$
Fraction with denominator of 100	$\frac{50}{100}$					
Percent	50%					

$\frac{1}{3} = \underline{?}\%$

$$\begin{array}{r} .33\frac{1}{3} \\ 3\overline{)1.00} \\ -9 \\ \hline 10 \\ -9 \\ \hline 1 \end{array}$$

$\frac{1}{3} = 33\frac{1}{3}\%$

Change to a percent.

16. $\frac{1}{6}$ 17. $\frac{1}{8}$ 18. $\frac{3}{8}$ 19. $\frac{5}{8}$

20. $\frac{5}{6}$ 21. $\frac{7}{8}$ 22. $\frac{7}{16}$ 23. $\frac{5}{12}$

27% of 62 = ?

$$\begin{array}{r} 62 \\ \times .27 \\ \hline 434 \\ 124 \\ \hline 16.74 \end{array}$$

27% of 62 = 16.74

Compute.

24. 10% of 40 25. 20% of 45

26. 25% of 64 27. 75% of 96

28. 28% of 60 29. 64% of 128

CHAPTER CHALLENGE

Percy the percent man built this computer. To operate the computer, Percy dials the numbers in the first two windows and then turns the switch to "ON." The computer then flashes the answer in the third window.

One day the door to the middle window could not be raised. See if you can determine the second number that Percy dialed.

Hint: You can divide to find the number that was multiplied.

Find the missing number.

1. 50% of ▢ = 10
2. 25% of ▢ = 6
3. 20% of ▢ = 9
4. 10% of ▢ = 5
5. 75% of ▢ = 12
6. 100% of ▢ = 8
7. 40% of ▢ = 16
8. 60% of ▢ = 18
9. ★ 35% of ▢ = 14
10. ★ 65% of ▢ = 31.2

MAJOR CHECKUP
Standardized Format

Choose the correct letter.

1. Add.

$$\begin{array}{r} 3\frac{1}{3} \\ +2\frac{3}{4} \\ \hline \end{array}$$

a. 6
b. $6\frac{1}{12}$
c. $5\frac{1}{12}$
d. none of these

2. Subtract.

$$\begin{array}{r} 8\frac{1}{8} \\ -4\frac{1}{2} \\ \hline \end{array}$$

a. $3\frac{5}{8}$
b. $4\frac{5}{8}$
c. $4\frac{3}{8}$
d. none of these

3. $1\frac{3}{4} \div 2\frac{1}{3} =$

a. $\frac{3}{4}$
b. $\frac{4}{3}$
c. $1\frac{1}{6}$
d. none of these

4. $2\frac{1}{2} \div 1\frac{1}{2} =$

a. $\frac{3}{5}$
b. $\frac{5}{3}$
c. 2
d. none of these

5. Which number is greatest?

a. 52.6138
b. 526.138
c. 52613.8
d. 5261.38

6. 25.343 rounded to the nearest hundredth is

a. 25.3
b. 25.35
c. 25.34
d. none of these

7. 6 − 4.25 =

a. 1.75
b. 2.25
c. 2.75
d. none of these

8. Multiply.

$$\begin{array}{r} 5.26 \\ \times 34.1 \\ \hline \end{array}$$

a. 178.166
b. 179.366
c. 373.56
d. none of these

9. Change $\frac{5}{8}$ to a decimal.

a. 1.6
b. .675
c. .625
d. none of these

10. Divide.

$.48\overline{)10.32}$

a. 2.15
b. 21.5
c. 215
d. none of these

11. The area of the rectangle is

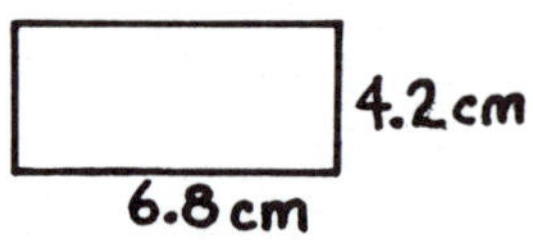

a. 28.56 square centimeters
b. 11 square centimeters
c. 22 square centimeters
d. none of these

12. What is the price of 2?

a. 92¢
b. 95¢
c. 93¢
d. none of these

12
Adding and Subtracting Integers

Positive and negative numbers

Imagine small objects that have electric charges. There are two kinds of charges—one is called **positive** and the other is called **negative.**

These charges are opposites. A positive charge and a negative charge together have no charge, or a 0 charge.

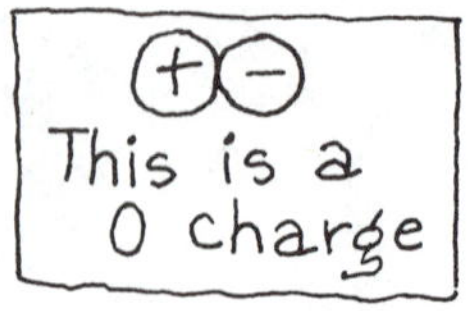

The total charge in this container is $^{+}4$ because there are 4 more positive charges than negative charges.

Here are other examples.

Positive and negative numbers are also used in other situations where opposites are involved.

EXAMPLE 1. Earning and spending money. $^{+}4$ stands for 4 dollars earned and $^{-}7$ stands for 7 dollars spent.

EXAMPLE 2. Traveling on an east-west road. $^{+}6$ stands for driving 6 kilometers east and $^{-}3$ stands for driving 3 kilometers west.

EXERCISES

Give the total charge in each container.

1. 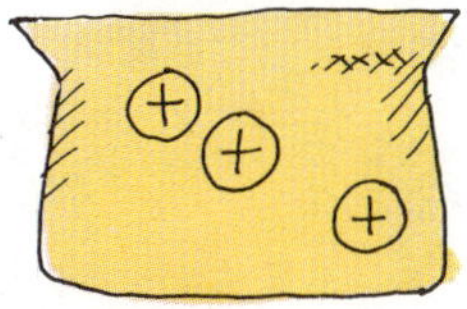

2. 

3. 

4. 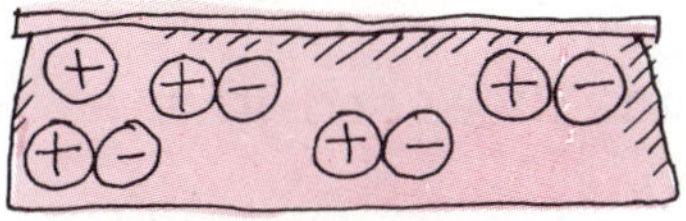

5. 

6. 

7. 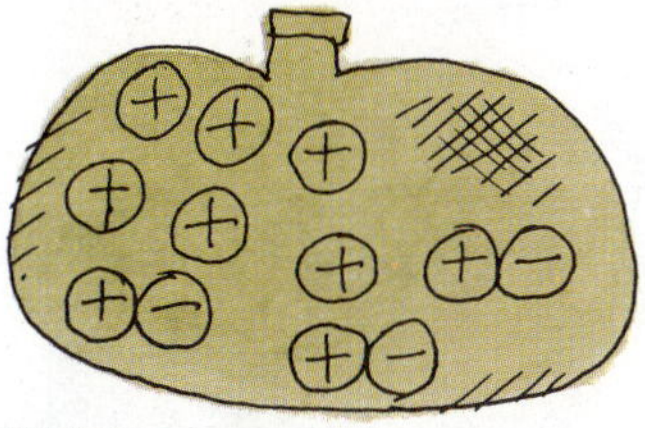

8. 

9. 

Answer.

10. If $^{+}8$ stands for earning 8 cents, what does $^{-}7$ stand for?

11. If $^{-}3$ stands for a trip of 3 kilometers south, what does $^{+}6$ stand for?

12. If $^{+}15$ stands for gaining 15 yards in football, what does $^{-}6$ stand for?

13. If $^{-}5$ stands for going down 5 floors in an elevator, what number stands for going up 8 floors?

14. If $^{-}28$ stands for a trip of 28 kilometers west, what number stands for a trip of 47 kilometers east?

What is the charge in each container? What is the total charge when the charges in the two containers are poured together?

15. 

16. 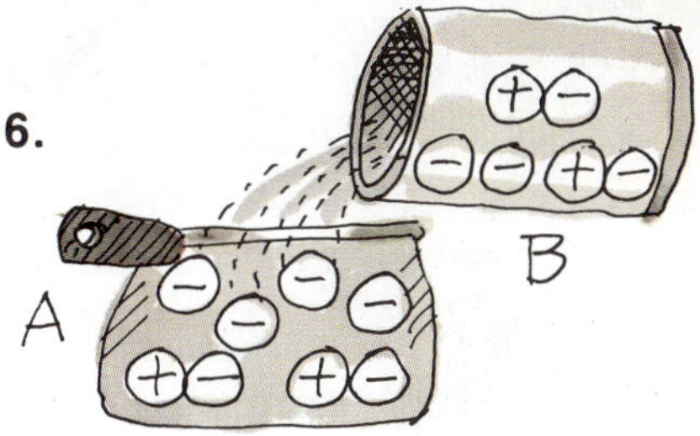

Adding integers

The numbers you used to stand for electrical charges in containers are called **integers.** To understand how to add integers, you can think about putting charges together.

If you wish to find the sum of $^{+}4$ and $^{-}3$, you can think about putting a $^{+}4$ charge together with a $^{-}3$ charge.

Notice that positive and negative charges neutralize each other as much as possible.

$^{+}4 + {}^{-}3$

$^{+}4 + {}^{-}3 = {}^{+}1$

Study these examples.

EXAMPLE 1.

$^{-}2 + {}^{-}3$

$^{-}2 + {}^{-}3 = {}^{-}5$

EXAMPLE 2.

$^{-}2 + {}^{+}2$

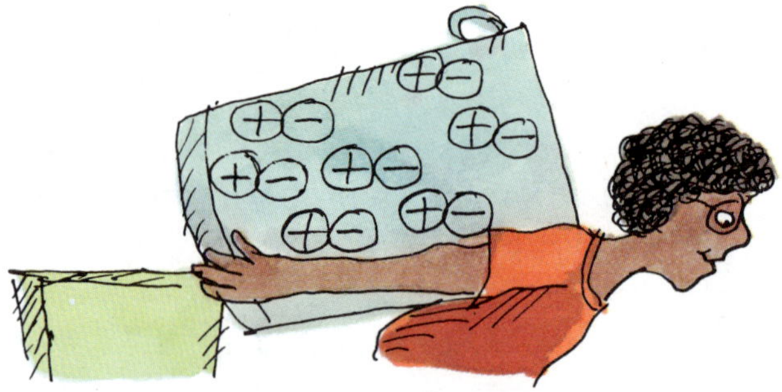

$^{-}2 + {}^{+}2 = 0$

$^{+}2$ and $^{-}2$ are called **opposites** because their sum is 0.

EXERCISES

Give each sum.

1.

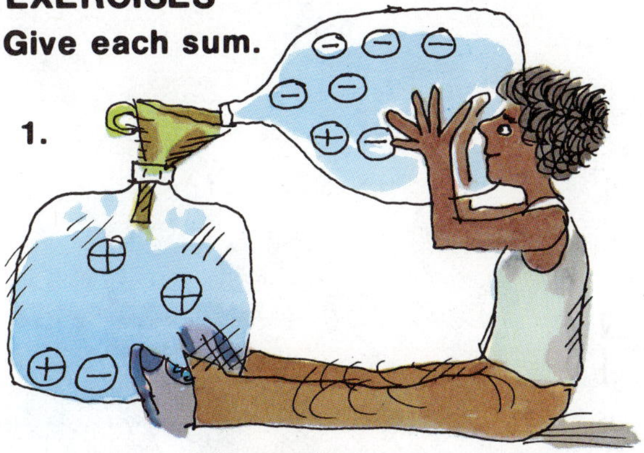

$^{+}2 + ^{-}5$

2.

$^{-}1 + ^{+}3$

3. $^{-}4 + ^{-}9$ **4.** $^{-}5 + ^{+}8$ **5.** $^{+}9 + ^{+}5$ **6.** $^{+}6 + ^{-}6$ (answer: 0)

7. $^{+}6 + ^{-}9$ **8.** $^{+}6 + ^{+}7$ **9.** $^{+}5 + ^{-}5$ **10.** $^{+}7 + ^{-}9$

11. $^{-}9 + ^{+}6$ **12.** $^{+}9 + ^{-}8$ **13.** $^{-}8 + ^{+}6$ **14.** $^{-}7 + ^{+}7$

15. $^{-}4 + ^{-}8$ **16.** $^{-}7 + ^{-}8$ **17.** $^{+}2 + ^{+}9$ **18.** $^{+}5 + ^{+}6$

19. $^{+}9 + ^{+}7$ **20.** $^{-}4 + ^{+}7$ **21.** $^{-}8 + ^{-}7$ **22.** $^{+}8 + ^{-}9$

23. $^{-}7 + 0$ (answer: $^{-}7$) **24.** $^{+}9 + ^{+}9$ **25.** $^{-}7 + ^{-}5$ **26.** $^{+}8 + ^{+}8$

27. $^{+}8 + ^{-}12$ **28.** $^{-}13 + ^{-}5$ **29.** $^{+}14 + ^{+}6$ **30.** $^{-}19 + ^{+}5$

31. $^{-}11 + ^{+}11$ **32.** $^{+}15 + ^{-}12$ **33.** $^{-}16 + ^{-}12$ **34.** $^{-}18 + ^{+}13$

35. $^{+}15 + ^{+}26$ **36.** $^{-}27 + ^{-}12$ **37.** $^{+}19 + 0$ **38.** $^{+}21 + ^{+}19$

39. $^{-}32 + ^{+}26$ **40.** $^{+}42 + ^{-}39$ **41.** $^{-}46 + ^{+}58$ **42.** $^{-}48 + ^{+}48$

Give the opposite of each integer.

43. $^{+}3$ **44.** $^{-}5$ **45.** $^{-}9$ **46.** $^{+}18$ **47.** $^{-}53$ **48.** 0

Complete.

49.

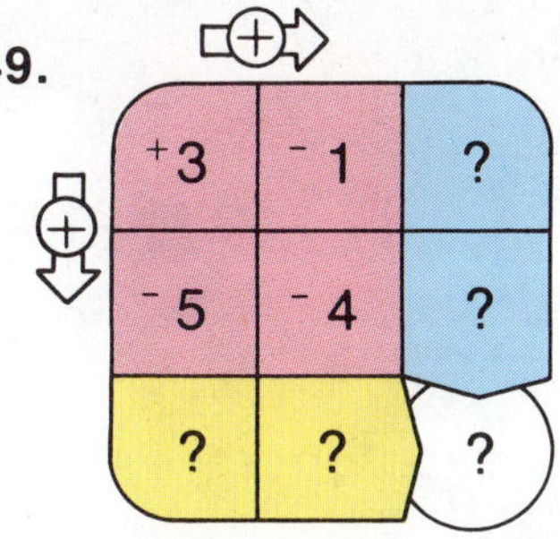

50.

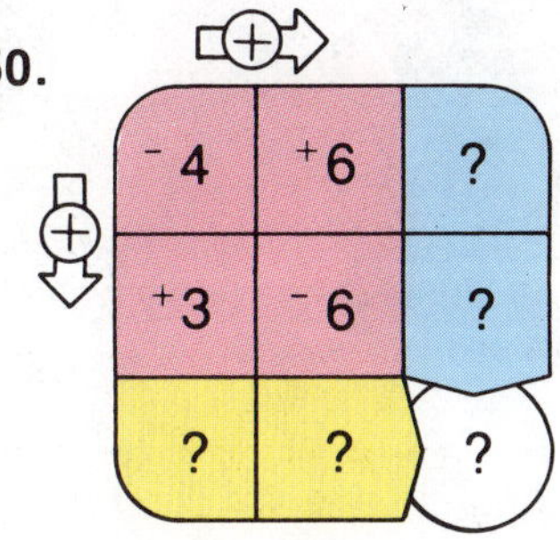

51.

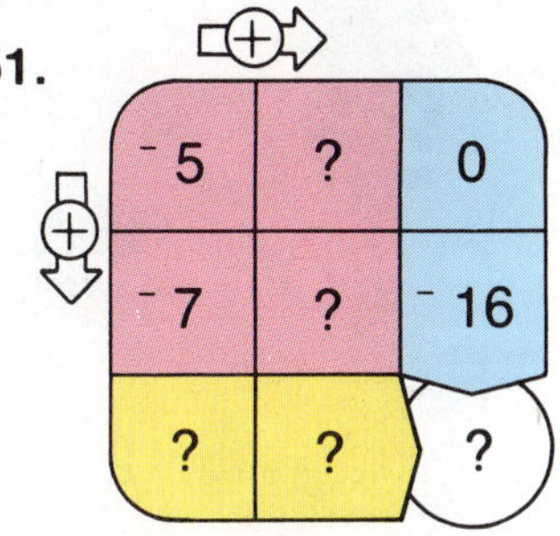

Subtracting integers

You can understand subtraction of integers by thinking about taking some charges out of a container.

If you wish to subtract ⁺2 from ⁻1, take 2 positive charges from a container that has a ⁻1 charge.

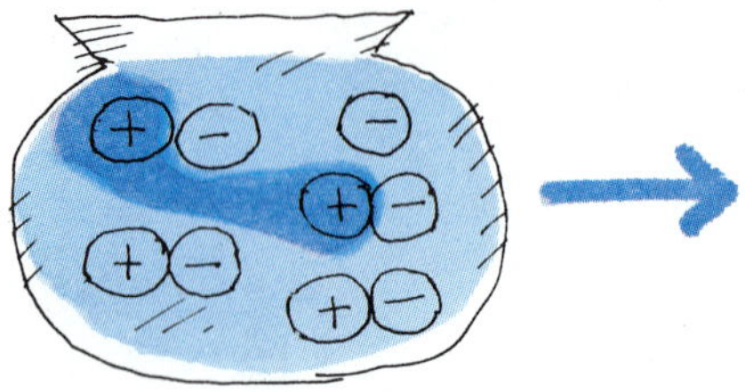

⁻1 − ⁺2 ⁻1 − ⁺2 = ⁻3

Study these examples.

EXAMPLE 1.

⁺4 − ⁺3 ⁺4 − ⁺3 = ⁺1

EXAMPLE 2.

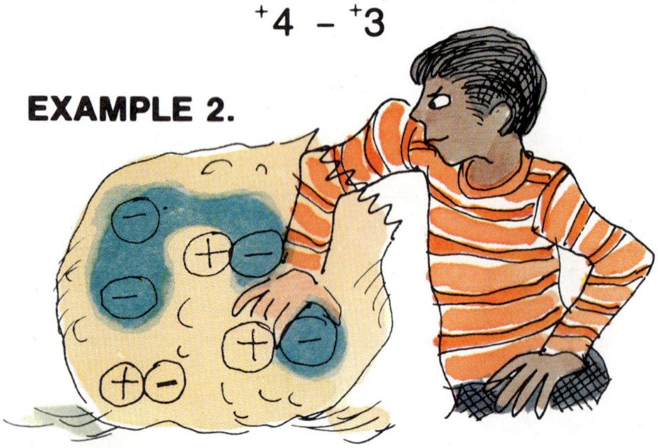

⁻2 − ⁻4 ⁻2 − ⁻4 = ⁺2

EXERCISES

Give each difference.

1.

$^{+}3 - {}^{-}2$

2.

$^{+}3 - {}^{+}4$

3. $^{+}7 - {}^{-}2$	**4.** $^{-}9 - {}^{+}5$	**5.** $^{-}3 - {}^{+}7$	**6.** $^{+}5 - {}^{+}2$
7. $^{+}1 - {}^{-}5$	**8.** $^{+}6 - {}^{+}9$	**9.** $^{-}4 - {}^{+}7$	**10.** $0 - {}^{-}5$
11. $^{+}12 - {}^{-}8$	**12.** $^{+}15 - {}^{-}4$	**13.** $^{+}12 - {}^{+}8$	**14.** $^{-}19 - {}^{+}7$
15. $^{-}9 - {}^{+}3$	**16.** $^{+}7 - {}^{-}6$	**17.** $^{+}12 - {}^{+}7$	**18.** $^{+}8 - {}^{+}17$
19. $^{-}16 - {}^{-}3$	**20.** $^{-}9 - {}^{-}9$	**21.** $^{-}6 - {}^{-}3$	**22.** $^{+}8 - {}^{-}13$

23. Give two ways to change this charge to $^{+}3$.

24. Give two ways to change this charge to $^{-}2$.

In exercises 23 and 24 you probably noticed this rule.

> Subtracting an integer is just the same as adding the opposite of the integer.
>
> $^{+}3 - {}^{-}2 = {}^{+}3 + {}^{+}2$

Subtract.

25. $^{+}8 - {}^{-}3$	**26.** $^{+}6 - {}^{-}4$	**27.** $^{+}9 - {}^{+}2$	**28.** $^{-}6 - {}^{-}8$
29. $^{-}9 - {}^{-}3$	**30.** $^{+}2 - {}^{+}8$	**31.** $^{-}5 - {}^{-}9$	**32.** $^{-}7 - {}^{+}1$
33. $^{+}7 - {}^{-}4$	**34.** $^{+}3 - {}^{-}9$	**35.** $^{-}9 - {}^{+}7$	**36.** $^{+}6 - {}^{+}5$
37. $^{+}6 - 0$	**38.** $^{+}5 - {}^{-}6$	**39.** $^{+}3 - {}^{+}4$	**40.** $^{+}21 - {}^{-}20$

Comparing integers

Here the integers are shown on the number line.

Numbers get smaller | Numbers get larger

This picture helps you compare integers.
To compare two integers, you can think of how the integers are ordered on the number line.

$^-7 < {}^+1$ $\qquad$ $^-4 < 0$

$^-7$ is less than $^+1$ $\qquad$ $^-4$ is less than 0

EXERCISES

<, =, or >?

1. $^-5$ ● $^-6$
2. $^+5$ ● $^+6$
3. $^-3$ ● 0
4. $^+8$ ● $^-5$
5. $^-8$ ● $^+5$
6. $^+8$ ● $^-8$
7. $^-16$ ● $^+16$
8. 0 ● $^-14$
9. $^-18$ ● $^-12$
10. $^+4 + {}^+2$ ● $^+7$
11. $^-3 + {}^-5$ ● $^-8$
12. $^+9 + {}^-3$ ● $^-6$
13. $^-7 + {}^+9$ ● 0
14. $^-9 + {}^-5$ ● $^-11$
15. $^+12 + {}^-12$ ● 0
16. $^+11 + {}^-6$ ● $^+5 + {}^+2$
17. $^-8 + {}^+3$ ● $^+7 + {}^-9$
18. $^-3 + {}^-6$ ● $^+2 + {}^-12$
19. $^+6 - {}^-3$ ● $^+4 + {}^+2$
20. $^-4 - {}^-2$ ● $^+3 - {}^-2$
21. $^+3 - {}^-4$ ● $^+3 + {}^-4$

22. $^-100$ ● $^+3$

23. $^-700$ ● $^+6$

24. $^-340$ ● $^-560$

25. $^-2$ ● $^-58$

26. $^+3$ ● $^-500$

27. $^+500$ ● $^-3$

Complete.

28. (⊕ across, ⊕ down)

$^-3$	$^+4$	?
$^-2$	$^+5$	?
?	?	?

29. (⊕ across, ⊕ down)

$^-8$	$^+2$	?
$^+4$	$^-3$	?
?	?	?

30. (⊕ across, ⊕ down)

$^+6$	$^-3$	?
$^-4$	$^-7$	?
?	?	?

31. (⊕ across, ⊕ down)

$^-6$	?	$^-4$
?	$^+2$	$^+5$
?	?	?

32. (⊕ across, ⊕ down)

$^-4$	?	$^+3$
$^-2$	?	$^+4$
?	?	?

33. (⊕ across, ⊕ down)

?	$^+2$	?
$^-3$	?	$^-4$
?	?	0

Follow each path.

34.

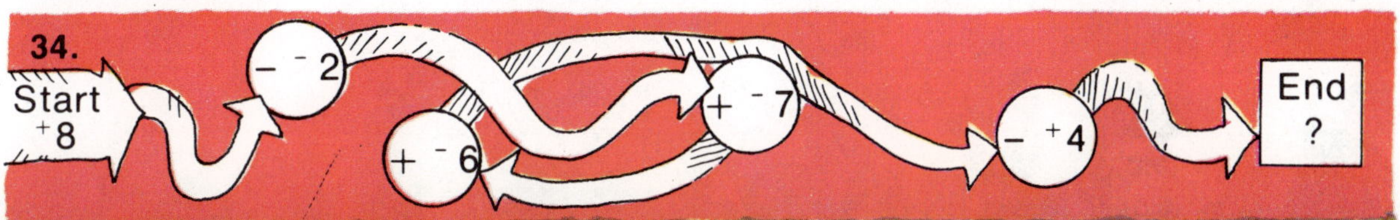

35.

36.

Practice exercises

Add.

1. $^{+}8 + {}^{+}9$
2. $^{+}8 + {}^{-}9$
3. $^{-}8 + {}^{+}9$
4. $^{-}8 + {}^{-}9$
5. $^{+}6 + {}^{+}5$
6. $^{+}6 + {}^{-}5$
7. $^{-}6 + {}^{+}5$
8. $^{-}6 + {}^{-}5$
9. $^{+}9 + {}^{+}4$
10. $^{+}6 + {}^{-}8$
11. $^{-}9 + {}^{-}6$
12. $^{-}8 + {}^{+}8$
13. $^{-}14 + {}^{+}6$
14. $^{-}15 + {}^{-}21$
15. $^{-}20 + {}^{+}20$
16. $^{+}12 + {}^{+}11$
17. $^{-}19 + {}^{-}19$
18. $^{+}11 + {}^{-}19$
19. $^{+}15 + {}^{+}10$
20. $^{+}20 + {}^{-}14$

Subtract.

21. $^{+}5 - {}^{+}2$
22. $^{+}5 - {}^{-}2$
23. $^{-}5 - {}^{+}2$
24. $^{-}5 - {}^{-}2$
25. $^{+}4 - {}^{+}7$
26. $^{-}4 - {}^{+}7$
27. $^{+}4 - {}^{-}7$
28. $^{-}4 - {}^{-}7$
29. $^{+}9 - {}^{+}3$
30. $^{-}8 - {}^{-}8$
31. $^{+}7 - {}^{-}4$
32. $^{-}3 - {}^{-}6$
33. $^{-}12 - {}^{+}16$
34. $^{+}18 - {}^{-}12$
35. $^{-}16 - {}^{-}16$
36. $^{+}16 - {}^{+}12$
37. $^{-}9 - 0$
38. $^{-}15 - {}^{-}22$
39. $^{+}15 - {}^{+}18$
40. $0 - {}^{-}25$

Compute.

41. $({}^{+}8 + {}^{+}2) + {}^{-}4$
42. $^{+}8 + ({}^{+}2 + {}^{-}4)$
43. $({}^{-}2 + {}^{-}5) - {}^{+}2$
44. $^{-}2 + ({}^{-}5 - {}^{+}2)$
45. $({}^{-}4 - {}^{-}5) - {}^{-}6$
46. $^{-}4 - ({}^{-}5 - {}^{-}6)$
47. $({}^{-}3 - {}^{+}7) + {}^{+}8$
48. $^{-}3 - ({}^{+}7 + {}^{+}8)$
49. $({}^{-}6 + {}^{-}9) - {}^{-}7$
50. $^{-}6 + ({}^{-}9 - {}^{-}7)$

Mr. Allen was put on a diet by his doctor. He kept this line graph of his weight.

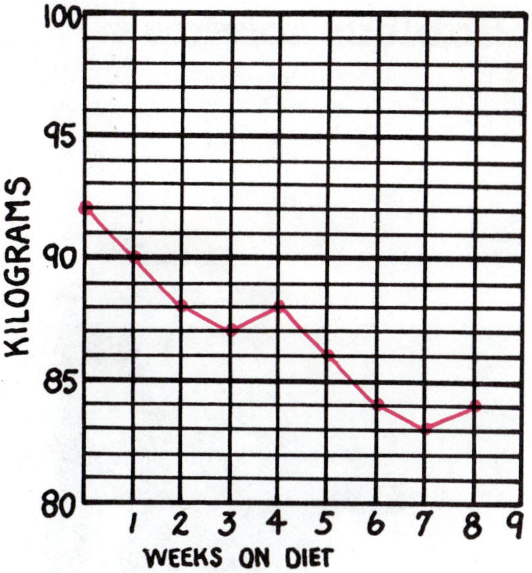

51. How much did he lose during the first week?

52. What was his weight at the end of the fifth week?

53. If $^{+}2$ represents a gain of 2 kilograms, what integer would be used to show the gain or loss of weight between

  a. week 1 and week 2?

  b. week 7 and week 8?

Christine used an integer to record the net change in cents of her money supply. For example, if during a day she ended up with 20¢ less, she would record $^{-}20$ for that day.

	DAY						
WEEK	SUN.	MON.	TUE.	WED.	THURS.	FRI.	SAT.
1	$^{+}20$	$^{-}30$	$^{+}50$	$^{+}80$	$^{-}20$	$^{-}10$	$^{-}30$
2	$^{+}30$	$^{+}50$	$^{-}30$	$^{+}80$	$^{-}30$	$^{-}20$	$^{-}40$
3	$^{+}30$	$^{+}75$	$^{-}25$	$^{+}25$	$^{+}35$	$^{+}70$	$^{-}30$
4	$^{+}20$	$^{-}15$	$^{+}75$	$^{-}30$	$^{-}5$	0	$^{+}65$
5	$^{-}15$	$^{-}40$	$^{+}80$	0	$^{+}90$	$^{-}35$	$^{+}20$
6	0	$^{+}35$	$^{-}25$	$^{-}5$	$^{+}60$	$^{+}60$	$^{+}90$

54. How did her money supply change on Wednesday of week 1?

55. On which day of week 4 was there no change in her money supply?

56. What was the net change during week 3?

Keeping Skills Sharp

1. $\begin{array}{r} 12 \\ -\ 3\frac{1}{2} \\ \hline \end{array}$

2. $\begin{array}{r} 15 \\ -\ 7\frac{4}{5} \\ \hline \end{array}$

3. $\begin{array}{r} 16\frac{1}{3} \\ -\ 8\frac{1}{2} \\ \hline \end{array}$

4. $\begin{array}{r} 17\frac{5}{8} \\ -\ 9\frac{3}{4} \\ \hline \end{array}$

5. $\begin{array}{r} 13\frac{2}{3} \\ -\ 5\frac{5}{6} \\ \hline \end{array}$

6. $\begin{array}{r} 14\frac{1}{4} \\ -\ 6\frac{7}{8} \\ \hline \end{array}$

Problem solving

1. If a football team can gain 10 yards or more in 4 or fewer downs (tries), they keep the football for another 4 downs. First compute the total yardage of each 4 downs. Then tell if the team keeps the ball for another 4 downs. A gain is given by a positive integer, and a loss is given by a negative integer.

YARDAGE

	DOWN 1	DOWN 2	DOWN 3	DOWN 4	TOTAL	KEEP BALL?
a	+3	-4	+6	-2	+3	NO
b	-1	+8	-2	+4		
c	-3	-5	+10	+7		
d	+5	0	+4	+8		
e	+9	-2	+1	+6		
f	+4	-5	-1	+7		
g	-7	+8	+5	-1		
h	+6	+3	0	+6		
i	-2	+7	+1	+4		
j	+8	-9	+7	+5		

YARDAGE	PLAYER						
	BROWN	DAVIS	GARCIA	JACKSON	MARTIN	PORTER	STEVENS
	+5	-4	+6	-4	+2	-3	-5
	+8	+6	+18	0	+2	+8	-1
	-2	+3	-2	+8	0	+2	+8
	+6	+1	-1	+10	+3	-4	0
	-3	+8	+9	-4	-4	+2	+9
	+2	-5	-3	+8	-5		+7
	+5	+4	+2		+6		+5
	+1	+2			+16		+4
	0	+1					-3
		-3					

2. This table shows the yards gained or lost each time a player carried the ball. Find the total yardage for each player. Who gained the most yardage? The least?

Graphing number pairs

Here is how to graph a number pair of integers. To graph ($^+4$, $^+3$), we start at 0, count 4 units to the right, and count 3 units up. To graph a number pair, we first move along the horizontal number line and then move parallel to the vertical number line.

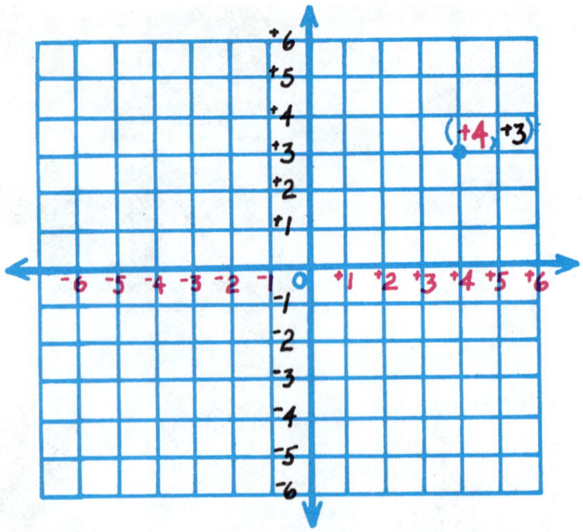

Study these examples.

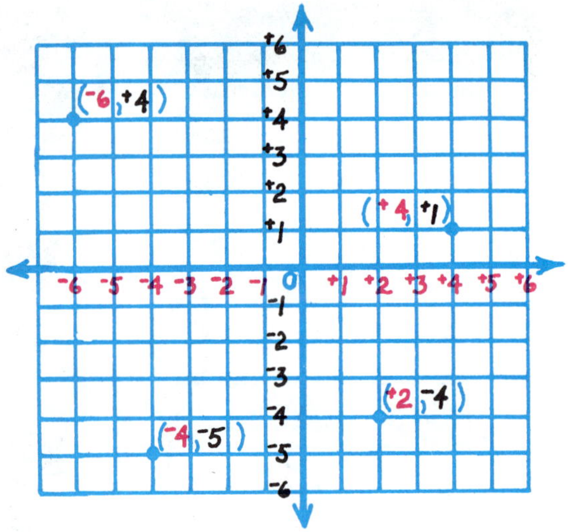

The number pair for a point is called the **coordinates** of the point. The coordinates of point A are ($^+2$, $^+4$). The coordinates of point B are ($^-2$, $^+2$). What are the coordinates of points C and D?

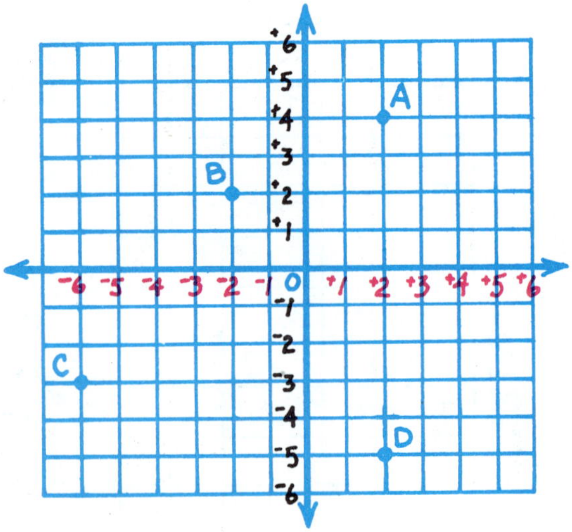

EXERCISES

Give the coordinates of each point.

1. A	**2.** F	**3.** C
4. E	**5.** H	**6.** K
7. L	**8.** B	**9.** I
10. G	**11.** J	**12.** D

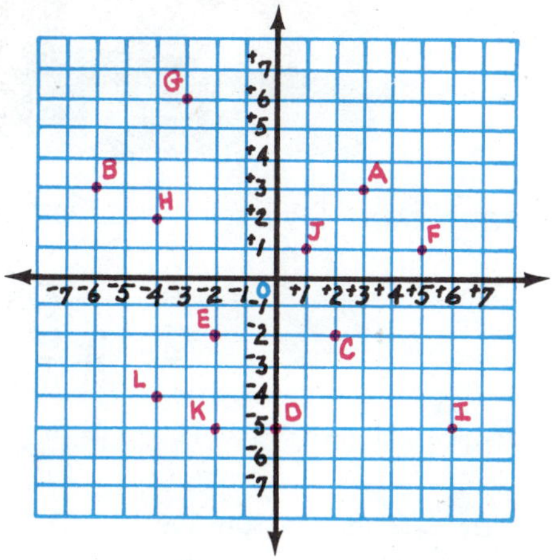

Get a piece of graph paper and graph these number pairs. Label each point with its coordinates.

13. $(^-3, ^+5)$	**14.** $(^-5, ^+1)$	**15.** $(^+7, ^+3)$
16. $(^-4, ^-2)$	**17.** $(0, 0)$	**18.** $(^+5, ^+3)$
19. $(^+4, ^+6)$	**20.** $(^+5, ^-5)$	**21.** $(^+3, ^-6)$
22. $(^-3, ^-3)$	**23.** $(^+5, 0)$	**24.** $(0, ^-6)$

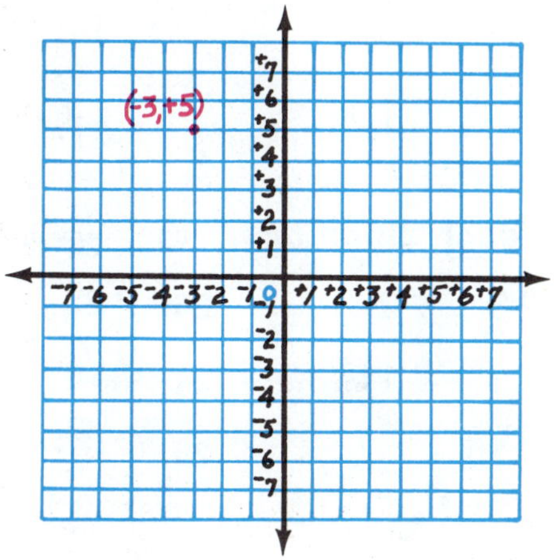

What picture do you get? Graph the points in the order given. Connect the points in the same order with line segments.

25. $(^+1, ^+5)$	**26.** $(^-1, ^+3)$	**27.** $(0, ^+3)$
28. $(^-2, ^+1)$	**29.** $(^-1, ^+1)$	**30.** $(^-3, ^-2)$
31. $(0, ^-2)$	**32.** $(0, ^-5)$	**33.** $(^+2, ^-5)$
34. $(^+2, ^-2)$	**35.** $(^+5, ^-2)$	**36.** $(^+3, ^+1)$
37. $(^+4, ^+1)$	**38.** $(^+2, ^+3)$	**39.** $(^+3, ^+3)$
40. $(^+1, ^+5)$		

CHAPTER CHECKUP

Give each sum. [pages 304–305, 309–310]

1. $^{+}6 + {}^{+}8$
2. $^{+}3 + {}^{-}9$
3. $^{-}6 + {}^{+}5$
4. $^{+}7 + {}^{-}7$
5. $^{-}8 + {}^{+}8$
6. $^{-}4 + {}^{-}9$
7. $^{+}8 + 0$
8. $^{+}6 + {}^{+}7$
9. $^{+}7 + {}^{-}9$
10. $^{-}9 + {}^{+}3$
11. $^{+}9 + {}^{+}5$
12. $^{-}9 + {}^{-}9$

<, =, or >? [pages 308–309]

13. $^{-}6$ ● $^{-}5$
14. $^{+}2$ ● $^{-}3$
15. 0 ● $^{-}8$
16. $^{+}5 + {}^{-}2$ ● $^{+}3$
17. $^{-}6 + {}^{+}6$ ● 0
18. $^{-}4 + {}^{-}4$ ● 0

Give each difference. [pages 306–307, 309–310]

19. $^{+}8 - {}^{+}2$
20. $^{+}2 - {}^{-}8$
21. $^{-}7 - {}^{+}4$
22. $^{-}4 - {}^{-}7$
23. $^{-}3 - {}^{-}11$
24. $^{-}16 - {}^{+}9$
25. $^{+}9 - {}^{+}14$
26. $^{-}10 - {}^{+}8$
27. $^{+}16 - {}^{-}7$
28. $^{+}6 - {}^{+}13$
29. $^{-}18 - {}^{-}9$
30. $0 - {}^{-}12$

Solve. [pages 302–303, 311–313]

Sandy used integers to keep a record of the net change in the number of golf balls she had. $^{+}2$ represented gaining 2 golf balls and $^{-}2$ represented losing 2 golf balls.

31. If she found 3 golf balls, what integer would she use?
32. During a week she had these daily net changes:
$^{+}2, {}^{-}1, {}^{-}3, {}^{+}2, {}^{+}1, {}^{+}4$
What was her net change for the week?

Give the coordinates of each point. [pages 314–315]

33. U
34. R
35. W
36. Z
37. X
38. T
39. S
40. V
41. Y

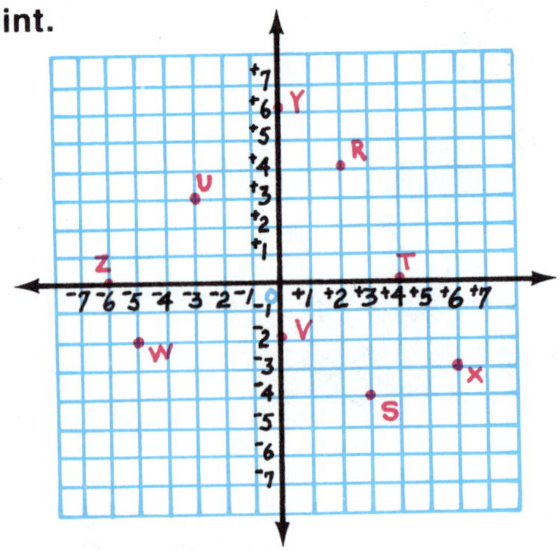

Project

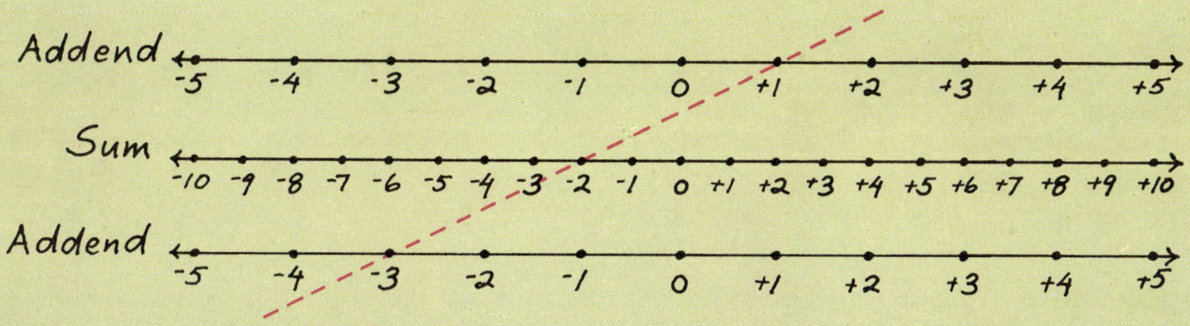

Example:

$^{+}1 + {}^{-}3 = {}^{-}2$

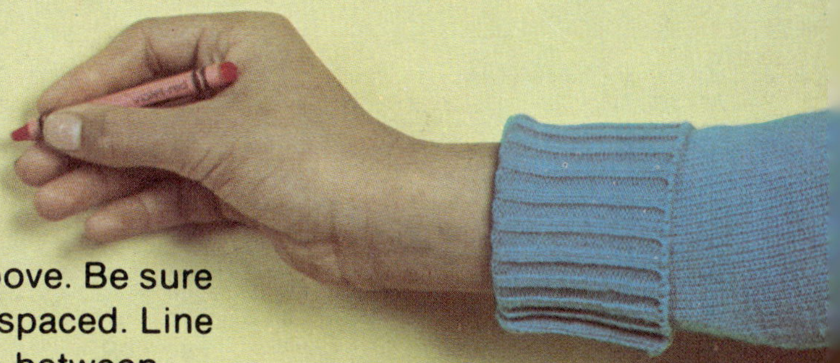

1. Draw 3 number lines as shown above. Be sure that they are parallel and equally spaced. Line up the 0's and make the distance between points on the "sum" number line half the distance between points on the "addend" number lines.

2. The example above shows how to use the number lines to find a sum.

Find the following sums on your number lines.

a. $0 + {}^{+}2$ **b.** ${}^{-}3 + {}^{+}1$ **c.** ${}^{-}2 + {}^{+}2$ **d.** ${}^{-}4 + {}^{-}1$

e. ${}^{-}3 + {}^{+}3$ **f.** ${}^{+}2 + {}^{+}3$ **g.** ${}^{-}4 + {}^{+}5$ **h.** ${}^{-}3 + 0$

3. Figure out how to use your number lines to subtract. *Hint:* Subtraction is finding a missing addend.

a. ${}^{+}3 - {}^{-}1$ **b.** ${}^{-}2 - {}^{+}3$ **c.** ${}^{-}3 - {}^{-}2$ **d.** ${}^{+}2 - {}^{-}3$

e. $0 - {}^{-}4$ **f.** ${}^{-}2 - {}^{-}3$ **g.** ${}^{+}3 - 0$ **h.** ${}^{-}4 - {}^{+}2$

CHAPTER REVIEW

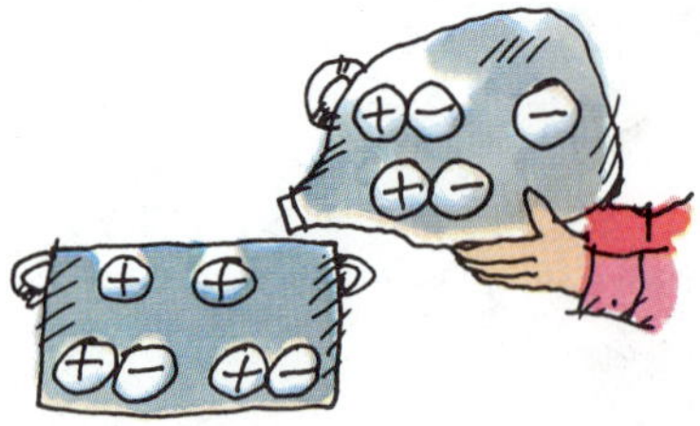

$^{+}2 + ^{-}1 = ^{+}1$

Give each sum.

1. $^{+}2 + ^{+}1$
2. $^{+}3 + ^{-}2$
3. $^{+}3 + ^{-}4$
4. $^{-}2 + ^{+}2$
5. $^{+}1 + ^{+}2$
6. $^{-}1 + ^{-}2$
7. $^{-}4 + ^{+}3$
8. $^{+}3 + ^{-}3$
9. $^{+}3 + 0$

< or >?

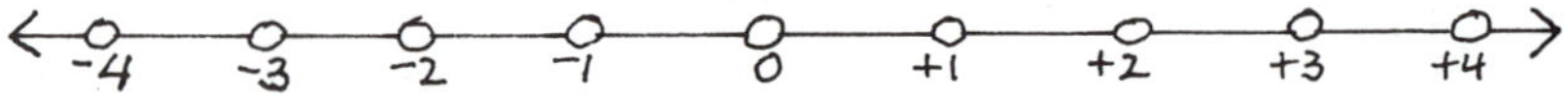

10. $^{-}3$ ● $^{+}1$
11. 0 ● $^{-}4$
12. $^{-}2$ ● $^{+}1$
13. 0 ● $^{+}3$

Give each difference.

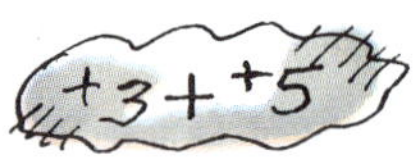

14. $^{+}8 - ^{+}2$
15. $^{+}3 - ^{-}5$
16. $0 - ^{-}5$
17. $^{-}5 - ^{+}5$
18. $^{+}4 - ^{+}6$
19. $^{-}6 - 0$
20. $^{+}8 - ^{-}5$
21. $^{-}6 - ^{-}5$
22. $^{+}9 - ^{-}11$

Give the coordinates of each point.

23. B
24. D
25. F
26. A
27. C
28. E

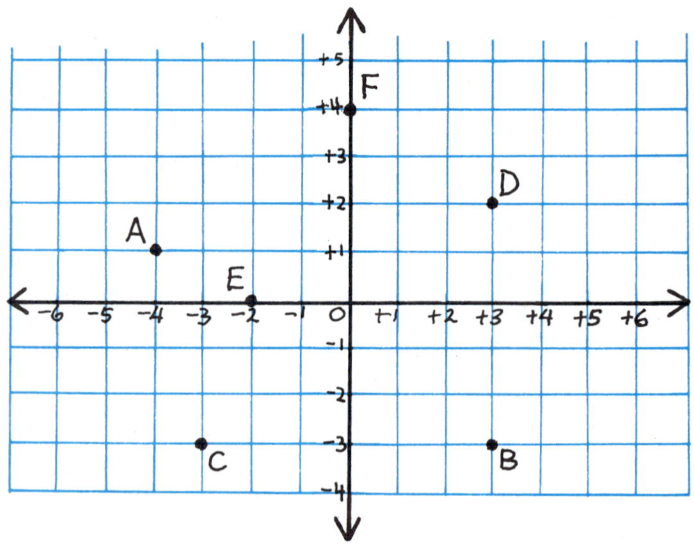

CHAPTER CHALLENGE

Here is a computer. To get the cards shown, the operator first dialed a rule **Add $^{-}2$**. Then input numbers were dialed. The machine added $^{-}2$ to each input number and printed the (input, output) number pair on a card. Notice that all the number pairs go with the rule **Add $^{-}2$**.

Here is how to graph the number pairs for the rule **Add $^{-}2$**.

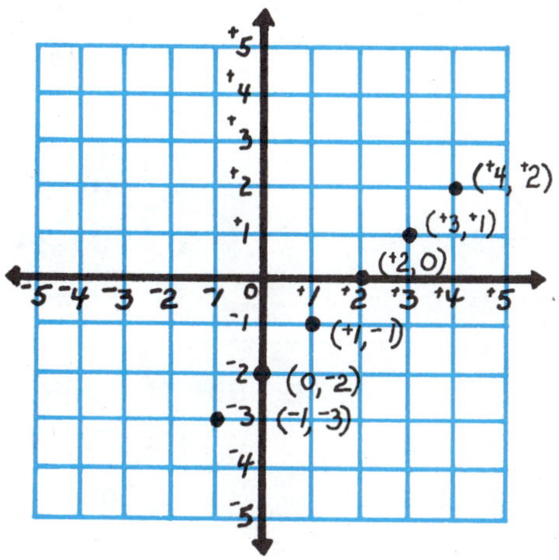

First copy and complete the cards for the given rule. Then graph the number pairs.

1. Add $^{+}3$	2. Subtract $^{+}2$	3. Subtract $^{-}2$
($^{-}2$, ?)	($^{-}2$, ?)	($^{-}2$, ?)
($^{-}1$, ?)	($^{-}1$, ?)	($^{-}1$, ?)
(0, ?)	(0, ?)	(0, ?)
($^{+}1$, ?)	($^{+}1$, ?)	($^{+}1$, ?)
($^{+}2$, ?)	($^{+}2$, ?)	($^{+}2$, ?)
($^{+}3$, ?)	($^{+}3$, ?)	($^{+}3$, ?)

MAJOR CHECKUP
Standardized Format

Choose the correct letter.

1. 526,385,000 rounded to the nearest ten thousand is

a. 530,000,000
b. 526,400,000
c. 526,390,000
d. none of these

2. Add.

$$\begin{array}{r} 9\frac{7}{10} \\ +4\frac{3}{4} \\ \hline \end{array}$$

a. $13\frac{9}{20}$
b. $14\frac{9}{20}$
c. $13\frac{4}{5}$
d. none of these

3. Subtract.

$$\begin{array}{r} 6\frac{1}{2} \\ -1\frac{5}{8} \\ \hline \end{array}$$

a. $5\frac{7}{8}$
b. $5\frac{1}{8}$
c. $4\frac{7}{8}$
d. none of these

4. $\frac{3}{8} \times \frac{4}{8} =$

a. $\frac{1}{2}$
b. $\frac{7}{24}$
c. $\frac{9}{32}$
d. none of these

5. $\frac{5}{3} \div \frac{3}{4} =$

a. $\frac{9}{20}$
b. $\frac{5}{4}$
c. $\frac{20}{9}$
d. none of these

6. Multiply.

$$\begin{array}{r} 46.8 \\ \times .43 \\ \hline \end{array}$$

a. 17.704
b. 20.124
c. 201.24
d. none of these

7. Divide.

$.73\overline{)9.125}$

a. 1.25
b. 125
c. 12.5
d. none of these

8. Find the volume.

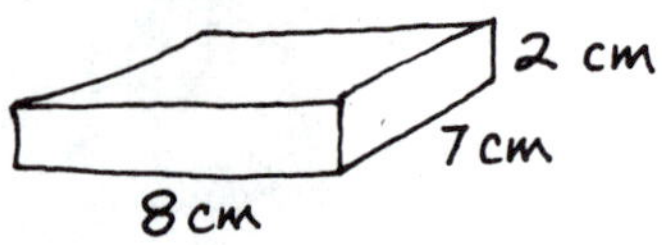

a. 56 cubic cm
b. 172 cubic cm
c. 17 cubic cm
d. none of these

9. The sale price is

a. \$2.83
b. \$7.08
c. \$2.84
d. none of these

10. $.6 =$

a. 6%
b. 60%
c. 600%
d. none of these

11. $\frac{2}{3} =$

a. $33\frac{1}{3}\%$
b. $16\frac{2}{3}\%$
c. $66\frac{2}{3}\%$
d. none of these

12. 25% of 84 =

a. 21
b. 210
c. 63
d. none of these

SKILL TEST

1	Adding any two numbers	582 + 163	5978 + 4169		421758 + 29694
2	Adding three or more numbers	38 46 + 75	298 46 343 + 156	5831 2703 568 + 2931	564 29671 432 + 85
3	Subtracting any two numbers	594 − 238	4832 − 1756		680031 − 246772
4	Multiplying any number by a 1-digit number	31 × 2	126 × 4	3764 × 7	28075 × 8
5	Multiplying any number by 10, 100, or 1000	57 × 10	381 × 100	526 × 1000	7284 × 100
6	Multiplying by tens, hundreds, or thousands	85 × 30	246 × 200	5218 × 400	7835 × 5000
7	Multiplying any number by a 2-digit number	75 × 23	258 × 48	2964 × 92	52718 × 76
8	Multiplying any number by a 3-digit number	216 × 314	528 × 205	3821 × 746	28034 × 187
9	Dividing a 2-digit or 3-digit number by a 1-digit number	$3\overline{)47}$	$5\overline{)298}$	$4\overline{)602}$	$8\overline{)311}$

	Skill				
10	Dividing any number by a 1-digit number	$7\overline{)2814}$	$8\overline{)29653}$	$5\overline{)172834}$	
11	Dividing any number by a 2- or 3-digit number	$56\overline{)52834}$	$179\overline{)71938}$	$527\overline{)749683}$	
12	Reducing a fraction to lowest terms	$\frac{7}{14} = \underline{?}$	$\frac{12}{9} = \underline{?}$	$\frac{32}{36} = \underline{?}$	$\frac{24}{56} = \underline{?}$
13	Comparing fractions < or >?	$\frac{1}{2}$ ● $\frac{3}{4}$	$\frac{3}{8}$ ● $\frac{1}{4}$	$\frac{3}{2}$ ● $\frac{2}{3}$	$\frac{7}{8}$ ● $\frac{5}{6}$
14	Adding fractions	$\frac{3}{7} + \frac{1}{7}$	$\frac{1}{2} + \frac{1}{4}$	$\frac{3}{4} + \frac{1}{3}$	$\frac{3}{2} + \frac{5}{8}$
15	Subtracting fractions	$\frac{5}{9} - \frac{2}{9}$	$\frac{3}{4} - \frac{1}{2}$	$\frac{3}{2} - \frac{2}{3}$	$\frac{5}{6} - \frac{4}{5}$
16	Changing a fraction to a mixed number	$\frac{5}{2} = \underline{?}$	$\frac{7}{4} = \underline{?}$	$\frac{10}{3} = \underline{?}$	$\frac{16}{7} = \underline{?}$
17	Changing a mixed number to a fraction	$1\frac{1}{3} = \underline{?}$	$2\frac{1}{2} = \underline{?}$	$3\frac{2}{3} = \underline{?}$	$5\frac{3}{4} = \underline{?}$
18	Adding mixed numbers	$2\frac{1}{2} + 1\frac{1}{3}$	$2\frac{1}{4} + 3\frac{2}{3}$	$3\frac{1}{4} + 4\frac{5}{8}$	$5\frac{2}{3} + 1\frac{5}{6}$
19	Subtracting mixed numbers	$6\frac{5}{8} - 2\frac{3}{8}$	$4\frac{3}{4} - 1\frac{2}{3}$	$8\frac{1}{3} - 4\frac{2}{3}$	$5\frac{1}{4} - 2\frac{7}{8}$

	Skill				
20	Multiplying fractions	$2 \times \frac{1}{3}$	$\frac{3}{7} \times \frac{7}{3}$	$\frac{2}{3} \times \frac{2}{5}$	$\frac{3}{4} \times \frac{8}{9}$
21	Finding a fraction of a number	$\frac{1}{3}$ of $24 = \underline{?}$	$\frac{3}{4}$ of $20 = \underline{?}$	$\frac{3}{2}$ of $12 = \underline{?}$	
22	Multiplying mixed numbers	$2\frac{1}{4} \times 3$	$1\frac{2}{3} \times 1\frac{2}{3}$	$2\frac{3}{4} \times 1\frac{1}{2}$	$3\frac{3}{8} \times 2\frac{1}{8}$
23	Dividing fractions	$\frac{4}{5} \div \frac{1}{5}$	$\frac{3}{4} \div 2$	$\frac{1}{3} \div \frac{2}{9}$	$\frac{4}{5} \div \frac{3}{2}$
24	Dividing mixed numbers	$2\frac{1}{2} \div 2$	$3\frac{1}{2} \div 1\frac{3}{4}$	$3\frac{1}{2} \div 2\frac{1}{4}$	$4\frac{1}{3} \div 1\frac{5}{8}$
25	Comparing decimals	< or >? 3.8 ● 3.9	5.02 ● 5.002	361.75 ● 36.175	
26	Adding decimals	$6.7 + 3.5$	$63.8 + 14.9$	$2.9 + 4.67$	$1.385 + .715$
27	Subtracting decimals	$4.8 - 2.1$	$3.75 - 1.69$	$20.041 - 9.375$	$14.6 - 2.83$
28	Multiplying decimals	$.6 \times .5$	$23 \times .7$	21.63×1.8	$5.385 \times .54$
29	Changing a decimal to a fraction in lowest terms	$.25 = ?$	$.8 = ?$	$.05 = ?$	$1.5 = ?$
30	Dividing decimals	$.6\overline{)2.4}$	$.15\overline{)32.10}$	$.025\overline{)6.25}$	
31	Changing a fraction to a percent	$\frac{1}{2} = ?$	$\frac{7}{10} = ?$	$\frac{5}{4} = ?$	$\frac{4}{5} = ?$
32	Changing a decimal to a percent	$.13 = ?$	$2.42 = ?$	$.9 = ?$	$.05 = ?$

33	Finding a percent of a number	50% of 24 = ?	75% of 48 = ?	150% of 16 = ?	
34	Adding integers	$^{+}6 + {}^{+}4$	$^{-}8 + {}^{+}5$	$^{-}11 + {}^{-}6$	$^{+}12 + {}^{-}4$
35	Subtracting integers	$^{+}5 - {}^{+}6$	$^{-}13 - {}^{+}8$	$^{+}9 - {}^{-}2$	$^{-}13 - {}^{-}6$
36	Comparing integers	< or >? $^{-}3$ ● $^{+}2$	$^{-}15$ ● $^{-}16$	$^{-}23$ ● $^{+}23$	0 ● $^{-}9$

EXTRA PRACTICE

Set 1 What digit is in the: 472,985

1. hundreds place?
2. ten thousands place?
3. ten place?
4. one thousands place?
5. hundred thousands place?
6. ones place?

Set 2 $<$ or $>$?

1. 352 ● 354	2. 480 ● 479	3. 600 ● 599
4. 4731 ● 4687	5. 6383 ● 6391	6. 4834 ● 4798
7. 26,428 ● 27,221	8. 6431 ● 22,314	9. 62,861 ● 9468
10. 86,463 ● 103,214	11. 99,999 ● 100,000	12. 16,874 ● 16,873
13. 298,341 ● 298,431	14. 843,261 ● 835,821	15. 687,342 ● 678,342

Set 3 Round to the nearest thousand.

1. 5348	2. 2931	3. 2987	4. 4632
5. 5099	6. 5499	7. 5500	8. 5501
9. 38,673	10. 39,751	11. 48,326	12. 297,512
13. 298,500	14. 372,041	15. 699,871	16. 899,500

Set 4 Give the standard numeral.

1. twelve million
2. two hundred fifty-eight million
3. seven million, two hundred forty-five thousand, four hundred fifty-two
4. six billion
5. ninety billion, ninety million, ninety thousand
6. fifteen trillion, six hundred billion, three million, forty thousand, six

Set 5 Add.

1.	2.	3.	4.	5.	6.	7.
8 +9	7 +6	4 +9	5 +9	3 +9	7 +5	7 +9

8.	9.	10.	11.	12.	13.	14.
9 +9	6 +6	6 +9	6 +5	6 +4	7 +7	8 +8

15.	16.	17.	18.	19.	20.	21.
8 +4	8 +9	8 +3	0 +9	8 +5	8 +6	5 +3

Set 6 Subtract.

1.	2.	3.	4.	5.	6.	7.
16 − 7	12 − 5	15 − 9	14 − 7	13 − 7	13 − 8	15 − 8

8.	9.	10.	11.	12.	13.	14.
12 − 4	11 − 3	13 − 9	18 − 9	14 − 9	17 − 8	11 − 2

15.	16.	17.	18.	19.	20.	21.
12 − 3	14 − 8	13 − 6	14 − 5	12 − 6	16 − 8	15 − 6

Set 7 Add.

1.	2.	3.	4.	5.
87 +46	274 +328	653 +497	8841 +2673	8294 +9277

6.	7.	8.	9.
673541 +285667	298734 +385499	289999 +579999	423651 +217688

Set 8 Add.

1.	2.	3.	4.	5.
345 258 +619	415 318 +567	967 431 +783	95 652 +833	827 258 +946

6.	7.	8.	9.	10.
486 297 341 +288	594 27 568 +493	2971 4831 2883 +1541	3074 291 68 +432	83416 29318 8764 +83516

Set 9 Subtract.

1. $\begin{array}{r} 87 \\ -\ 58 \\ \hline \end{array}$ 2. $\begin{array}{r} 93 \\ -\ 46 \\ \hline \end{array}$ 3. $\begin{array}{r} 70 \\ -\ 29 \\ \hline \end{array}$ 4. $\begin{array}{r} 45 \\ -\ 19 \\ \hline \end{array}$ 5. $\begin{array}{r} 827 \\ -\ 486 \\ \hline \end{array}$ 6. $\begin{array}{r} 937 \\ -\ 239 \\ \hline \end{array}$

7. $\begin{array}{r} 673 \\ -\ 488 \\ \hline \end{array}$ 8. $\begin{array}{r} 2794 \\ -\ 1395 \\ \hline \end{array}$ 9. $\begin{array}{r} 4836 \\ -\ 2867 \\ \hline \end{array}$ 10. $\begin{array}{r} 9351 \\ -\ 4672 \\ \hline \end{array}$ 11. $\begin{array}{r} 3814 \\ -\ 1835 \\ \hline \end{array}$

12. $\begin{array}{r} 67542 \\ -\ 18673 \\ \hline \end{array}$ 13. $\begin{array}{r} 375416 \\ -\ 158488 \\ \hline \end{array}$ 14. $\begin{array}{r} 973541 \\ -\ 654938 \\ \hline \end{array}$ 15. $\begin{array}{r} 876541 \\ -\ 293882 \\ \hline \end{array}$

Set 10 Subtract.

1. $\begin{array}{r} 603 \\ -\ 215 \\ \hline \end{array}$ 2. $\begin{array}{r} 500 \\ -\ 328 \\ \hline \end{array}$ 3. $\begin{array}{r} 901 \\ -\ 493 \\ \hline \end{array}$ 4. $\begin{array}{r} 802 \\ -\ 573 \\ \hline \end{array}$ 5. $\begin{array}{r} 504 \\ -\ 201 \\ \hline \end{array}$

6. $\begin{array}{r} 2004 \\ -\ 917 \\ \hline \end{array}$ 7. $\begin{array}{r} 6108 \\ -\ 3249 \\ \hline \end{array}$ 8. $\begin{array}{r} 8011 \\ -\ 2735 \\ \hline \end{array}$ 9. $\begin{array}{r} 5000 \\ -\ 4731 \\ \hline \end{array}$ 10. $\begin{array}{r} 8000 \\ -\ 1977 \\ \hline \end{array}$

11. $\begin{array}{r} 50301 \\ -\ 2943 \\ \hline \end{array}$ 12. $\begin{array}{r} 60000 \\ -\ 9731 \\ \hline \end{array}$ 13. $\begin{array}{r} 800000 \\ -\ 512345 \\ \hline \end{array}$ 14. $\begin{array}{r} 70304 \\ -\ 59391 \\ \hline \end{array}$ 15. $\begin{array}{r} 800000 \\ -\ 387999 \\ \hline \end{array}$

Set 11 Multiply.

1. $\begin{array}{r} 9 \\ \times 5 \\ \hline \end{array}$ 2. $\begin{array}{r} 9 \\ \times 6 \\ \hline \end{array}$ 3. $\begin{array}{r} 3 \\ \times 8 \\ \hline \end{array}$ 4. $\begin{array}{r} 9 \\ \times 8 \\ \hline \end{array}$ 5. $\begin{array}{r} 7 \\ \times 5 \\ \hline \end{array}$ 6. $\begin{array}{r} 5 \\ \times 8 \\ \hline \end{array}$ 7. $\begin{array}{r} 7 \\ \times 6 \\ \hline \end{array}$

8. $\begin{array}{r} 8 \\ \times 4 \\ \hline \end{array}$ 9. $\begin{array}{r} 7 \\ \times 7 \\ \hline \end{array}$ 10. $\begin{array}{r} 9 \\ \times 9 \\ \hline \end{array}$ 11. $\begin{array}{r} 9 \\ \times 4 \\ \hline \end{array}$ 12. $\begin{array}{r} 5 \\ \times 6 \\ \hline \end{array}$ 13. $\begin{array}{r} 8 \\ \times 7 \\ \hline \end{array}$ 14. $\begin{array}{r} 8 \\ \times 8 \\ \hline \end{array}$

15. $\begin{array}{r} 4 \\ \times 7 \\ \hline \end{array}$ 16. $\begin{array}{r} 6 \\ \times 6 \\ \hline \end{array}$ 17. $\begin{array}{r} 3 \\ \times 9 \\ \hline \end{array}$ 18. $\begin{array}{r} 8 \\ \times 6 \\ \hline \end{array}$ 19. $\begin{array}{r} 9 \\ \times 7 \\ \hline \end{array}$ 20. $\begin{array}{r} 5 \\ \times 9 \\ \hline \end{array}$ 21. $\begin{array}{r} 5 \\ \times 5 \\ \hline \end{array}$

Set 12 Give the least common multiple.

1. 3, 9 2. 4, 6 3. 5, 4 4. 7, 3 5. 6, 12

6. 6, 8 7. 8, 5 8. 8, 10 9. 5, 9 10. 3, 18

11. 4, 12 12. 9, 6 13. 3, 12 14. 7, 14 15. 1, 8

16. 5, 2 17. 8, 12 18. 6, 10 19. 4, 10 20. 18, 12

Set 13 Multiply.

1. $\begin{array}{r} 57 \\ \times 2 \\ \hline \end{array}$ 2. $\begin{array}{r} 38 \\ \times 3 \\ \hline \end{array}$ 3. $\begin{array}{r} 45 \\ \times 5 \\ \hline \end{array}$ 4. $\begin{array}{r} 72 \\ \times 4 \\ \hline \end{array}$ 5. $\begin{array}{r} 86 \\ \times 7 \\ \hline \end{array}$

6. $\begin{array}{r} 923 \\ \times 8 \\ \hline \end{array}$ 7. $\begin{array}{r} 486 \\ \times 7 \\ \hline \end{array}$ 8. $\begin{array}{r} 395 \\ \times 4 \\ \hline \end{array}$ 9. $\begin{array}{r} 804 \\ \times 9 \\ \hline \end{array}$ 10. $\begin{array}{r} 903 \\ \times 6 \\ \hline \end{array}$

11. $\begin{array}{r} 5113 \\ \times 4 \\ \hline \end{array}$ 12. $\begin{array}{r} 6874 \\ \times 9 \\ \hline \end{array}$ 13. $\begin{array}{r} 8003 \\ \times 9 \\ \hline \end{array}$ 14. $\begin{array}{r} 57031 \\ \times 8 \\ \hline \end{array}$ 15. $\begin{array}{r} 58326 \\ \times 6 \\ \hline \end{array}$

Set 14 Multiply.

1. $\begin{array}{r} 37 \\ \times 10 \\ \hline \end{array}$ 2. $\begin{array}{r} 50 \\ \times 10 \\ \hline \end{array}$ 3. $\begin{array}{r} 623 \\ \times 10 \\ \hline \end{array}$ 4. $\begin{array}{r} 487 \\ \times 10 \\ \hline \end{array}$ 5. $\begin{array}{r} 500 \\ \times 10 \\ \hline \end{array}$

6. $\begin{array}{r} 672 \\ \times 100 \\ \hline \end{array}$ 7. $\begin{array}{r} 480 \\ \times 100 \\ \hline \end{array}$ 8. $\begin{array}{r} 8031 \\ \times 100 \\ \hline \end{array}$ 9. $\begin{array}{r} 7238 \\ \times 1000 \\ \hline \end{array}$ 10. $\begin{array}{r} 9000 \\ \times 1000 \\ \hline \end{array}$

Set 15 Multiply.

1. $\begin{array}{r} 48 \\ \times 20 \\ \hline \end{array}$ 2. $\begin{array}{r} 76 \\ \times 50 \\ \hline \end{array}$ 3. $\begin{array}{r} 39 \\ \times 80 \\ \hline \end{array}$ 4. $\begin{array}{r} 421 \\ \times 70 \\ \hline \end{array}$ 5. $\begin{array}{r} 683 \\ \times 90 \\ \hline \end{array}$

6. $\begin{array}{r} 604 \\ \times 300 \\ \hline \end{array}$ 7. $\begin{array}{r} 956 \\ \times 800 \\ \hline \end{array}$ 8. $\begin{array}{r} 403 \\ \times 200 \\ \hline \end{array}$ 9. $\begin{array}{r} 856 \\ \times 300 \\ \hline \end{array}$ 10. $\begin{array}{r} 675 \\ \times 800 \\ \hline \end{array}$

11. $\begin{array}{r} 777 \\ \times 7000 \\ \hline \end{array}$ 12. $\begin{array}{r} 421 \\ \times 600 \\ \hline \end{array}$ 13. $\begin{array}{r} 395 \\ \times 400 \\ \hline \end{array}$ 14. $\begin{array}{r} 6713 \\ \times 30 \\ \hline \end{array}$ 15. $\begin{array}{r} 8514 \\ \times 2000 \\ \hline \end{array}$

Set 16 Multiply.

1. $\begin{array}{r} 53 \\ \times 21 \\ \hline \end{array}$ 2. $\begin{array}{r} 48 \\ \times 23 \\ \hline \end{array}$ 3. $\begin{array}{r} 97 \\ \times 42 \\ \hline \end{array}$ 4. $\begin{array}{r} 86 \\ \times 13 \\ \hline \end{array}$ 5. $\begin{array}{r} 92 \\ \times 63 \\ \hline \end{array}$

6. $\begin{array}{r} 248 \\ \times 72 \\ \hline \end{array}$ 7. $\begin{array}{r} 305 \\ \times 68 \\ \hline \end{array}$ 8. $\begin{array}{r} 421 \\ \times 93 \\ \hline \end{array}$ 9. $\begin{array}{r} 572 \\ \times 86 \\ \hline \end{array}$ 10. $\begin{array}{r} 893 \\ \times 99 \\ \hline \end{array}$

11. $\begin{array}{r} 2831 \\ \times 62 \\ \hline \end{array}$ 12. $\begin{array}{r} 4328 \\ \times 84 \\ \hline \end{array}$ 13. $\begin{array}{r} 9761 \\ \times 39 \\ \hline \end{array}$ 14. $\begin{array}{r} 87782 \\ \times 74 \\ \hline \end{array}$ 15. $\begin{array}{r} 49836 \\ \times 26 \\ \hline \end{array}$

Set 17 Multiply.

1. $\begin{array}{r} 753 \\ \times 123 \\ \hline \end{array}$
2. $\begin{array}{r} 293 \\ \times 240 \\ \hline \end{array}$
3. $\begin{array}{r} 641 \\ \times 374 \\ \hline \end{array}$
4. $\begin{array}{r} 795 \\ \times 458 \\ \hline \end{array}$
5. $\begin{array}{r} 443 \\ \times 443 \\ \hline \end{array}$
6. $\begin{array}{r} 926 \\ \times 288 \\ \hline \end{array}$
7. $\begin{array}{r} 502 \\ \times 665 \\ \hline \end{array}$
8. $\begin{array}{r} 674 \\ \times 324 \\ \hline \end{array}$
9. $\begin{array}{r} 839 \\ \times 550 \\ \hline \end{array}$
10. $\begin{array}{r} 753 \\ \times 700 \\ \hline \end{array}$
11. $\begin{array}{r} 6515 \\ \times 305 \\ \hline \end{array}$
12. $\begin{array}{r} 2066 \\ \times 204 \\ \hline \end{array}$
13. $\begin{array}{r} 3541 \\ \times 1206 \\ \hline \end{array}$
14. $\begin{array}{r} 4500 \\ \times 3418 \\ \hline \end{array}$
15. $\begin{array}{r} 7293 \\ \times 2964 \\ \hline \end{array}$

Set 18 Divide.

1. $8\overline{)64}$
2. $9\overline{)72}$
3. $6\overline{)36}$
4. $9\overline{)27}$
5. $7\overline{)49}$
6. $7\overline{)28}$
7. $8\overline{)48}$
8. $3\overline{)18}$
9. $9\overline{)81}$
10. $9\overline{)45}$
11. $9\overline{)36}$
12. $7\overline{)21}$
13. $7\overline{)42}$
14. $8\overline{)56}$
15. $4\overline{)32}$
16. $4\overline{)24}$
17. $6\overline{)54}$
18. $8\overline{)40}$
19. $7\overline{)63}$
20. $8\overline{)72}$

Set 19 Give the greatest common factor.

1. 14, 18
2. 40, 16
3. 32, 24
4. 24, 18
5. 54, 36
6. 24, 9
7. 63, 45
8. 28, 14
9. 27, 24
10. 9, 36
11. 42, 28
12. 20, 15
13. 32, 56
14. 12, 36
15. 45, 15
16. 15, 35
17. 18, 36
18. 14, 49
19. 21, 28
20. 24, 16

Set 20 Give the prime factorization.

1. 10
2. 8
3. 12
4. 15
5. 24
6. 32
7. 36
8. 48
9. 52
10. 64
11. 70
12. 80
13. 84
14. 96
15. 100
16. 120
17. 128
18. 144

Set 21 Divide.

1. 3)82   2. 5)60   3. 4)84   4. 6)93   5. 8)77

6. 3)268   7. 5)725   8. 4)964   9. 5)327   10. 6)981

11. 9)683   12. 8)593   13. 7)674   14. 9)213   15. 7)893

Set 22 Divide.

1. 5)3682   2. 4)9102   3. 7)3471   4. 5)8526   5. 9)3042

6. 3)5006   7. 6)3089   8. 9)2714   9. 8)36510   10. 5)29783

11. 8)42065   12. 6)78302   13. 7)91706   14. 4)283451   15. 8)761745

Set 23 Compute the average. Round to the nearest tenth.

1. 3, 5, 8, 8, 10, 11, 14
2. 5, 6, 9, 9, 11, 12
3. 32, 38, 46, 49, 50
4. 26, 32, 34, 41, 42, 42, 46
5. 68, 69, 70, 71, 72
6. 39, 43, 56, 57, 65
7. 45, 45, 45, 45, 45, 45,
8. 83, 86, 86, 89, 90, 91
9. 126, 124, 132, 128, 130
10. 153, 162, 147, 165, 147

Set 24 Divide.

1. 15)542   2. 16)706   3. 42)653   4. 39)802   5. 43)217

6. 23)7821   7. 18)3748   8. 46)7403   9. 70)6508   10. 58)5921

11. 92)59708   12. 84)35214   13. 76)60058   14. 95)93774   15. 42)32671

Set 25 Divide.

1. $259\overline{)38251}$	2. $307\overline{)79604}$	3. $521\overline{)36525}$	4. $846\overline{)74238}$
5. $938\overline{)69752}$	6. $578\overline{)43869}$	7. $651\overline{)86534}$	8. $742\overline{)92761}$
9. $853\overline{)73824}$	10. $926\overline{)69385}$	11. $509\overline{)71005}$	12. $642\overline{)83742}$

Set 26 Copy and complete.

1. $\frac{1}{2} = \frac{?}{8}$	2. $\frac{3}{4} = \frac{?}{20}$	3. $\frac{2}{3} = \frac{?}{18}$	4. $\frac{3}{2} = \frac{?}{18}$
5. $\frac{5}{6} = \frac{?}{30}$	6. $\frac{6}{5} = \frac{?}{30}$	7. $\frac{3}{4} = \frac{?}{28}$	8. $\frac{2}{3} = \frac{?}{27}$
9. $\frac{5}{7} = \frac{?}{42}$	10. $\frac{3}{5} = \frac{?}{15}$	11. $\frac{4}{5} = \frac{?}{30}$	12. $\frac{5}{8} = \frac{?}{32}$

Set 27 Reduce to lowest terms.

1. $\frac{10}{12}$	2. $\frac{9}{12}$	3. $\frac{8}{12}$	4. $\frac{6}{12}$	5. $\frac{4}{12}$	6. $\frac{3}{12}$	7. $\frac{2}{12}$
8. $\frac{12}{18}$	9. $\frac{24}{20}$	10. $\frac{24}{28}$	11. $\frac{32}{40}$	12. $\frac{25}{30}$	13. $\frac{30}{50}$	14. $\frac{9}{6}$
15. $\frac{12}{15}$	16. $\frac{18}{12}$	17. $\frac{10}{16}$	18. $\frac{18}{24}$	19. $\frac{24}{32}$	20. $\frac{36}{30}$	21. $\frac{42}{48}$

Set 28 $<$ or $>$?

1. $\frac{3}{5} \bigcirc \frac{2}{5}$	2. $\frac{4}{9} \bigcirc \frac{5}{9}$	3. $\frac{3}{2} \bigcirc \frac{1}{2}$	4. $\frac{5}{4} \bigcirc \frac{5}{7}$
5. $\frac{3}{8} \bigcirc \frac{3}{4}$	6. $\frac{3}{5} \bigcirc \frac{1}{3}$	7. $\frac{2}{5} \bigcirc \frac{1}{2}$	8. $\frac{3}{7} \bigcirc \frac{1}{4}$
9. $\frac{3}{7} \bigcirc \frac{1}{2}$	10. $\frac{5}{7} \bigcirc \frac{3}{4}$	11. $\frac{2}{3} \bigcirc \frac{3}{4}$	12. $\frac{5}{6} \bigcirc \frac{7}{9}$

Set 29 **Add. Reduce answers to lowest terms.**

1. $\frac{2}{9} + \frac{2}{9}$
2. $\frac{3}{10} + \frac{3}{10}$
3. $\frac{3}{8} + \frac{1}{8}$
4. $\frac{1}{4} + \frac{1}{2}$
5. $\frac{1}{2} + \frac{3}{4}$
6. $\frac{1}{3} + \frac{1}{2}$
7. $\frac{1}{4} + \frac{3}{8}$
8. $\frac{7}{3} + \frac{2}{2}$
9. $\frac{3}{8} + \frac{3}{4}$
10. $\frac{2}{5} + \frac{5}{2}$
11. $\frac{3}{4} + \frac{5}{8}$
12. $\frac{2}{3} + \frac{4}{5}$
13. $\frac{2}{3} + \frac{5}{9}$
14. $\frac{3}{2} + \frac{5}{8}$

Set 30 **Subtract. Reduce answers to lowest terms.**

1. $\frac{1}{2} - \frac{1}{4}$
2. $\frac{1}{2} - \frac{1}{3}$
3. $\frac{2}{3} - \frac{1}{4}$
4. $\frac{2}{5} - \frac{1}{4}$
5. $\frac{3}{2} - \frac{2}{3}$
6. $\frac{1}{2} - \frac{2}{5}$
7. $\frac{2}{3} - \frac{1}{8}$
8. $\frac{3}{5} - \frac{5}{10}$
9. $\frac{4}{5} - \frac{3}{10}$
10. $\frac{3}{8} - \frac{1}{6}$
11. $\frac{5}{8} - \frac{1}{4}$
12. $\frac{5}{6} - \frac{2}{3}$
13. $\frac{2}{3} - \frac{5}{8}$
14. $\frac{11}{12} - \frac{3}{8}$

Set 31 **Divide. Write your answer as a mixed number.**

1. $6\overline{)38}$
2. $7\overline{)40}$
3. $5\overline{)37}$
4. $5\overline{)42}$
5. $7\overline{)60}$
6. $9\overline{)70}$
7. $9\overline{)74}$
8. $6\overline{)45}$
9. $8\overline{)53}$
10. $9\overline{)60}$
11. $5\overline{)49}$
12. $8\overline{)46}$
13. $7\overline{)53}$
14. $8\overline{)52}$
15. $6\overline{)46}$
16. $9\overline{)46}$
17. $6\overline{)55}$
18. $9\overline{)80}$
19. $7\overline{)39}$
20. $8\overline{)60}$

Set 32 **Change to mixed numbers.**

1. $\frac{5}{2}$
2. $\frac{9}{2}$
3. $\frac{15}{2}$
4. $\frac{7}{3}$
5. $\frac{8}{3}$
6. $\frac{17}{3}$
7. $\frac{7}{4}$
8. $\frac{9}{4}$
9. $\frac{21}{4}$
10. $\frac{7}{5}$
11. $\frac{9}{5}$
12. $\frac{11}{5}$
13. $\frac{19}{5}$
14. $\frac{17}{6}$
15. $\frac{14}{6}$
16. $\frac{13}{4}$
17. $\frac{19}{2}$
18. $\frac{23}{8}$
19. $\frac{21}{6}$
20. $\frac{25}{3}$
21. $\frac{32}{10}$

Set 33 **Add or subtract.**

1. $3\frac{1}{3} + 4\frac{1}{3}$
2. $2\frac{1}{4} + 3\frac{1}{8}$
3. $3\frac{1}{3} + 5\frac{1}{2}$
4. $2\frac{1}{4} + 5\frac{1}{2}$
5. $4\frac{1}{6} + 3\frac{3}{4}$
6. $2\frac{3}{8} + 4\frac{1}{3}$
7. $8\frac{7}{8} - 2\frac{3}{8}$
8. $5\frac{3}{4} - 3\frac{1}{2}$
9. $9\frac{5}{8} - 6\frac{1}{4}$
10. $7\frac{5}{6} - 1\frac{2}{3}$
11. $8\frac{3}{5} - 6\frac{1}{4}$
12. $8\frac{1}{3} - 3\frac{1}{6}$

Set 34 **Add.**

1. $5\frac{1}{2} + 3\frac{3}{4}$
2. $5\frac{1}{4} + 4\frac{3}{4}$
3. $7\frac{3}{8} + 9\frac{3}{4}$
4. $7\frac{1}{3} + 5\frac{7}{9}$
5. $6\frac{3}{8} + 7\frac{5}{8}$
6. $8\frac{3}{4} + 5\frac{5}{8}$
7. $6\frac{3}{4} + 8\frac{5}{6}$
8. $7\frac{5}{8} + 7\frac{5}{8}$
9. $8\frac{2}{3} + 6\frac{2}{5}$
10. $4\frac{1}{6} + 2\frac{8}{9}$
11. $8\frac{3}{4} + 8\frac{3}{5}$
12. $9\frac{3}{5} + 6\frac{1}{2}$

Set 35 **Subtract.**

1. $8\frac{1}{5} - 2\frac{4}{5}$
2. $3\frac{1}{6} - 1\frac{5}{6}$
3. $7\frac{1}{3} - 3\frac{2}{3}$
4. $8\frac{1}{4} - 5\frac{3}{4}$
5. $7\frac{3}{8} - \frac{5}{8}$
6. $9\frac{1}{3} - \frac{2}{3}$
7. $7 - 3\frac{1}{2}$
8. $8 - 4\frac{3}{4}$
9. $9 - 5\frac{1}{3}$
10. $9 - 6\frac{5}{6}$
11. $4 - 3\frac{1}{8}$
12. $6 - 1\frac{5}{8}$

Set 36 **Give each product.**

1.  $\frac{1}{4} \times \frac{1}{3} = \underline{?}$
2. 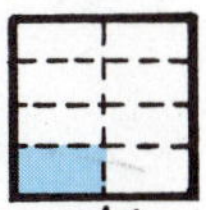 $\frac{1}{4} \times \frac{1}{2} = \underline{?}$
3. 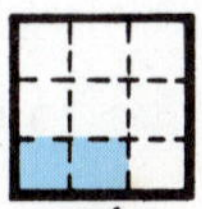 $\frac{1}{3} \times \frac{2}{3} = \underline{?}$
4. 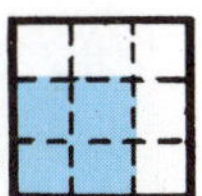 $\frac{2}{3} \times \frac{2}{3} = \underline{?}$
5. 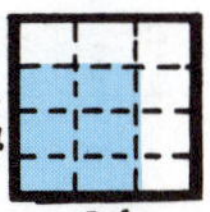 $\frac{3}{4} \times \frac{2}{3} = \underline{?}$
6. 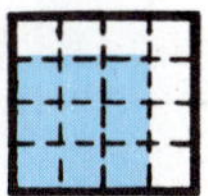 $\frac{3}{4} \times \frac{3}{4} = \underline{?}$

Set 37 **Multiply.**

1. $\frac{3}{4} \times \frac{5}{8}$ 2. $\frac{1}{2} \times \frac{1}{4}$ 3. $\frac{3}{4} \times \frac{1}{3}$ 4. $\frac{4}{5} \times \frac{5}{8}$

5. $\frac{3}{4} \times \frac{2}{9}$ 6. $\frac{2}{3} \times \frac{2}{3}$ 7. $\frac{5}{6} \times \frac{3}{5}$ 8. $\frac{7}{8} \times \frac{4}{7}$

9. $\frac{5}{2} \times \frac{3}{10}$ 10. $3 \times \frac{1}{2}$ 11. $6 \times \frac{1}{3}$ 12. $9 \times \frac{2}{3}$

13. $\frac{4}{5} \times 3$ 14. $\frac{3}{2} \times \frac{2}{3}$ 15. $\frac{4}{3} \times \frac{7}{10}$ 16. $\frac{4}{5} \times \frac{4}{5}$

Set 38 **Complete.**

1. $\frac{2}{3}$ of 30 = ? 2. $\frac{3}{5}$ of 15 = ? 3. $\frac{3}{8}$ of 40 = ? 4. $\frac{1}{3}$ of 15 = ?

5. $\frac{5}{6}$ of 30 = ? 6. $\frac{3}{4}$ of 36 = ? 7. $\frac{4}{3}$ of 18 = ? 8. $\frac{5}{3}$ of 18 = ?

9. $\frac{5}{9}$ of 18 = ? 10. $\frac{2}{3}$ of 24 = ? 11. $\frac{3}{2}$ of 18 = ? 12. $\frac{3}{4}$ of 24 = ?

13. $\frac{1}{4}$ of 16 = ? 14. $\frac{5}{8}$ of 24 = ? 15. $\frac{4}{3}$ of 18 = ? 16. $\frac{4}{5}$ of 30 = ?

Set 39 **Multiply. Reduce answers to lowest terms.**

1. $3\frac{1}{2} \times 2$ 2. $4\frac{1}{4} \times 8$ 3. $5\frac{1}{2} \times 3$ 4. $6\frac{1}{4} \times 6$

5. $3\frac{1}{8} \times 7$ 6. $2\frac{5}{8} \times 7$ 7. $4\frac{3}{8} \times 6$ 8. $2\frac{3}{4} \times 6$

9. $2\frac{1}{2} \times 3\frac{1}{2}$ 10. $2\frac{1}{3} \times 4\frac{1}{2}$ 11. $5\frac{3}{4} \times 2\frac{1}{2}$ 12. $3\frac{2}{3} \times 4\frac{1}{3}$

13. $3\frac{1}{6} \times 2\frac{3}{4}$ 14. $5\frac{1}{2} \times 4\frac{3}{4}$ 15. $3\frac{1}{6} \times 2\frac{2}{3}$ 16. $5\frac{3}{8} \times 4\frac{1}{5}$

Set 40 **Divide.**

1. $\frac{5}{8} \div \frac{1}{8}$ 2. $\frac{4}{7} \div \frac{2}{7}$ 3. $\frac{1}{3} \div \frac{4}{5}$ 4. $\frac{3}{8} \div \frac{3}{4}$

5. $\frac{2}{5} \div \frac{4}{5}$ 6. $\frac{5}{4} \div \frac{1}{2}$ 7. $\frac{5}{8} \div \frac{5}{4}$ 8. $\frac{5}{8} \div \frac{3}{4}$

9. $\frac{5}{9} \div \frac{2}{3}$ 10. $\frac{3}{8} \div \frac{2}{5}$ 11. $\frac{3}{8} \div \frac{1}{2}$ 12. $\frac{2}{3} \div \frac{4}{5}$

13. $\frac{7}{9} \div \frac{2}{3}$ 14. $\frac{4}{7} \div \frac{1}{2}$ 15. $\frac{3}{2} \div \frac{5}{4}$ 16. $\frac{7}{8} \div \frac{3}{4}$

Set 41 **Divide.**

1. $3\frac{1}{2} \div 2$
2. $5\frac{1}{4} \div 3$
3. $4\frac{3}{4} \div 2$
4. $4\frac{1}{6} \div 5$
5. $3\frac{3}{4} \div 5$
6. $2\frac{7}{8} \div 4$
7. $3\frac{2}{5} \div 7$
8. $4\frac{3}{8} \div 4$
9. $2\frac{1}{2} \div 2\frac{1}{2}$
10. $3\frac{1}{2} \div 1\frac{3}{4}$
11. $7\frac{1}{2} \div 2\frac{1}{2}$
12. $6\frac{3}{4} \div 3\frac{1}{2}$
13. $2\frac{7}{8} \div 3\frac{1}{4}$
14. $4\frac{5}{8} \div 2\frac{2}{3}$
15. $3\frac{5}{6} \div 2\frac{1}{3}$
16. $5\frac{3}{4} \div 2\frac{2}{3}$

Set 42 **Give a decimal for the number of shaded squares.**

1. 
2. 
3. 
4. 
5. 
6. 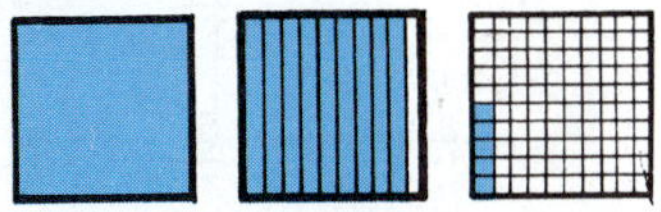
7. 
8. 
9. 

Set 43 **Tell what the red digit stands for.**

1. 3.0**8**
2. 3.**8**0
3. 5.9**2**1
4. 7.43**6**
5. **5**.803
6. 2.97**4**
7. 3**8**.21
8. 7.5**6**9
9. 8.14**9**
10. 63.5**9**
11. **6**6.666
12. 6**6**.666
13. 66.**6**66
14. 66.6**6**6
15. 66.66**6**

Set 44 **In 95,263.71084 what digit is in the:**

1. ones place?
2. tenths place?
3. hundredths place?
4. thousandths place?
5. tens place?
6. ten thousandths place?
7. hundred thousandths place?
8. ten thousands place?
9. thousands place?
10. hundreds place?

Set 45 <, =, or >?

1. 0.7 ◯ 0.70	2. 0.5 ◯ 0.05	3. .007 ◯ 0.70
4. 1 ◯ .99	5. .09 ◯ .088	6. .257 ◯ .256
7. 3.5281 ◯ 3.5279	8. .5263 ◯ 5.263	9. 7.5281 ◯ 7.4999

Set 46 Copy and complete.

	Number	Rounded to nearest .1	Rounded to nearest .01	Rounded to nearest .001
1.	1.8952	?	?	?
2.	2.6347	?	?	?
3.	2.9352	?	?	?
4.	3.7510	?	?	?
5.	8.3965	?	?	?

Set 47 Add.

1. 2.93 +4.67	2. 5.08 +0.39	3. .374 +.296	4. 7.65 +3.82	5. 18.9 +6.7
6. 3.624 +5.839	7. 7.097 +5.658	8. 539.8 +27.5	9. 4.636 +3.294	10. 58.49 +6.97
11. .20 .716 +.384	12. 2.5 6.49 +3.85	13. 21.7 35.6 + 9.3	14. 4.56 0.97 +6.7	15. .417 .93 +.261

Set 48 Subtract.

1. 3.81 − 2.67	2. 5.03 − 2.84	3. 70.2 − 36.9	4. 59.3 − 26.7	5. .841 − .729
6. 4.6 − 2.83	7. 7.2 − 5.94	8. 6 − 4.25	9. 8 − 3.94	10. .27 − .186
11. 52.4 − 1.87	12. 4.28 − .164	13. 3.45 − 1.283	14. 8.29 − 2.163	15. 9.43 − 5.286

Set 49 First estimate the product. Then multiply.

1. 3.8×2.1
2. 2.6×1.8
3. 5.9×6.3
4. 9.4×2.9
5. 8.2×3.2
6. 2.78×16
7. $39.5 \times .33$
8. 4.82×8.1
9. 9.31×60
10. 67.4×2.9
11. 73.5×5.0
12. 68.2×6.4
13. 9.27×5.3
14. 8.04×8.2
15. 5.38×4.4

Set 50 Multiply.

1. 5.8×2.9
2. $.26 \times 1.4$
3. $7.4 \times .35$
4. 9.3×6.4
5. $.52 \times 18$
6. 3.81×16
7. $25.3 \times .33$
8. 5.64×8.1
9. 78.22×60
10. 9.113×2.9
11. 53.84×1.96
12. 956.5×5.34
13. $81.34 \times .273$
14. $7.026 \times .582$
15. $51.88 \times .649$

Set 51 Divide.

1. $3\overline{)348}$
2. $7\overline{)25.9}$
3. $8\overline{)43.2}$
4. $9\overline{)6.75}$
5. $4\overline{)936}$
6. $2\overline{).938}$
7. $6\overline{)5.22}$
8. $5\overline{)385}$
9. $3\overline{)6.24}$
10. $7\overline{)84.7}$
11. $12\overline{)1.44}$
12. $23\overline{)71.3}$
13. $11\overline{).198}$
14. $36\overline{)10.08}$
15. $47\overline{).2491}$
16. $12\overline{).3324}$
17. $42\overline{)162.12}$
18. $65\overline{)35.295}$

Set 52 Divide.

1. $5\overline{)8.1}$
2. $4\overline{)6.3}$
3. $5\overline{).74}$
4. $2\overline{).95}$
5. $4\overline{)4.2}$
6. $8\overline{)18}$
7. $5\overline{)32}$
8. $8\overline{)74}$
9. $4\overline{)59}$
10. $8\overline{)75}$
11. $4\overline{)3.2}$
12. $5\overline{)43.1}$
13. $8\overline{)75.6}$
14. $4\overline{)39.8}$
15. $2\overline{)74.2}$
16. $16\overline{)3}$
17. $32\overline{)5}$
18. $16\overline{)6}$
19. $64\overline{)8}$
20. $32\overline{)4}$

Set 53 **Copy and complete.**

	1.	2.	3.	4.	5.	6.	7.	8.	9.	10.
Decimal	.2	.9	.25	.5	.75	?	?	?	?	?
Fraction	?	?	?	?	?	$\frac{4}{5}$	$\frac{3}{2}$	$\frac{3}{8}$	$\frac{5}{8}$	$\frac{13}{20}$

Set 54 **Multiply or divide.**

1. 85.347×10
2. 85.347×100
3. 85.347×1000
4. $529.8 \div 10$
5. $529.8 \div 100$
6. $529.8 \div 1000$
7. 74.36×10
8. 917.8×10
9. $.5063 \times 100$
10. 17.094×1000
11. $528.36 \div 10$
12. $9217.8 \div 1000$
13. $32.916 \div 10$
14. 5.810×1000
15. $391.65 \div 100$

Set 55 **Divide.**

1. $.7\overline{)3.836}$
2. $.6\overline{)26.34}$
3. $.9\overline{)5.067}$
4. $.8\overline{)45.84}$
5. $.03\overline{)1.473}$
6. $.5\overline{)36.05}$
7. $.04\overline{)2.656}$
8. $.03\overline{)9.630}$
9. $.4\overline{).6400}$
10. $.6\overline{)350.4}$
11. $.18\overline{)6705}$
12. $.16\overline{)18.72}$
13. $.05\overline{)2.870}$
14. $1.2\overline{)37.62}$
15. $.008\overline{)6.680}$
16. $.64\overline{).31040}$

Set 56 **How much for**

1. 3 for 79¢
   a. 1?
   b. 2?
2. 4 for 78¢
   a. 1?
   b. 3?
3. 3 for $1.00
   a. 1?
   b. 2?
4. 5 for 88¢
   a. 2?
   b. 3?
5. 3 for $1.39
   a. 1?
   b. 4?
6. 4 for $1.79
   a. 3?
   b. 5?

Set 57 **Compute the discount and the sale price.**

1. $2.75 $\frac{1}{3}$ off
2. $1.69 $\frac{1}{2}$ off
3. $5.45 $\frac{1}{4}$ off
4. $3.85 $\frac{1}{5}$ off
5. $6.29 $\frac{1}{3}$ off
6. $7.49 $\frac{1}{4}$ off

Set 58 **Compute the unit price to three decimal places.**

1. 8 oz for $.92
2. 10 oz for $1.53
3. 7 oz for $1.95
4. 3 pounds for $.76
5. 2 pounds for $1.19
6. 4 pounds for $3.27

Set 59 **Copy and complete.**

	Cash Price	Down Payment	Monthly Payment	Number of Payments	Installment Price	Interest
1.	$49	$10.92	$3.50	12		
2.	$85	$28.20	$5.30	12		
3.	$314	$26.56	$14.07	24		
4.	$512	$120.70	$29.50	15		
5.	$788	$113.96	$42.70	18		
6.	$900	$218.40	$70.00	12		

Set 60 **Copy and complete.**

1. $\frac{1}{2} = \frac{?}{8}$
2. $\frac{1}{3} = \frac{?}{15}$
3. $\frac{3}{4} = \frac{?}{12}$
4. $\frac{0}{5} = \frac{?}{20}$
5. $\frac{5}{8} = \frac{15}{?}$
6. $\frac{3}{5} = \frac{12}{?}$
7. $\frac{3}{5} = \frac{30}{?}$
8. $\frac{1}{3} = \frac{6}{?}$
9. $\frac{2}{3} = \frac{?}{6}$
10. $\frac{5}{6} = \frac{30}{?}$
11. $\frac{1}{2} = \frac{?}{6}$
12. $\frac{3}{2} = \frac{18}{?}$
13. $\frac{5}{3} = \frac{?}{12}$
14. $\frac{2}{3} = \frac{6}{?}$
15. $\frac{5}{2} = \frac{15}{?}$
16. $\frac{6}{5} = \frac{?}{30}$

Set 61 **Change to a percent.**

1. $\frac{15}{100}$ 2. $\frac{95}{100}$ 3. $\frac{5}{100}$ 4. $\frac{68}{100}$ 5. $\frac{20}{100}$

6. $\frac{13}{100}$ 7. $\frac{37}{100}$ 8. $\frac{200}{100}$ 9. $\frac{52}{100}$ 10. $\frac{95}{100}$

11. $\frac{100}{100}$ 12. $\frac{10}{100}$ 13. $\frac{23}{100}$ 14. $\frac{175}{100}$ 15. $\frac{1}{100}$

Set 62 **Change to a percent.**

1. $\frac{7}{20}$ 2. $\frac{3}{4}$ 3. $\frac{3}{5}$ 4. $\frac{1}{4}$ 5. $\frac{6}{25}$ 6. $\frac{11}{20}$ 7. $\frac{9}{10}$

8. $\frac{5}{20}$ 9. $\frac{1}{2}$ 10. $\frac{8}{40}$ 11. $\frac{7}{10}$ 12. $\frac{6}{8}$ 13. $\frac{3}{12}$ 14. $\frac{18}{20}$

15. $\frac{3}{6}$ 16. $\frac{9}{25}$ 17. $\frac{9}{20}$ 18. $\frac{4}{8}$ 19. $\frac{9}{15}$ 20. $\frac{9}{12}$ 21. $\frac{12}{25}$

Set 63 **Copy and complete.**

	1.	2.	3.	4.	5.	6.	7.	8.	9.	10.
Fraction	$\frac{1}{3}$	$\frac{1}{6}$	$\frac{2}{3}$	$\frac{3}{8}$	$\frac{5}{8}$	$\frac{1}{16}$	$\frac{5}{16}$	$\frac{5}{6}$	$\frac{7}{8}$	$\frac{13}{16}$
Decimal										
Percent										

Set 64 **Complete.**

1. 50% of 10 = ? 2. 10% of 20 = ? 3. 100% of 6 = ?

4. 20% of 45 = ? 5. 8% of 50 = ? 6. 40% of 60 = ?

7. 60% of 140 = ? 8. 125% of 36 = ? 9. 10% of 80 = ?

10. 125% of 68 = ? 11. 50% of 28 = ? 12. 6% of 200 = ?

13. 100% of 42 = ? 14. 200% of 42 = ? 15. 75% of 96 = ?

16. 40% of 180 = ? 17. 25% of 108 = ? 18. 300% of 18 = ?

Set 65 **Give each sum.**

1. $^{+}5 + ^{+}9$
2. $^{+}9 + ^{-}5$
3. $^{-}8 + 0$
4. $^{-}5 + ^{-}6$
5. $^{+}9 + ^{+}3$
6. $^{-}9 + ^{-}5$
7. $^{-}6 + ^{-}6$
8. $^{+}9 + 0$
9. $^{-}15 + ^{+}11$
10. $^{-}11 + ^{-}3$
11. $^{+}8 + ^{-}9$
12. $^{-}9 + ^{+}13$
13. $^{+}11 + ^{-}11$
14. $^{-}1 + ^{-}10$
15. $^{+}7 + ^{+}9$
16. $^{-}14 + ^{+}14$
17. $^{+}13 + ^{+}5$
18. $^{-}19 + 0$
19. $^{-}13 + ^{+}13$
20. $^{+}13 + ^{-}20$
21. $^{-}16 + ^{+}16$
22. $0 + 0$
23. $^{+}12 + ^{-}5$
24. $^{+}11 + ^{+}9$

Set 66 **Give each difference.**

1. $^{+}6 - ^{+}4$
2. $^{+}6 - ^{-}4$
3. $^{+}7 - 0$
4. $^{-}5 - ^{+}7$
5. $0 - ^{-}6$
6. $^{-}50 - ^{-}3$
7. $^{-}3 - ^{+}9$
8. $^{+}12 - ^{+}5$
9. $^{-}2 - ^{-}6$
10. $^{-}9 - ^{+}8$
11. $^{+}8 - ^{+}3$
12. $^{-}8 - ^{-}2$
13. $^{+}3 - ^{-}9$
14. $0 - 0$
15. $^{+}12 - ^{-}3$
16. $^{-}2 - ^{-}9$
17. $^{+}3 - ^{+}8$
18. $^{-}4 - ^{+}11$
19. $^{-}9 - ^{-}4$
20. $^{+}4 - ^{-}6$
21. $^{-}4 - ^{+}6$
22. $^{+}13 - ^{-}6$
23. $^{+}6 - ^{+}14$
24. $^{-}1 - ^{-}5$

Set 67 $<$ or $>$?

1. $^{+}5$ ● $^{+}4$
2. $^{-}5$ ● $^{+}4$
3. $^{-}5$ ● $^{-}4$
4. $^{+}5$ ● $^{-}4$
5. $^{+}9$ ● $^{-}8$
6. $^{+}6$ ● $^{+}5$
7. $^{-}9$ ● $^{-}8$
8. $^{-}5$ ● $^{+}4$
9. $^{-}14$ ● $^{+}16$
10. $^{+}9$ ● $^{-}12$
11. $^{+}1$ ● 0
12. 0 ● $^{-}1$
13. $^{+}12$ ● $^{-}13$
14. $^{+}12$ ● $^{+}13$
15. $^{-}12$ ● $^{+}13$
16. $^{-}12$ ● $^{-}13$
17. $^{-}18$ ● $^{+}19$
18. $^{+}32$ ● $^{+}33$
19. $^{-}16$ ● $^{-}15$
20. $^{+}12$ ● $^{-}12$

Set 68 **Graph these number pairs. Label each point with its coordinates.**

1. $(0, 0)$
2. $(^{-}5, ^{+}5)$
3. $(^{-}4, ^{+}5)$
4. $(^{-}6, ^{-}6)$
5. $(^{-}6, ^{-}4)$
6. $(0, ^{+}8)$
7. $(^{-}6, ^{+}6)$
8. $(^{-}4, ^{+}6)$
9. $(^{+}9, ^{-}3)$
10. $(^{+}8, 0)$
11. $(^{+}6, ^{+}6)$
12. $(^{-}8, 0)$
13. $(0, ^{-}8)$
14. $(^{+}6, ^{-}6)$
15. $(^{-}3, ^{+}8)$

EXTRA PROBLEM SOLVING

Set 1

Mileage Chart

	Amarillo	Birmingham	Charleston	Dallas	Houston	Jackson	Louisville	Mobile	Tampa	Tulsa
Amarillo, Texas	—	967	1265	358	596	758	1019	944	1431	361
Birmingham, Alabama	967	—	552	645	639	241	364	241	539	647
Charleston, West Virginia	1265	552	—	1035	1144	770	246	793	848	905
Dallas, Texas	358	645	1035	—	243	404	819	590	1077	257
Houston, Texas	596	639	1144	243	—	406	928	478	967	478
Jackson, Mississippi	758	241	770	404	406	—	564	186	673	527
Louisville, Kentucky	1019	364	246	819	928	564	—	605	835	659
Mobile, Alabama	944	241	793	590	478	186	605	—	489	713
Tampa, Florida	1431	539	848	1077	967	673	835	489	—	1177
Tulsa, Oklahoma	361	647	905	257	478	527	659	713	1177	—

1. What is the distance from Dallas to Jackson?

2. What city is 967 miles from Houston?

3. How far is a Houston–Tulsa–Louisville trip?

4. How far is a Tampa–Mobile–Charleston trip?

5. Amarillo is how much farther from Houston than from Dallas?

6. Tampa is how much farther from Birmingham than from Mobile?

7. How much farther is a Mobile–Jackson–Houston trip than a Mobile–Houston trip?

8. How far is a Birmingham–Houston–Dallas–Amarillo trip?

9. How much farther is a Jackson–Houston trip than a Jackson–Dallas trip?

10. Suppose that you have driven 397 miles of a Charleston–Amarillo trip. How far are you from Amarillo?

11. Suppose that you have driven the first 249 miles of a Charleston–Jackson–Mobile trip. How much farther do you have to go?

12. Suppose that you are on a Mobile–Birmingham–Charleston trip and have 117 miles to go to reach Birmingham. How much farther do you have to go?

Set 2

1. Jefferson Elementary School had $487.52 to spend for playground equipment. Their Bike-a-Thon raised $379.68 more. How much did they have then?

2. Dennis got $1.43 for each kilometer he rode. How much did he get for riding 9 kilometers?

3. Joan had a total pledge of $1.29 per kilometer. How much did she earn for riding 7 kilometers?

4. Anita had pledges of 50¢, 26¢, 38¢, 19¢, and 45¢ for each kilometer that she rode. How much did she get per kilometer?

5. Anita rode a total of 8 kilometers. How much money did she raise? (See exercise 4.)

6. Bob had pledges of $.28 and $.19 per kilometer. How much did he get for riding 8 kilometers?

7. Alan had pledges of $.25, $.20, and $.38 per kilometer. How much did he get for riding 6 kilometers?

8. Alice had total pledges of $1.65 per kilometer. How much more would she earn for riding 9 kilometers than for 6 kilometers?

9. Tomas had total pledges of $1.78 per kilometer. Ann had total pledges of $2.15 per kilometer. How much more would Ann raise for riding 7 kilometers?

10. Alex had total pledges of $3.15 and rode 6 kilometers. Betty had total pledges of $2.64 and rode 8 kilometers. Who raised more money?

11. Each rider was given a $1.75 gift certificate to Bob's Burgers. Patricia bought 2 hot dogs for 69¢ each. How much money did she have left?

12. David wanted to buy a Super Burger for $1.59 and a large shake for $1.05. How much more money does he need with his $1.75 gift certificate?

Set 3

1. Sarah delivers 63 papers each morning. How many papers does she deliver in a week?
2. Sarah collected $.85 a week from each of her 63 customers. How much did she collect?
3. John delivers 118 papers each morning. How many papers does he deliver in a 365-day year?
4. John collects $.80 a week from each of his 118 customers. How much does he collect each week?
5. If Sarah delivers 63 papers each morning and John delivers 118, how many more papers does John deliver in a week?
6. Sarah and John are paid $.32 per customer per week. How much does each earn in a week? (See exercise 5.)

7. How much more per week does John earn than Sarah? (See exercise 6.)

8. One week, 9 of John's 118 customers were on vacation. If John earns $.32 per customer, how much did he earn that week?

9. If Sarah is paid $.32 a customer, how much more would she earn if she got 15 new customers?

10. John would like to have 150 customers. If he made $.32 per customer per week, how much would he make in 4 weeks?

11. One week Sarah collected $.85 from 56 of her customers. Twelve of her customers gave her a $.25 tip. What was the total collected?

12. The newspaper decided to give each carrier a weekly raise of $.04 per customer. How much raise will John get delivering 118 papers?

Set 4

The pictograph shows the number of students in Willow Elementary School who watched television each day of one week.

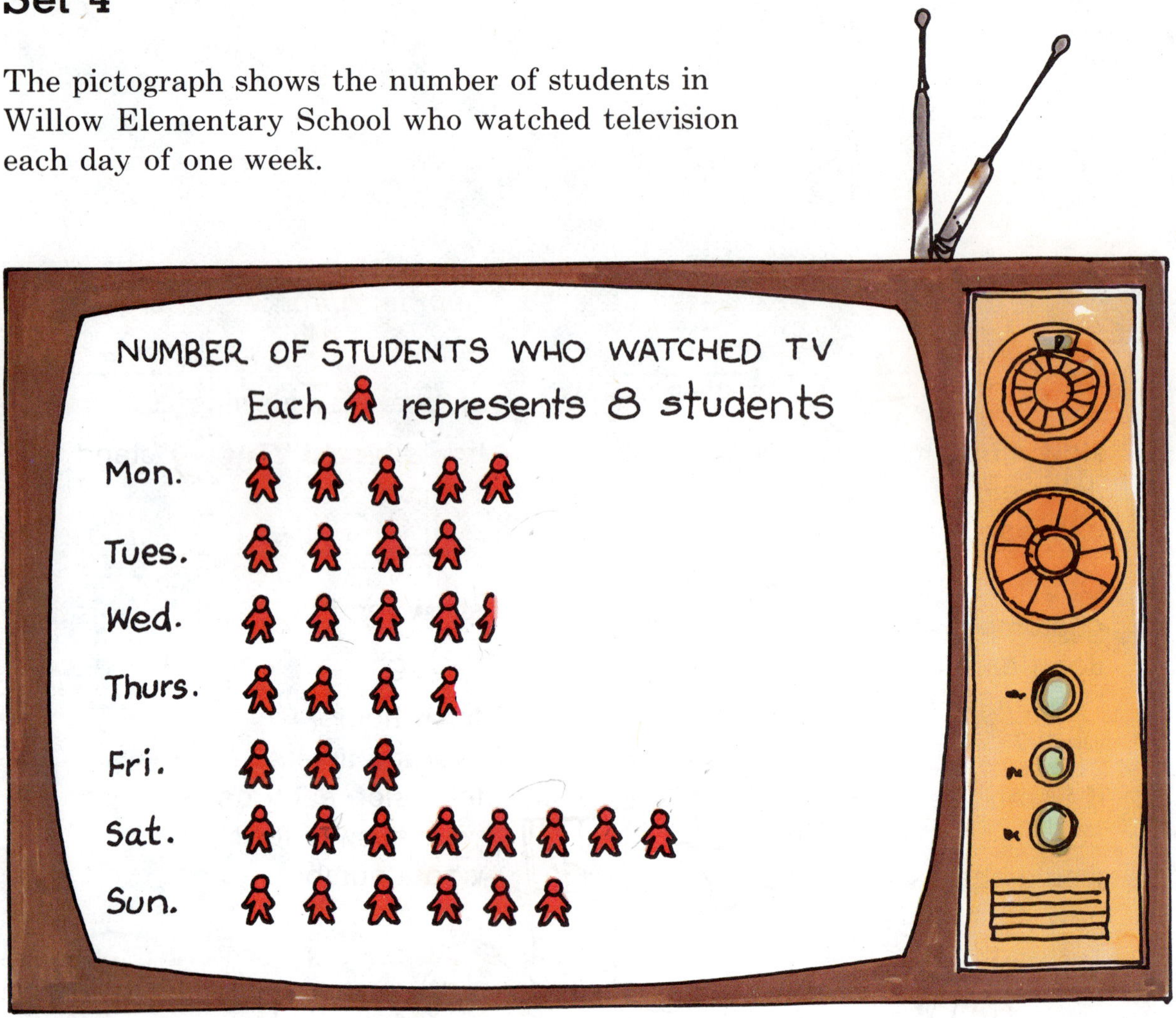

1. How many students does each represent?

2. What day did the greatest number of students watch television?

3. How many watched on Monday? On Saturday?

4. How many more watched on Saturday than on Friday?

5. About how many students watched on Wednesday? On Thursday?

6. Suppose it was found that 96 students watched television on a particular day. How many should be drawn for that day?

7. On Monday of the next week, 153 watched a special. If represented 9 students, how many did they draw?

8. If 1 stands for 9 viewers, how many viewers would 17 stand for?

9. On Tuesday of the next week, 105 students watched television. If represented 9 students, how many did they draw?

10. Another graph was made by letting 1 stand for 6 hours of television viewing. How many hours would 27 stand for?

11. If 1 stands for 6 hours of viewing, how many should be drawn for 156 hours of viewing?

12. Thirty-four were used to show the hours viewed Monday through Friday. If 1 stood for 6 hours, what was the average time spent watching television each day? Round your answer to the nearest whole number.

Set 5

World's Tallest Buildings

NAME	*STORIES*	*HEIGHT IN FEET*
Sears Roebuck Building	110	1450
World Trade Center	110	1353
Empire State Building	102	1250
Standard Oil Building	80	1136
John Hancock Center	100	1107
Chrysler Building	77	1046
60 Wall Tower	67	950
Bank of Manhattan	71	900
RCA Building	70	850
U. S. Steel Building	64	841
One Chase Manhattan Plaza	60	813

1. How many of the buildings listed are over 1000 feet in height?

2. Which building is 307 feet taller than the Chrysler Building?

3. Which building is 609 feet shorter than the tallest building?

4. The Empire State Building is how much taller than One Chase Manhattan Plaza?

5. Which two buildings are most nearly the same height?

6. What is the average distance from one floor to the next in the U. S. Steel building? Round your answer to the nearest whole number.

7. What is the average distance from one floor to the next in the John Hancock Center? Round your answer to the nearest whole number.

8. About how high above the street is a window on the 42nd floor of the John Hancock Center? (*Hint:* Use your answer from exercise 7.)

9. What is the average distance from one floor to the next in the Standard Oil Building? Round your answer to the nearest whole number.

10. About how high above the street is a window on the 25th floor of the Standard Oil Building? Round your answer to the nearest whole number. (*Hint:* Use your answer from exercise 9.)

11. What is the average height of the 5 tallest buildings? Round your answer to the nearest whole number.

12. What is the average number of stories of the buildings listed? Round your answer to the nearest whole number.

Set 6

1. Shopping

   2 milks: $.97 each
   Flour: $2.59

   How much for all?

2. Shopping

   Had $20
   Spent $13.74

   How much was left?

3. Baked Cookies

   Brown sugar: $\frac{1}{2}$ cup

   White sugar: $\frac{1}{4}$ cup

   How much sugar?

4. Baked Cakes

   Needed: $\frac{3}{4}$ cup of sugar

   Had: $\frac{3}{8}$ cup of sugar

   How much more was needed?

5. Cookies

   Cost: $2.88
   Number baked: 6 dozen

   How much did each cookie cost to make?

6. Cakes

   Number sold: 16
   Price: $2.50 each

   How much for the cakes?

7. Pies

Cost: $1.27 each
Selling price: $2.15

How much profit on 18 pies?

8. Brownies

Number sold: 235
Price: $.25 each

How much was the total sale?

9. Dinner Rolls

Price: $.06 each
Number sold: 12 dozen

How much for the dinner rolls?

10. Cupcakes

Morning: sold $\frac{2}{3}$

Afternoon: sold $\frac{1}{4}$

What fraction were sold?

11. Making Change

Had $10
Bought:
3 cakes—$2.50 each
2 loaves of date bread—$1.05 each

How much change?

12. Making Change

Had $8.50
Bought:
24 cupcakes—$.15 each
18 brownies—$.25 each

How much change?

Set 7

1. A 2-person canoe rents for \$11.50, and a 3-person canoe rents for \$14.70. Which is cheaper per person? How much cheaper?

2. A party of 6 decided to go canoeing. A 2-person canoe rents for \$11.50, and a 3-person canoe rents for \$14.70. How much would they save by renting two 3-person canoes?

3. Laura spent \$5.75 to rent a canoe, \$3.25 for a lunch, and \$1.95 for suntan lotion. How much did she spend in all?

4. Greg bought 3 rolls of film for \$2.85 each. He gave the clerk \$10. How much change did he get?

5. Jerry spent $\frac{2}{3}$ of his money to rent a canoe and $\frac{1}{4}$ of his money for lunch. What fraction of his money did he have left?

6. There were 3 people in each of 9 canoes and 2 people in each of 6 canoes. How many people in the group?

7. The group left at 8:30 A.M. and planned to be back at 5:00 P.M. How many hours would that be?

8. Julian was to sit in the bow (front) of his canoe for the first $1\frac{1}{4}$ hours. After $\frac{3}{4}$ of an hour they had just gone through White Water Bend. How much more time did he have in the bow of the canoe?

9. The group canoed $12\frac{1}{4}$ miles before lunch and $8\frac{1}{2}$ miles after lunch. How far did they canoe altogether?

10. The first part of the canoe trip took $4\frac{1}{2}$ hours. The second part took $3\frac{3}{4}$ hours. How much longer was the first part?

11. The canoers going the farthest without tipping over got a $5 prize. The first upset was at the $3\frac{3}{4}$-mile marker. The winner upset at the $17\frac{1}{4}$-mile marker. How much farther did the winning canoe go without upsetting?

12. The 39 people returned by station wagons. If 9 people could ride in a wagon, how many wagons were needed?

Set 8

1. Skis generally rent for \$6.75. On Wednesday they can be rented for $\frac{1}{3}$ off. How much do they cost on Wednesday?
2. Single lift tickets cost \$8.50. Lisa bought a book of 12 tickets for \$75. How much did she save by buying a book instead of 12 single tickets?
3. Ken rented skis for \$6.75 and bought a lift ticket for \$8.50. How much did he have left from \$20?
4. Bear Slope was $2\frac{1}{2}$ miles and Mule Slope was $1\frac{3}{4}$ miles. How much longer was Bear Slope?
5. Ken's first run down Bear Slope took $8\frac{1}{4}$ minutes. His second run was $1\frac{1}{2}$ minutes shorter. What was the time of his second run?
6. Lisa skied $2\frac{3}{4}$ hours in the morning and $3\frac{3}{4}$ hours in the afternoon. How many hours did she ski that day?

7. Ken skied $2\frac{1}{2}$ hours in the morning and 4 hours in the afternoon. How many hours longer did he ski in the afternoon?

8. Lisa skied Mule Slope 9 times. If each run was $1\frac{3}{4}$ miles, how far did she ski?

9. Ken skied Bear Slope ($2\frac{1}{2}$ miles) 6 times and Mule Slope ($1\frac{3}{4}$ miles) 5 times. How far did he ski in all?

10. The lift could pick up 2 people every 5 seconds. How many people could be picked up in one minute? In an hour?

11. At one time Lisa counted 44 people ahead of her in the lift line. How long did Lisa have to wait if the lift picked up 2 people every 5 seconds?

12. If the lift picked up 2 people every 5 seconds, how many minutes would it take to pick up 456 people?

Set 9

1. Nate bought a jogging suit for \$23.50, shoes for \$27.65, and socks for \$2.35. What was the total cost?

2. Allison had \$32.50. She bought a suit for \$27.65 and socks for \$2.85. How much money did she have left?

3. Rosa bought a jogging suit for $\frac{2}{5}$ off. The regular price was \$32.80. How much was the sale price?

4. Nate jogged for $16\frac{1}{2}$ minutes, and Fred jogged for $10\frac{3}{4}$ minutes. How many more minutes did Nate jog?

5. Kay averaged $5\frac{1}{2}$ minutes per kilometer. How long did it take her to jog 4 kilometers?

6. Brian jogs 6 kilometers a day. How many kilometers does he jog in a year (365 days)?

7. Allison jogged 4 kilometers in $24\frac{3}{4}$ minutes. What was her average time per kilometer?

8. The record for a certain course was 15 minutes and 15 seconds. Jan ran the course in 21 minutes. How many minutes over the course record was she?

9. One week Fred jogged $152\frac{1}{2}$ minutes. During the next week he increased his jogging time by $18\frac{3}{4}$ minutes. How many minutes did he jog the second week?

10. Allison's doctor told her that she burned up about 65 calories per kilometer. If she jogs 4 kilometers a day, how many calories does she burn up in a week?

11. The first time Brian jogged 8 kilometers, it took him an hour. A year later he could do it in $\frac{2}{3}$ of the time. How many minutes was that?

12. When he's at rest, Nate's heart rate is 72 beats a minute. When he jogs, his heart beats $1\frac{3}{4}$ times as fast. What is his heart rate when he's jogging?

Set 10

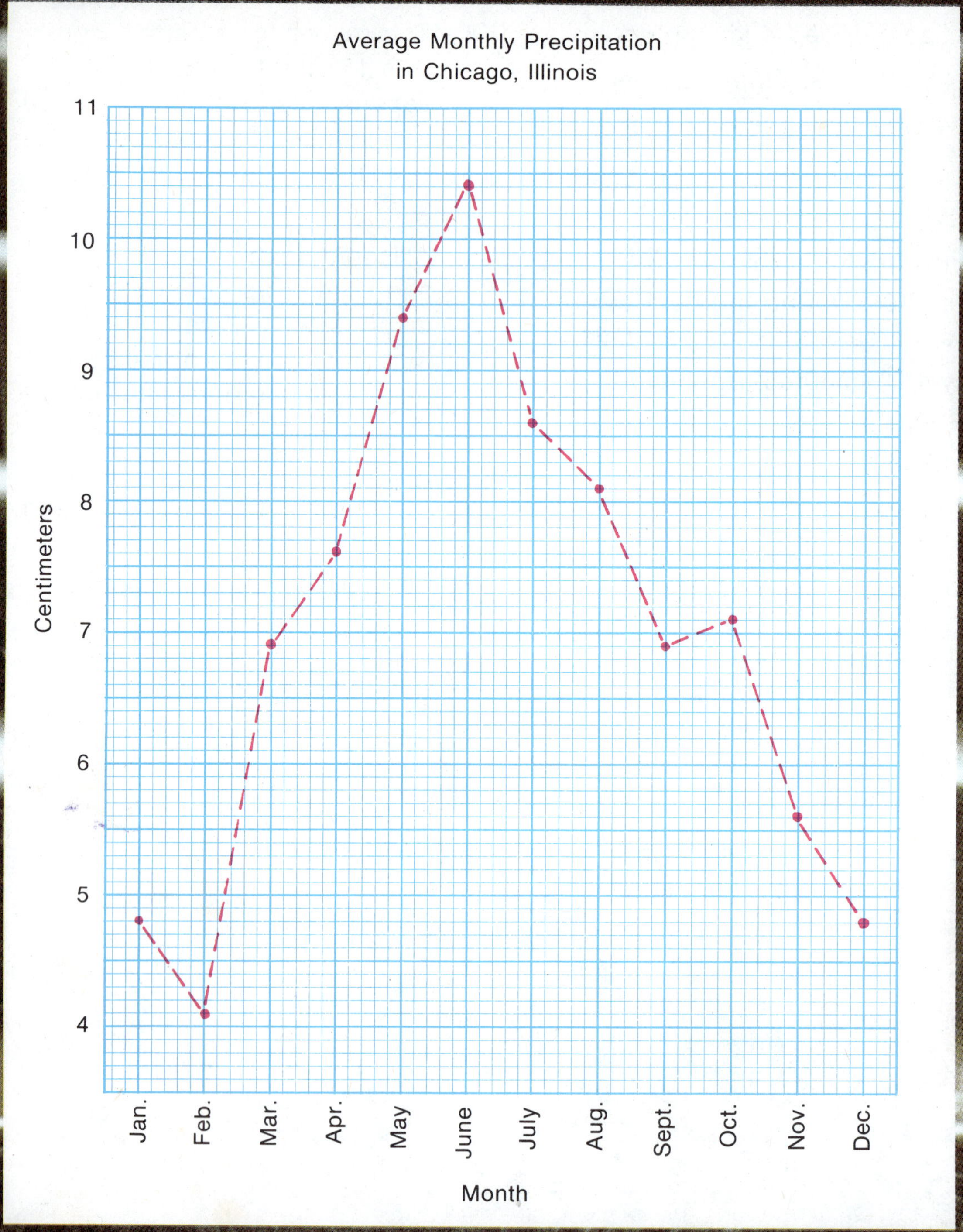

Average Monthly Precipitation
in Chicago, Illinois
Centimeters
11
10
9
8
7
6
5
4
Jan.
Feb.
Mar.
Apr.
May
June
July
Aug.
Sept.
Oct.
Nov.
Dec.
Month

1. Which month has the most precipitation? The least?
2. Which month has the same amount as March?
3. Which two-month period has more precipitation, January and February or March and April? How much more?
4. During how many months is the precipitation less than 6.0 cm? More than 8.5 cm?
5. During which two-month period is there the most precipitation?
6. What is the total precipitation during the first 3 months of the year?
7. What is the total precipitation during the last 3 months of the year?
8. How much more precipitation is there in the last 3 months than in the first 3 months?
9. How much more precipitation is there in July than in January?
10. What is the difference in precipitation of the month having the most and the month having the least?
11. What is the difference in precipitation of the first 6 months and the last 6 months?
12. What is the total precipitation for the year?

Set 11

The table shows the power used by some of the electrical devices found in the Jackson home.

Electrical Device	*Power Used in Kilowatts*
Light bulbs	.015–.300
Radio	.100
Television	.225
Refrigerator	.250
Vacuum cleaner	.500
Washer	.425
Toaster	.800
Air conditioner	1.275
Clothes dryer	4.200
Stove	Up to 22.500

1. The number of kilowatt-hours (kW·h) is found by multiplying the kilowatts used by the number of hours the device is used. If the television was left on for 4 hours, how many kW·h would that be?
2. How many kW·h would be needed to run the air conditioner for 24 hours?
3. How many kW·h would be needed to run the washer for 3.4 hours?
4. Which takes more electricity, having the radio on for 6 hours or running the television for 4 hours? How many kW·h more?
5. The rate the Jacksons are charged is $.12 per kW·h. How much will it cost them to run their air conditioner for 24 hours? (See exercise 2.) Round answer to nearest cent.

6. Suppose the washer takes 30 minutes to wash a load. How many kW·h would that be? At $.12 per kW·h, what would the cost be? Round your answer to the nearest cent.

7. At $.12 per kW·h, how much would it cost to run the vacuum cleaner for 5 hours?

8. At $.12 per kW·h, how much would it cost to run the air conditioner from 10:30 A.M. to 8:30 P.M.?

9. It takes 30 minutes to wash a load of clothes and 45 minutes to dry the load. At $.12 per kW·h, how much would it cost to wash and dry a load of clothes? Round your answer to the nearest cent.

10. At $.12 per kW·h, how much would it cost to bake a cake for 45 minutes if the stove uses 8.500 kw of power? Round your answer to the nearest cent.

Set 12

1. Teresa's dog cost $35. She also paid $7.50 for shots and $3.75 for a license. What was the total cost?

2. Her dog weighs 75 pounds. It is to be fed 25 calories per pound each day. How many calories per day is that?

3. A dog that weighs 15 pounds should be fed 38 calories per pound each day. How many calories is that per day?

4. Should a dog that weighs twice as much be fed twice as much? (*Hint:* Compare answers to exercises 2 and 3.)

5. A 50-pound dog should be fed 27 calories per pound per day. How many calories is that?

6. Soft (moist-type) dog food has 1350 calories per pound. How much of it should you feed a 50-pound dog each day? (*Hint:* Use answer from exercise 5.)

7. A 9-pound dog needs 43 calories per pound each day. An 18-pound dog needs 36 calories per pound each day. Which requires more calories, two 9-pound dogs or one 18-pound dog? How much more?

8. One pound of ration-type canned dog food has 500 calories. A 40-pound dog requires 1160 calories a day. How many pounds of dog food are needed? Round your answer to the nearest tenth pound.

9. A 16.5-pound dog requires 610.5 calories per day. How many calories per pound is that?

10. A 13.5-pound dog needs 526.5 calories per day. How many calories per pound is that?

11. Teresa's dog is 4.5 years old. The vet told her that to find the equivalent age in "human years" she should multiply by 6.7. How old is her dog in "human years"?

12. A 50-year-old human is equivalent in age to a 7.9-year-old dog. Using these ages, what is the average number of "human years" that is equivalent to 1 "dog year"?

Set 13

1. Payments

   $103.58 a month
   12 months

   What is the cost per year?

2. Miles Driven

   First year: 9085.7
   Second year: 11,135

   How many more miles were driven the second year?

3. Filling the Tank

   Tank holds: 20.5 gallons
   Filled tank with: 13.8 gallons

   How much was in the tank before it was filled?

4. Finding the Mileage

   Drove: 289 miles
   Used: 14 gallons

   How many miles per gallon? Round your answer to the nearest tenth.

5. Servicing Car

   Filter: $5.65
   5 quarts of oil: $1.35 each

   What is the total cost?

6. Maximum Range

   Miles per gallon: 21.6
   Gallons: 20.5

   How far can the car travel on a full tank of gasoline?

7. Buying Gasoline

Price: $1.43 per gallon
Amount: 6.8 gallons

What is the total price? Round your answer to the nearest cent.

8. Tune-up

Labor: $27.50
6 spark plugs: $1.65 each

What is the total cost?

9. Driving Time

Distance: 210 miles
Average Speed: 52.5 miles per hour

How many hours did it take?

10. Operating Cost

Total cost: $1660
Miles driven: 11,135

What was the average cost per mile? Give your answer to the nearest cent.

11. Finding Distance Driven

Odometer Reading—
Before trip: 18,236.5
After trip: 19,078.1

How many miles were driven?

12. Finding the Mileage

Odometer Reading—
Before trip: 21,253.8
After trip: 21,511.8
Gallons used: 13.5 gallons

How many miles per gallon? Round your answer to the nearest tenth.

Set 14

1. Roberto bought a pair of jeans for \$18.75, a belt for \$7.45, and a shirt for \$12.69. What was the total cost?
2. Alan bought 2 pairs of jeans for \$21.65 each. What was the total cost?
3. Leon bought 3 T-shirts for \$5.69 each and 2 pairs of jeans for \$18.45 each. What was the total cost?
4. Mandy bought some jeans that regularly sold for \$22.20 for $\frac{1}{3}$ off. What was the sale price?
5. Susan spent $\frac{1}{2}$ of her money on blue jeans and $\frac{1}{3}$ of her money on a blouse. What fraction of her money did she spend?
6. Kerry bought 3 pairs of jeans for \$42.16. What was the average price? Round your answer to the nearest cent.

7. All T-shirts were on sale for $\frac{1}{4}$ off regular price. What percent off is that?

8. Bob had \$20. He wanted to buy some \$24 jeans for $\frac{1}{3}$ off and a \$7.50 shirt. How much more money did he need?

9. A certain kind of T-shirt regularly sells for \$5.80. How much will 4 cost if they are marked $\frac{1}{4}$ off regular price?

10. What is the total price of a \$6 T-shirt marked $\frac{1}{4}$ off and a \$6 belt marked $\frac{1}{3}$ off?

11. One week, \$24.60 jeans were on sale at $\frac{1}{3}$ off. The next week they were marked $\frac{1}{2}$ off. How much cheaper were they the second week?

12. Which costs less, a \$21.60 pair of jeans on sale for $\frac{1}{3}$ off or a \$19.40 pair of jeans on sale for $\frac{1}{4}$ off? How much less?

Set 15

Johnson Grain Company

No. 57054	Date April 5, 1981
Name	Howard Thomas
Price per bushel	$4.25
Dockage	2.0%
Weight (in pounds)	
Gross	14660
Tare	6540
Net	8120
Dockage	162
Adjusted net weight	7958
Bushels	132.63
Net price	$563.68

Use the ticket above to answer the following questions.

1. What was the price per bushel?

2. Because the wheat contained too much moisture, the net weight was reduced (docked) by what percent?

3. What was the gross weight (the weight of the loaded truck)?

4. What was the tare weight (the weight of the empty truck)?

5. What was the net weight (the weight of the wheat)?

6. To find the pounds of dockage, 2.0% of the net weight was computed. How many pounds was the dockage?

7. The dockage was subtracted from the net weight to get the adjusted net weight. What was the adjusted net weight?

8. To find the number of bushels, the adjusted net weight was divided by 60. How many bushels were there?

9. The number of bushels was multiplied by $4.25 (price per bushel) to find the net price. What was the net price?

Complete each of these tickets.

10.

Johnson Grain Company

No. 57183 Date April 11, 1981
Name John Dow
Price per bushel $4.06
Dockage 3.4%
Weight (in pounds)
Gross 18593
Tare 9654
Net ______
Dockage ______
Adjusted net weight ______
Bushels ______
Net price ______

11.

Johnson Grain Company

No. 57261 Date April 15, 1981
Name Jack Brent
Price per bushel $4.19
Dockage 2.7%
Weight (in pounds)
Gross 16073
Tare 5846
Net ______
Dockage ______
Adjusted net weight ______
Bushels ______
Net price ______

Set 16

1. Season tickets for all 8 home games cost $78.80. What was the cost per ticket?

2. Adult tickets were reduced by 10%. How much was an adult ticket if one usually sold for $12.50?

3. About $\frac{2}{3}$ of the seats were filled. If the stadium holds 72,000, about how many were at the game?

4. During the season, the quarterback had thrown 135 passes and completed 81. What percent had he completed?

5. The place kicker had made 18 out of 24 field goals attempted. What percent had he made? What percent had he missed?

6. The starting fullback had carried the ball 62 times and had averaged 4.5 yards per carry. How many yards had he gained?

7. During the season one halfback gained 483 yards in 94 carries. How many yards did he average per carry? Round your answer to the nearest tenth.

8. For the first two carries of the game, "gains" of $^{+}18$ yards and $^{-}4$ yards were made. What was the net yardage?

9. The yardage gained or lost during the next 4 carries was $^{+}6$, $^{-}2$, $^{+}8$, and $^{-}3$. What was the net yardage?

10. During the first half the home team had $^{+}96$ yards rushing and $^{+}47$ yards passing. What was the total?

11. During the game the quarterback completed 8 out of 12 passes. Did he complete more or less than 60% of his passes?

12. One back gained 82 yards in 16 carries. Another gained 67 yards in 12 carries. Which back had the better average?

Glossary

acute angle An angle whose measure is less than 90°.

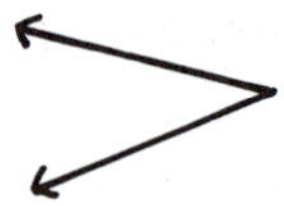

addend A number used in an addition problem.

angle A figure formed by two rays with the same endpoint.

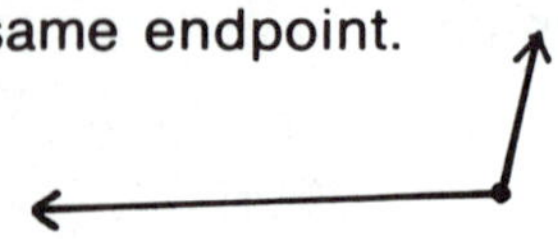

area The number of unit squares that cover a figure.

associative property of addition Changing the grouping of the addends does not change the sum.

$$(9 + 4) + 6 = 9 + (4 + 6)$$

associative property of multiplication Changing the grouping of the factors does not change the product.

$$(7 \times 25) \times 4 = 7 \times (25 \times 4)$$

average The average of 4, 5, 5, 7, and 9 is 6. To find the average, add the numbers and divide by the number of numbers.

bisect To cut into halves. The segment is bisected.

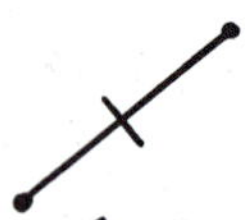

The angle is bisected.

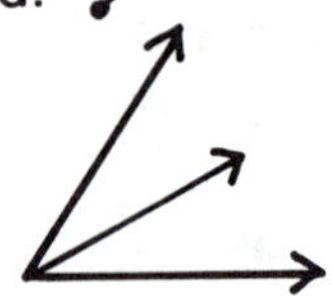

Celsius temperature (°C) The metric temperature scale in which 0°C is the freezing point of water and 100°C is the boiling point of water.

centimeter A metric unit of length. One centimeter is one hundredth of a meter.

circle A curved figure with all points a given distance from the center.

circumference The distance around a circle.

common denominator A common denominator for $\frac{1}{2}$ and $\frac{1}{3}$ is 6, because $\frac{1}{2} = \frac{3}{6}$ and $\frac{1}{3} = \frac{2}{6}$. A common denominator is a common multiple of the denominators of two fractions.

common factor 2 is a common factor of 4 and 6, because 2 is a factor of 4 and a factor of 6.

common multiple 30 is a common multiple of 5 and 6, because it is a multiple of 5 and a multiple of 6.

commutative property of addition Changing the order of the addends does not change the sum.

$$23 + 89 = 89 + 23$$

commutative property of multiplication Changing the order of the factors does not change the product.

$$19 \times 54 = 54 \times 19$$

congruent figures Figures that have the same size and shape.

coordinates A pair of numbers that locates a point on a grid.

corresponding parts In congruent figures, the parts that fit are called corresponding parts.

cube A rectangular solid with all edges the same length.

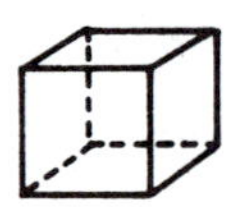

decimal A number written in our place-value system with a decimal point before the tenths place.

decimeter A metric unit of length. One decimeter is one tenth of a meter.

degree A unit for measuring angles. This is a 1° (1 degree) angle.

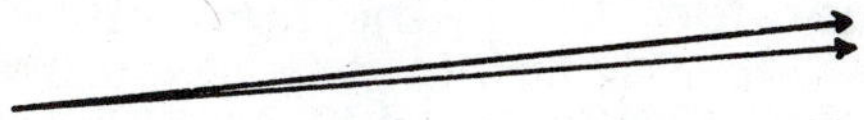

denominator In $\frac{2}{3}$, the denominator is 3.

diameter The distance across a circle through its center.

difference The answer to a subtraction problem.

discount An amount subtracted from the regular price of an item.

distributive property of multiplication A product can be written as the sum of two products.

$$3 \times (10 + 2) = (3 \times 10) + (3 \times 2)$$

down payment The first amount paid when buying on an installment plan.

equation A sentence with an equals sign, such as $3 \times 9 = 27$.

equilateral triangle A triangle with all sides congruent.

equivalent fractions Fractions for the same number. $\frac{1}{2}$, $\frac{2}{4}$, and $\frac{3}{6}$ are equivalent fractions.

even number A multiple of 2.

exponent An exponent tells the number of factors.

$$2^3 = \underbrace{2 \times 2 \times 2}_{\text{3 factors}} \quad (3 = \text{exponent})$$

factors Numbers used in a multiplication problem.

$$\begin{array}{rl} 8 & \leftarrow \text{factor} \\ \times 6 & \leftarrow \text{factor} \\ \hline 48 & \leftarrow \text{product} \end{array}$$

finance charge Buying an item on an installment plan costs more than paying cash. The difference is called the finance charge.

gram A metric unit of weight (mass). One gram is one thousandth of a kilogram.

graph A picture used to show numerical information.

hectare A metric unit of area. One hectare is 10,000 square meters.

hexagon A plane figure with six sides.

installment buying A way of buying expensive items. You pay part of the cost (the down payment) when you get the item and then agree to pay a certain amount each month for a certain number of months.

integers The numbers ..., $^{-}5$, $^{-}4$, $^{-}3$, $^{-}2$, $^{-}1$, 0, $^{+}1$, $^{+}2$, $^{+}3$, $^{+}4$, $^{+}5$, ...

interest A payment for the use of money.

isosceles trapezoid A trapezoid with 2 congruent sides.

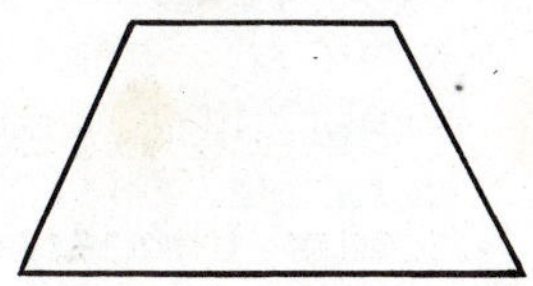

isosceles triangle A triangle with two congruent sides.

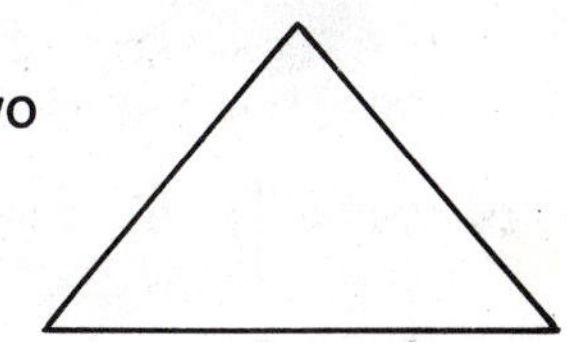

kilogram A unit of weight (mass) in the metric system. A kilogram is 1000 grams.

kilometer A unit of length in the metric system. A kilometer is 1000 meters.

kite A quadrilateral with 2 pairs of congruent sides.

least common denominator The least common multiple of the denominators of two or more fractions.

least common multiple The least (smallest) common multiple of two or more numbers. The least common multiple of 6 and 15 is 30.

line of symmetry If a figure can be folded along a line so the two parts of the figure match, the fold line is a line of symmetry.

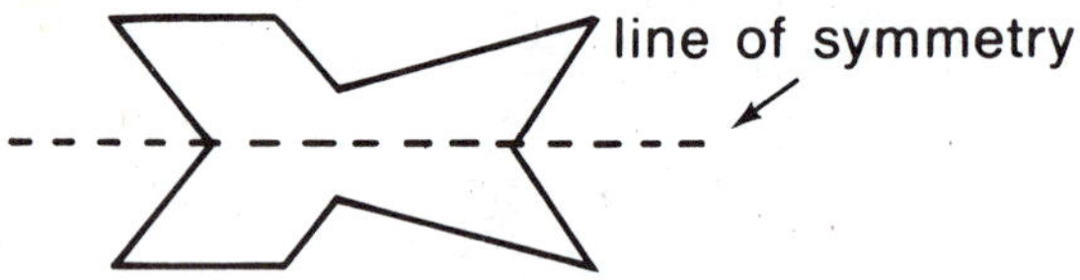

liter A unit of volume in the metric system.

meter A unit of length in the metric system. A meter is 100 centimeters.

metric system An international system of measurement that uses meter, liter, gram, and Celsius temperature.

milligram A metric unit of weight (mass). One milligram is one thousandth of a gram.

milliliter A metric unit of volume. One milliliter is one thousandth of a liter.

millimeter A metric unit of length. One millimeter is one thousandth of a meter.

mixed number A number that has a whole-number part and a fraction part. $2\frac{3}{4}$ is a mixed number.

multiple A product. 0, 4, 8, 12, 16, 20, and so on, are multiples of 4.

numerator In $\frac{2}{3}$, the numerator is 2.

obtuse angle An angle whose measure is greater than 90°.

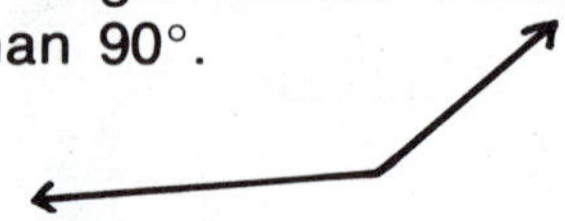

odd number A whole number that is not divisible by 2. The numbers 1, 3, 5, 7, 9, 11, and so on, are odd.

ordinal number The numbers *first*, *second*, *third*, *fourth*, *fifth*, and so on, are ordinal numbers. They tell the order of objects.

parallel lines Lines in a plane that do not cross.

parallelogram A quadrilateral with opposite sides parallel.

percent (%) Percent means per hundred. 5% is a percent. It equals $\frac{5}{100}$.

perimeter The distance around a figure. The sum of the lengths of the sides.

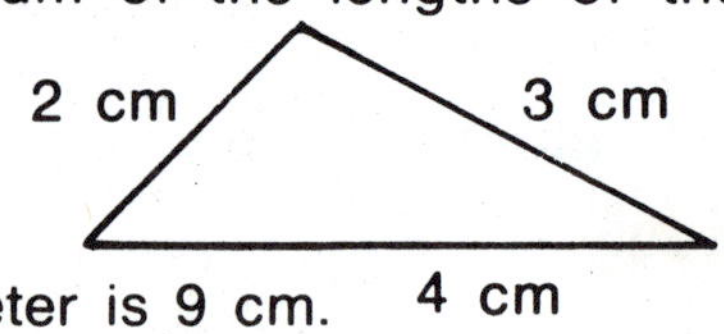

The perimeter is 9 cm.

perpendicular lines Lines that intersect to form right angles.

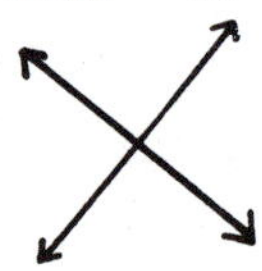

place value A system for writing numbers in which the value of a digit is determined by its position.

prime factorization Writing a composite number as a product of prime numbers. The prime factorization of 18 is $2 \times 3 \times 3$.

prime number 2, 3, 5, 7, 11, 13, and so on, are prime numbers. They cannot be obtained by multiplying smaller whole numbers.

principal An amount of money on which interest is paid.

probability A measure of the chances that an event will occur.

quadrilateral A plane figure with four sides.

quotient The answer to a division problem.

radius The distance from the center of a circle to the circle.

ratio A comparison of two quantities by division. In a quadrilateral, the ratio of sides to diagonals is 4 to 2, 4:2, or $\frac{4}{2}$.

ray A part of a line that has one endpoint. This is ray *AB*.

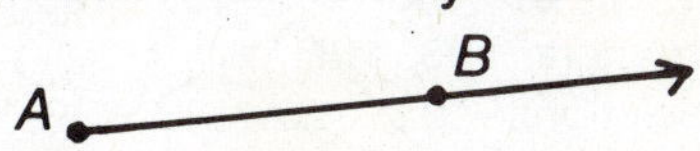

reciprocal Two numbers are reciprocals if their product is 1.

$$\frac{3}{4} \times \frac{4}{3} = 1$$

reciprocals

rectangle A parallelogram with four right angles.

rectangular solid A box whose faces are all rectangles. A rectangular solid has length, width, and height.

rhombus A parallelogram with 4 congruent sides.

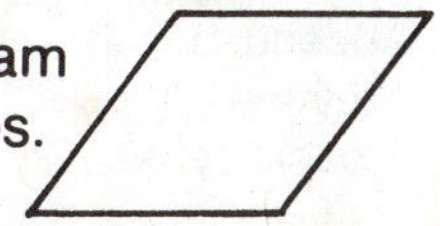

right angle An angle whose measure is 90°.

round To replace a number by another one that is easier to use. You round a number to the nearest ten by choosing the nearest multiple of ten. (5 is rounded up.)

13 → 10 27 → 30 45 → 50

You round a number to the nearest hundred by choosing the nearest multiple of one hundred.

487 → 500 1238 → 1200 550 → 600

sales tax A tax paid when you make a purchase.

scalene triangle A triangle with no congruent sides.

segment Part of a line that has two endpoints.

square A rectangle with four congruent sides.

symmetry A figure has symmetry if it can be folded so the two parts of the figure match.

trapezoid A quadrilateral with two parallel sides.

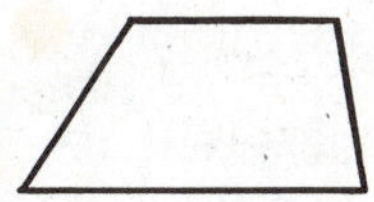

vertex The point at the "corner" of an angle, plane figure, or solid figure.

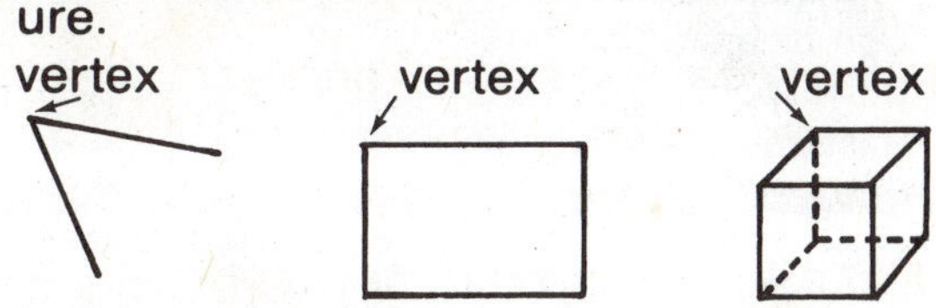

whole number Any of the numbers 0, 1, 2, 3, 4, and so on.

Index

3 4 5 6 7 8 9 0